Core Mathematics

In memory of my grandmother Agnes

Core Mathematics

George N. Frempong

Printed by CreateSpace

www.CreateSpace.com/TITLEID

Available from Amazon.com and other retail outlets

Contents

Preface

This book is one of two volumes written for students following the Core Curriculum, and is ideal for students preparing for Cambridge IGCSE, GCE O Level Mathematics and the West African School Certificate Examination.

This book has been developed from materials used in teaching mathematics at Accra High School for twenty eight years.

I have tried to present the concepts in forms that are easy to understand. Each of the sections offers a step-by-step explanation of the concepts together with many worked examples. The Try this exercises give you the opportunity to practise after every example. The many exercises allow you to have as much practice as possible. Every chapter ends with a test. The tests have been provided to help you track your progress and also help you retain mastery of the topics.

It will not be possible to look back and recall the origins of most of the materials in this book. I am very grateful to all whose work I have benefited from.

1

Sets

1.1 Sets

The word set is used to describe a collection of objects. For example, a collection of books, a group of students, a list of countries are sets.

Definition of Sets

A set is a well–defined collection of objects. Each object is called an element or a member of the set.

A set is properly defined if for any given object you can determine whether or not the object is a member of the set. For example, the set of teachers in your school is a well defined set because you can determine whether a particular teacher is a member or not a member of the set.

Sets can be defined in two ways:

1. Listing the elements

A set can be defined by listing all its elements between curly brackets { }. The symbol { } is read 'the set of'. For example, the set that contains the elements 1, 2, 2, 3, 1 and 1 may be written as {1, 2, 3}, {2, 1, 3} or {3, 1, 2}. Notice that 1 and 2 are listed only once and there is a comma after each element except the last. The elements can be listed in any order.

2. Describing the elements

A set can be defined by describing its elements. For example, the set of numbers divisible by 2 is described as the set of even numbers.

Set-builder notation

The set of real numbers less than 12 cannot be defined by listing all its members because there are an infinite number of them. In such cases, it is more convenient to define the set by stating a property of its elements. The set of real numbers less than 12 can be written as $\{x: x \in R, x < 12\}$.

This is read as 'the set of all x such that x is a real number less than 12'.

The colon : means such that. Sometimes the vertical bar | is used instead of the colon. This notation of a set is called the set – builder notation.

Naming Sets

We usually use capital letters, such as A, B, C etc to represent the name of a set. For example, we normally name the set of real numbers by the letter R. The set of natural numbers less than 6 can be written as N = {1,2, 3, 4, 5}.

The elements of sets are usually denoted by lower case letters such as a, b, c etc.

Example

Describe the following set by listing its element

Set A is the set of prime numbers less than 10

The prime numbers less than 10 are 2, 3, 5, and 7.

Thus, A = {2, 3, 5, 7}

Try this 1

Describe the following set by listing its element

Set A is the set of natural numbers between 11 and 15

Example

Write set B = {1, 2, 3, 4, 5, 6, 7, 8} in set-builder notation

Since set B consists of natural numbers less than 9, we can write
B = $\{x : x \in N \ and \ x < 9\}$

Try this 2

Write set A = {2, 4, 6, 8} in set-builder notation

Example

List the members of the set A = $\{x : x \in N \ and \ 2 < x \leq 6\}$

A = {3, 4, 5, 6}

Try this 3

List the members of set B = {x: x is an even number between 12 and 15}

Indicating Membership of a Set

The symbol $\in$ is used to indicate that a set contains a given element. For example, if x is an element of set A, we write $x \in A$. The symbol $\in$, is read 'is a member of' or ' is an element of' or 'belongs to'.

The symbol $\notin$. is read 'is not a member of' or 'is not an element of' or 'does not belong to'.

For example, if set $A = \{1,2,3,4\}$, then $1 \in A$ and $6 \notin A$.

Example

Write in symbol the statement below

2 is a real number

The set of real numbers is denoted by R. Thus, we write $2 \in R$

Try this 4

Write in symbol the statement:

-2 is not a natural number.

Exercise 1.1(a)

1. Describe the following sets by listing the elements between curly brackets

   (a) The set of positive integers less than 7

   (b) The set of prime numbers between 21 and 30

   (c) The set of factors of 12

   (d) The set of prime factors of 50

   (e) The set of multiples of 2 less than 16

2. Rewrite the following set using the set-builder notation

   (a) The set of integers greater than 12

   (b) The set of odd natural numbers

   (c) The set of multiples of 5

   (d) The set of factors of 20

3. Rewrite the following sets by listing the elements between curly brackets

   (a) $\{x : x$ is a multiple of 2 greater than 6 but less than 18$\}$

   (b) $\{x : x$ is a factor of 36$\}$

   (c) $\{x : x$ is an integer and $10 < x \leq 16\}$

(d) $\{x: x$ is even and $7 < x < 15\}$

4. Using the symbol $\in$ or $\notin$ and the letters representing the sets write the following statements in symbols

(a) 3 is not an even number   E = {even numbers}

(b) 8 is a multiple of 2   M = {multiples of 2}

(c) 7 is not an alphabet   A = {alphabet}

(d) 6 is a factor of 12   F = {factors of 12}

(e) 4 is not a prime number   P = {prime numbers}

5. If $A = \{a, b, c, d\}$ and $B = \{a, c, e\}$, complete the following statements by inserting the symbols $\in$ and $\notin$

(a) a A (b) e A (c) e B

(d) d B (e) c B

Empty Set

A set that has no element is called the empty set. For example, the set of university students who are five years old is empty. The empty set is also called the null set, and is denoted by the symbol { } or Ø. It is important to note that the empty set is not written as {Ø}. The set {Ø} contains the element Ø.

Example

State whether or not the set of integers that are both odd and even is an empty set.

There is no integer that is both odd and even. Thus, the set is an empty set

Try this 5

State whether or not the set of points common to two parallel lines is an empty set.

Exercise 1.1(b)

State whether the following sets are empty sets or not

1. $\{x : x$ is an integer and $3x + 2 = 7\}$

2. {integers between 4 and 5}

3. {even prime numbers}

4. {prime factors of 4}

5. {odd numbers divisible by 2}

Equal Sets

Two sets A and B are equal, written $A = B$ if they contain exactly the same elements. For example, if set $A = \{x : x \in N \; and \; x < 4\}$ and set B = {1, 2, 3} then A = B.

Cardinal Number

The cardinal number of a set A, written $n(A)$, is the number of elements in set A. The cardinal number of set $A = \{a, b, c, d, e, f\}$ is 6, since set A has 6 elements. Hence, we write $n(A) = 6$. Notice that, since the empty set has no elements $n(\emptyset) = 0$.

Try this 6

State the cardinal number of the set $B = \{x : x \in N \, and \, 4 \leq x < 9\}$

Equivalent Sets

Two sets A and B are said to be equivalent if they contain the same number of elements. For example, if set A = $\{0, 1, 2\}$ and set B = $\{a, b, c\}$, then set A and B are equivalent since each set has 3 elements.

Try this 7

State whether or not the set A = $\{a\}$ and set B = $\{even \, prime \, numbers\}$ are equivalent.

Exercise 1.1(c)

In Exercises 1 – 6, state whether the following pairs of sets are equal or not.

1. {factors of 6}   {1, 2, 3, 6}

2. {multiples of 2 less than 10}   {2, 4, 6, 8}

3. {multiples of 3}   {odd natural numbers}

4. {prime numbers less than 12}   {3, 5, 7, 11}

5. {0}   { }

6. {even natural numbers}   {multiples of 2}

In Exercises 7 – 11, state whether the following pairs of sets are equivalent or not.

7. {factors of 12}   {a, b, c, d, e, f}

8. {odd numbers between 2 and 6}   {a, b, c}

9. {multiples of 7 less than 28}   {0, 1, 3}

10 $\{x: x$ is an integer and $1 < x < 5\}$ {even prime numbers}

11. $\{x: 4x = 12\}$ {0}

Universal Set

We usually use sets whose elements are contained in a larger set, called the universal set. For example, the set of natural numbers contains prime numbers, odd numbers and even numbers. The set of natural numbers can serve as the universal set for say the set of odd numbers. Generally, the universal set, denoted by $\cup$ or ξ is a set that contains the entire element for any specific discussion.

When a universal set is given, only the elements in the universal set may be considered. If for example, the universal set is {1, 2, 3, 4, 5, 6, 7, 8, 9, 10} then the set of multiples of 2 is {2, 4, 6, 8, 10}.

Try this 8

Given that $\xi = \{1, 2, 3, 4, 5, 6, 7, 8\}$ list the elements of the set $A = \{\text{multiples of } 3\}$

Exercise 1.1(d)

For each of the given pair of sets choose the set that can be considered as the universal set

1. {Multiples of 2}   {multiples of 2 less than 20}

2. {Positive integers}   {integers}

3. {Equilateral triangles}   {triangles}

4. {Parallelograms}   {quadrilaterals}

5. {Even numbers}   {integers}

6. {Factors of 36}   {factors of 12}

Finite and Infinite Set

The number of elements of a set can be finite or infinite. A set is said to be finite if it is either empty or the number of its elements can be counted, and infinite if there is no end in counting its element. For example, the set $B = \{a, b, c, d, e, f, g\}$ is a finite set but the set of natural numbers is an infinite set. A large finite set such as the set of natural numbers from 1 to 100 can be listed as $\{1, 2, 3, 4, 5, \cdots 100\}$. The dots are used to indicate the missing elements.

An infinite set may be described by listing few elements followed by three dots. For example, the set of natural numbers may be written as $\{1, 2, 3, 4, 5 \ldots\}$. The three dots after 5 indicate that the elements in the set continue in the same pattern. For instance, the set also contains the numbers, 100, 101, 102, and so on

Try this 9

State whether the following sets are finite or infinite

(a) $A = \{x : x \in R, 1 < x < 2 \}$ (b) $B = \{x : x \text{ is a factor of } 6\}$

Exercise 1.1(e)

State whether the following sets are finite or infinite

1. {Integers}

2. {Multiples of 5 less than 20}

3. {Prime numbers less than 3}

4. {Rational numbers between 5 and 6}

5. $\{x : x \text{ is an integer and } x > 8\}$

6. {Factors of 54}

Subsets

If every element of set A is also an element of set B, then set A is said to be a subset of set B, written $A \subset B$. We can also write $B \supset A$, read as 'B contains A'.

Consider the set of students in your class. We can divide this set into different groups. For example, the girls may form one group. We may have the group of boys who are in the school football team or the group of students who failed the mathematics test. Each of these groups is a subset of the set of students in your class. Notice that, a particular set may have several subsets.

The empty set has no elements and is a subset of every set, including itself, and in particular every set is a subset of itself.

If A and B are two sets then

1. $\emptyset \subset A$

2. $A \subset A$

3. A = B if and only if A $\subset$ B and B $\subset$ A

Proper Subsets

A set A is called a proper subset of set B, if there is at least one element of set B that is not in set A. For example, if A = {1, 2, 3} and B = {1, 2, 3, 4, 5}, then set A is a proper subset of set B.

Example

Determine whether set A is a subset of set B.

(a) A = {1, 2, 3, 4}   B = {1, 2, 3, 4, 5, 6}

(b) A = {3, 4, 5, 8}   B = {1, 2, 3, 4, 8}

(a) Every element of set A is in set B. Hence, set A is a subset

   of set B.

(b) Not all the elements of set A are in set B. The number 5 is not

   in set B; hence set A is not a subset of set B.

Try this 10

Determine whether set A is a subset of set B

(a) A = {x: x is an even prime number}   B = {1, 2, 3, 4, 5}

(b) A = {2, 3, 4, 5}   B = {2, 3}

Number of Subsets

A subset of a given set may contain any number of elements or all the elements of the set. Recall that the empty set is a subset of every set, and that every set is a subset of itself. Hence the empty set has one subset and a set containing one element has two subsets, the empty set and the set itself.

You can find all the subset of a given set by making a list that contains the empty set, all subsets with one element, all subsets with two elements, and so on as illustrated in Table 1.1.

Table 1.1

Sets	Number of elements	Subsets	Number of subsets
{ }	0	{ }	$1 = 2^0$
{a}	1	{ }, {a}	$2 = 2^1$
{a, b}	2	{ }, {a}, {b}, {a, b}	$4 = 2^2$
{a, b, c}	3	{ }, {a}, {b}, {c}, {a, b} {a, c}, {b, c}, {a, b, c}	$8 = 2^3$
{a, b, c, d}	4	{ }, {a}, {b}, {c}, {d}, {a, b} {a, c}, {a, d}, {b, c}, {b, d}, {c, d}, {a, b, c}, {a, b, d}, {a, c, d}, {b, c, d}, {a, b, c, d}	$16 = 2^4$

You can see from Table 1.1 that a set with n elements has 2^n number of subsets.

For example, a set with 5 elements has 2^5, i.e. 32 subsets

Try this 11

(a) List all the possible subset of the set A = {2, 3, 5}

(b) Find the number of subsets of a set with 9 elements

Exercise 1.1(f)

1. Given P = {1, 2, 3, 4, 5, 6}, Q = {2, 4, 6} and R = {1, 2, 4, 6}, fill the space between the pair of sets with the symbol $\subset$ or $\supset$

(a) P   Q   (b) Q   R   (c) P   R

2. If A = {5, 6, 7}, how many subsets does set A have? List all the

   possible subsets

3. If P = {1, 2, 3, 4, 5}, how many subsets of set P have two

   elements?

4. If a set has 128 subsets, how many elements have the set?

5. The number of elements of set A is equal to half the number of the subsets of the set B = {0, 1, 2}. How many element has set A?

1.2 Operations on sets.

There are three basic set operations; union, intersection and complement.

Union of two sets

The union of sets A and B, written A ∪ B, is the set containing all the elements that are members of set A or set B or both sets. A ∪ B is read "A union B"

Example

If A = {1, 2, 4} and B = {2, 3, 5}, find A ∪ B

A ∪ B = {1, 2, 3, 4, 5}

Try this 12

If A = {a, c, d} and B = {a, b, d, e, f}, find A ∪ B and B ∪ A

Example

1f A = {multiples of 2 less than 10} and B = {factors of 6}, find A ∪ B.

First list the members of set A and set B.

$$A = \{2, 4, 6, 8\}$$

$$B = \{1, 2, 3, 6\}$$

$$\therefore \ A \cup B = \{1, 2, 3, 4, 6, 8\}$$

Try this 13

If A = {x: x is an integer and 5 < x < 9} and B = {multiples of 2 less than 12}, find A ∪ B and B ∪ A.

Some basic properties of union are:

If A and B are two sets, then

1. $A \cup B = B \cup A$

2. $(A \cup B) \cup C = A \cup (B \cup C)$

3. $A \subset (A \cup B)$

4. $A \cup \emptyset = A$

5. $A \cup A = A$

6. $A \subset B$ if and only if $A \cup B = B$

Exercise 1.2(a)

1. If P = {1, 3, 4, 6} and Q = {2, 3, 5}, find $P \cup Q$.

2. If A = {1, 2, 4, 5} and B = {1, 3, 4, 6, 8}, find $A \cup B$ and $B \cup A$

3. Given A = {4, 5, 8, 11, 12}, B = {3, 7, 9, 11} and C = {1, 2, 5, 6, 10}, find $A \cup (B \cup C)$

4. Given A = {2, 4, 6, 8}, B = {1, 2, 3, 6} and C = {3, 6, 9}, find $(A \cup B) \cup C$ and $A \cup (B \cup C)$

5. If $P = \{x : 0 < x < 6\}$ and $Q = \{x : 3 < x \le 8\}$, where $x \in$ {integers}, find $P \cup Q$.

6. If P = {prime numbers less than 10} and Q = {odd numbers less

than 10}, find $P \cup Q$.

7. Given A = {$x: x$ is a multiple of 2 and $x < 12$} and B = {$x: x$ is a

factor of 20}, find $A \cup B$

Intersection of two sets

The intersection of sets A and B, written A ∩ B, is the set containing all the elements that are common to both set A and set B. A ∩ B is read "A intersect B".

Example

If A = {1, 3, 4} and B = {1, 2, 4, 5, 6}, find A ∩ B

A ∩ B = {1, 4}

Try this 14

Given that A = {2, 3, 5, 6} and B = {1, 3, 4, 6, 8}, find A ∩ B.

Example

If A = {$x: x$ is a multiple of 3 less than10} and B = {factors of 12}, find A ∩ B.

First, list the elements of set A and set B.

A = {3, 6, 9} and B = {1, 2, 3, 4, 6, 12}, hence A ∩ B = {3, 6}

Try this 15

If A = {even numbers less than 12} and B = {prime numbers less than 8}, find A ∩ B and B ∩ A.

Some basic properties of intersections are:

If A and B are two sets, then

1. $A \cap B = B \cap A$

2. $(A \cap B) \cap C = A \cap (B \cap C)$

3. $A \cap B \subset A$

4. $A \cap A = A$

5. $A \cap \emptyset = \emptyset$

6. A $\subset$ B if and only if $A \cap B = A$

Notice that the operations of union and intersection each obey the associative law and the commutative law. Together they also obey the distributive law.

Disjoint sets

Two sets A and B are said to be disjoint if they have no elements in common.

i.e. $A \cap B = \emptyset$.

For example, if A = {2, 4, 6, 8} and B = {3, 5, 7, 9}, then set A and set B are disjoint set since $A \cap B = \emptyset$.

Try this 16

State whether A = {even numbers} and B = {prime numbers} are disjoint set or not.

Exercise 1.2(b)

1. If P = {a, f} and Q = {f, g}, find $P \cap Q$

2. Given A = {1, 2} and B = {1, 2, 3, 4}, find $A \cap B$

3. If A = {1, 3, 4, 6}, B = {1, 2, 5, 7} and C = {2, 3, 8}, find

$(A \cap B) \cap C$

4. Given A = {1, 2, 3, 4, 5, 6, 7, 8, 9, 10}, B = {2, 4, 6, 8} and

C = {3, 6, 9}, find $(A \cap B) \cap C$ and $A \cap (B \cap C)$

5. Given A = {$x: x \in Z, 2 < x < 8$} and B = {factors of 6}, find

$A \cap B$

6. Given A = {multiples of 2},   Q = {multiples of 3} and

R = {factors of 12}, find $P \cap Q \cap R$

7. Given A = {$x: x \in Z, x > 4$}, B = {$x: x \in Z, x < 15$} and

C = {factors of 36}, find $A \cap B \cap C$

8. If P = {factors of 20} and Q = {factors of 36}, find

(a) $P \cap Q$ (b) the hcf of 20 and 36

9. Given A = {1, 3, 5, 6}, B = {2, 3, 4, 8} and C = {1, 2, 4, 6},

find $A \cup (B \cap C)$ and $(A \cup B) \cap (A \cup C)$

10. Given A = {1, 2, 3, 4, 5, 6}, B = {2, 3} and C = {1, 2, 5},

find $A \cap (B \cup C)$ and $(A \cap B) \cup (A \cap C)$

11. State whether the following pair of sets are disjoint or not disjoint

   (a) {even numbers}   {odd numbers}

   (b) {even prime numbers}   {prime factors of 6}

   (c) {triangles}   {polygons}

   (d) {multiples of 3}   {factors of 10}

   (e) {vowels} $\{x, y, z\}$

Complement

The complement of set A, written A', is the set of all the elements in the universal set that are not in set A.

For example, given that $\xi = \{1, 2, 3, 4, 5, 6\}$ and A = $\{1, 3, 5\}$, then the complement of A is A' = $\{2, 4, 6\}$

Try this 17

Given that $\xi = \{1, 2, 3, 4, 5, 6, 7, 8, 9, 10\}$ and A= $\{1, 3, 4\ 7, 10\}$, find A'.

Some basic properties of complements are:

If A is a set, then

1. $A \cup A' = \xi$

2. $A \cap A' = \emptyset$

3. $(A')' = A$

4. $\xi' = \varnothing$ and $\varnothing' = \xi$

Exercise 1.2(c)

1. Given $\xi = \{1, 2, 3, 4, 5, 6, 7, 8\}$ and A = $\{1, 3, 5, 7\}$, find A'

2. Given $\xi = \{1, 2, 3, 4, 5, 6, 7, 8\}$ and A = $\{1, 3, 6, 8\}$, find (A')'

3. Given $\xi = \{x : x \in Z \text{ and } 0 < x \leq 12\}$ and A = {factors of 12},

   find A'

4. Given $\xi = \{x : 1 \leq x \leq 10\}$ and A = $\{x : x \geq 6\}$, where $x \in$

   {iintegers}, find A'

5. Given $\xi = \{1, 3, 5, 7, 9, 11\}$, A = $\{3, 5, 9, 11\}$ and B = $\{3, 7, 9,$

   $11\}$, find $A' \cap B$ and $A \cap B'$

6. Given $\xi = \{a, b, c, d, e, f\}$, A = $\{a, b, e\}$ and B = $\{b, c\}$, find

   $(A \cup B)'$ and $A' \cap B'$

7. Given $\xi = \{a, b, d, f\}$, P = $\{a, d\}$ and Q = $\{d, f\}$, find $(P \cap Q)'$

   and $P' \cup Q'$

8. Given $\xi = \{1, 2, 3, 5\}$, find ξ' and $\varnothing'$

9. Given $\xi = \{1, 2, 3, 4, 5, 6, 7, 8, 9, 10\}$, A = $\{1, 3, 4, 7, 9. 10\}$ and

   B = $\{2, 4, 5, 7, 9\}$ find $(A' \cup B)'$ and $A \cap B'$

10. Given ξ = {1, 2, 3, 4, 5, 6, 7, 8, 9, 10}, A = {1, 2, 4, 5, 9. 10}

   and B = {3, 4, 6, 7, 8} find $(A \cap B')'$ and $A' \cup B$

11. Given ξ = {1, 2, 3, 4, 5, 6, 7, 8, 9, 10}, A = {1, 3, 4, 6, 7. 10},

   B = {1, 3, 4, 5, 8} and C = {2, 5, 7, 9, 10} find $(A \cap B)' \cup C'$

12. Given ξ = {1, 2, 3, 4, 5, 6}, P = {1, 3, 6}, Q = {2, 3, 5} and

   R = {1, 4, 6}, find $P \cap (Q \cup R)'$

1. 3 Venn Diagrams

Relationship between sets can be express visually by drawing diagrams called Venn diagrams. A Venn diagram usually consists of a rectangle, representing the universal set, and circles within the rectangle, representing subsets of the universal sets.

The Venn diagrams below illustrate some relationships of sets. In Figure 1.1, the set A is within the rectangle, so set A is a subset of the universal set.

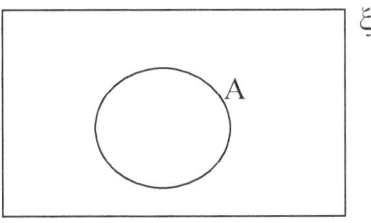

Figure 1.1

The union of set A and set B can be illustrated with a Venn diagram as shown in Figure 1.2. The shaded area contains the elements in the union of the sets.

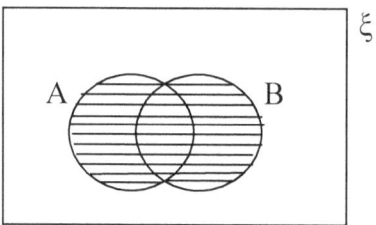

Figure 1.2

The intersection of set A and set B can be illustrated with a Venn diagram as shown in Figure 1.3.   The shaded area contains the elements in the intersection of the sets.

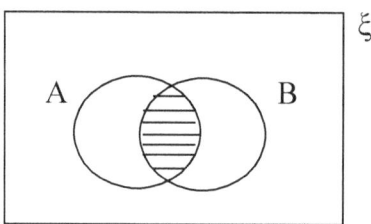

Figure 1.3

Try this 18

Draw a Venn diagram that illustrates the relationship described

(a). $A \subset B$ (b). A' (c). Set A and set B are disjoint

Classification

Two Sets

Two overlapping sets divide the universal set into four regions as shown in Figure 1.4

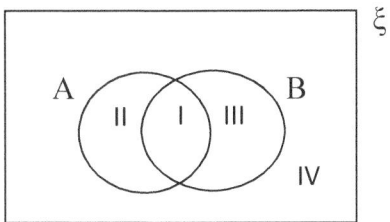

Figure 1.4

The region I contains all the elements which are in both set A and B. The region II, contains all the elements in set A which are not in set B. The region III contains all the elements in set B which are not in set A. The region IV contains all the elements which are not in set A or set B or both.

The various regions are named in terms of set A and B as follows;

I: $A \cap B$ II: $A \cap B'$ III: $A' \cap B$ IV: $A' \cap B'$ or $(A \cup B)'$

Example

For the sets ξ, A and B, draw a Venn diagram and place the elements in the proper regions

ξ = {1, 2, 3, 4, 5, 6, 7, 8}

A = {1, 4, 6, 7}

B = {2, 3, 4, 7, 8}

First determine the intersection of sets A and B. Since 4 and 7 are common to both sets, we place these elements in region I. Now place in region II, the elements in set A that have not been placed in region I. Similarly, place in region III the elements in set B that have not been placed in region I. Finally place those elements in ξ that are not in either sets in region IV.

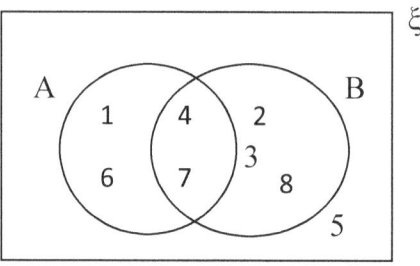

Figure 1.5

Try this 19

For the sets ξ, *A* and *B*, draw a Venn diagram and place the elements in the proper regions

ξ = {3, 6, 9, 12, 15, 18, 21, 24, 27, 30}

A = {6, 12, 18, 27, 30}

B = {3, 12, 15, 18, 24}

Three Sets

Three overlapping sets A, B and C divide the universal set into eight regions as indicated in Figure 1.6.

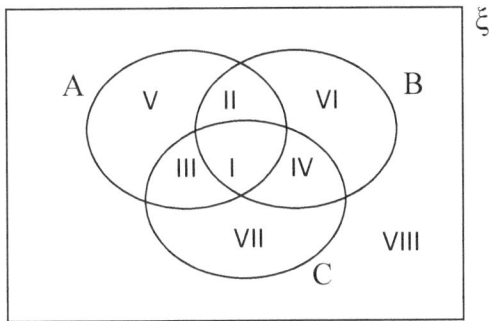

Figure 1.6

The names of the regions in terms of sets A, B and C are as follows:

I: $A \cap B \cap C$ II: $A \cap B \cap C'$ III: $A \cap B' \cap C$

IV: $A' \cap B \cap C$ V: $A \cap B' \cap C'$ VI: $A' \cap B \cap C'$

VII: $A' \cap B' \cap C$ VIII: $A' \cap B' \cap C'$ or $(A \cup B \cup C)'$

Example

For the sets ξ, A, B and C, draw a Venn diagram and place the elements in the proper regions

$\xi = \{1, 2, 3, 4, 5, 6, 7, 8, 9, 10, 11, 12\}$

$A = \{2, 3, 4, 5, 6, 12\}$

$B = \{1, 2, 3, 4, 8\}$

$C = \{1, 2, 3, 5, 6, 10\}$

When placing the elements in the regions, we generally start with region I and work outward. First find the intersection of all three sets. Since the elements 2 and 3 are in all three sets, place 2 and 3 in region I. Next complete region II by determining the intersection of sets A and B. Notice that $A \cap B$, consist of regions I and II. Placed those elements in $A \cap B$, not placed in region I. Similarly, placed those elements in $A \cap C$ and $B \cap C$ not placed in region I, in region III and region IV respectively. Now complete set A by placing in region V elements in set A that have not previously been placed in regions I, II or III. Similarly, place in regions VI and VII, those elements in sets B and C respectively that have not previously been placed. Finally, place the remaining elements in region VIII. The Venn diagram is shown in Figure 1.7

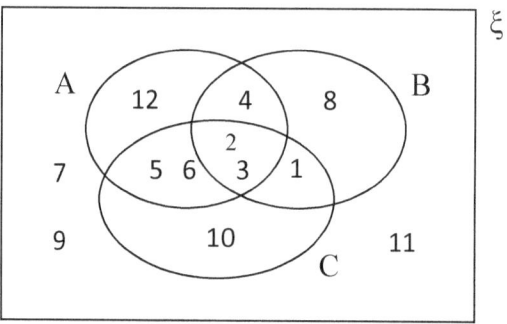

Figure 1.7

Try this 20

For the sets ξ, A, B and C, draw a Venn diagram and place the elements in the proper regions

$\xi = \{1, 2, 3, 4, 5, 6, 7, 8, 9\}$

$A = \{2, 4, 6, 7, 8\}$

$B = \{1, 3, 4, 6, 8\}$

$C = \{1, 2, 4, 5, 8\}$

Exercise 1.3

1. Name the shaded region in each Venn diagram

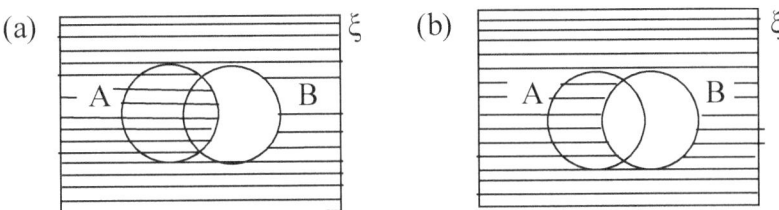

(a)   (b)

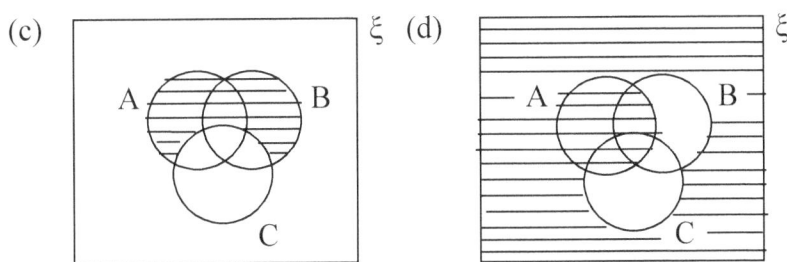

(c)   (d)

2.

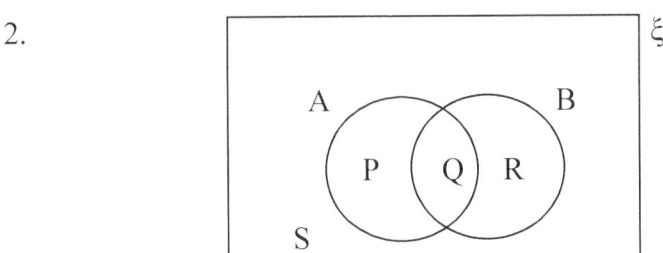

Name the regions (a) R and (b) S in terms of A and B

3. $P = \{a, b, c, d\}$ and $Q = \{a, c, e, f\}$ are subsets of $\xi =$

$\{a, b, c, d, e, f, g\}$

(a) Draw a Venn diagram to illustrate the given information

(b) From your diagram find

(i) $(P \cap Q') \cup (P' \cap Q)$ (ii) $(P \cup Q)'$

4.

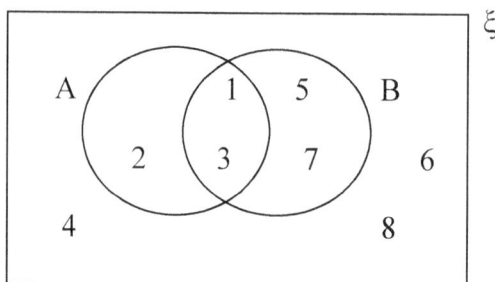

Using the information in the Venn diagram above list the following sets

(a) A (b) A' (c) B (d) B' (e) $A \cap B$ (f) $(A \cap B)'$

(g) $A \cup B$ (h) $(A \cup B)'$ (i) $A' \cup B'$ (j) $A' \cap B'$

Which two pairs of the above sets are equal?

5. Given the sets

$\xi = \{1, 2, 3, 4, 5, 6, 7, 8, 9, 10\}$

$A = \{3, 4, 5, 7, 8, 9\}$

$B = \{1, 3, 4, 7\}$

$C = \{3, 6, 9, 10\}$

(a) Draw a Venn diagram of the three sets and show all the members of each set

(b) Using your diagram find

(i) $(A \cup B) \cap C'$ (ii) $A' \cap B' \cap C'$

6.

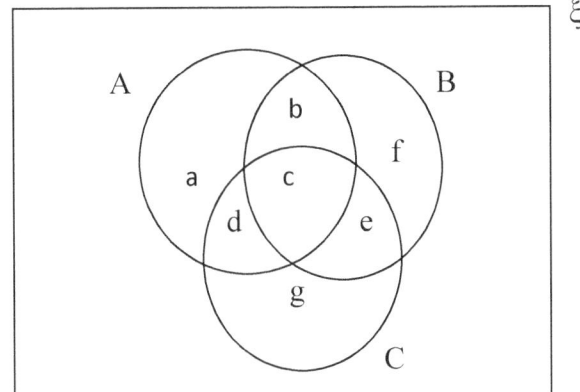

List the elements in the regions that correspond to

(a) $A \cap (B \cup C)$ (b) $(A \cup B) \cap (A \cup C)$

7. The Venn diagram below shows three intersecting sets P, Q and

R

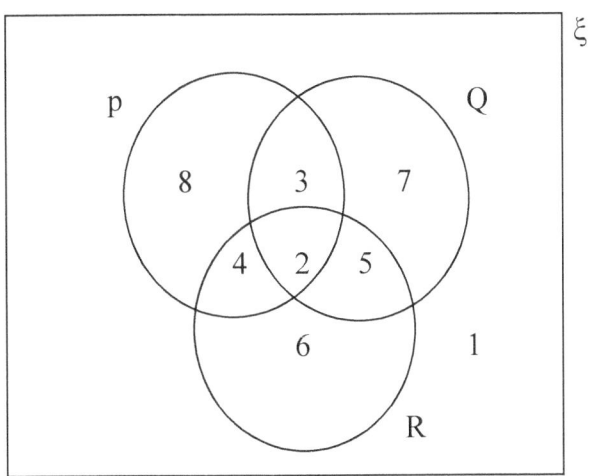

(i) What region represents the set {5, 6, 7}

(ii) Find $P' \cap Q'$

8. P, Q and R are sets such that $P \subset Q$ and $Q \subset R$. Illustrate this information in a Venn diagram

1.4 Using Venn Diagrams to Solve Problems

Venn diagrams can be used to solve problems containing two or three sets of elements.

Two Set Problems

Consider two overlapping sets A and B. If we add the elements in set A and set B, we are counting the elements common to both sets twice. Hence, if two sets have some elements in common, then the number of elements in the union of the sets is

$$n(A \cup B) = n(A) + n(B) - n(A \cap B)$$

This is called the cardinal number formula for the union of sets.

Example

In a class of 18 girls, 10 study Mathematics and 14 study Physics. How many girls study both subjects?

First we illustrate the information on a Venn diagram as shown in Figure 1.8. M represents the set of girls who study mathematics, P those who study Physics, and x the number of girls who study both subjects.

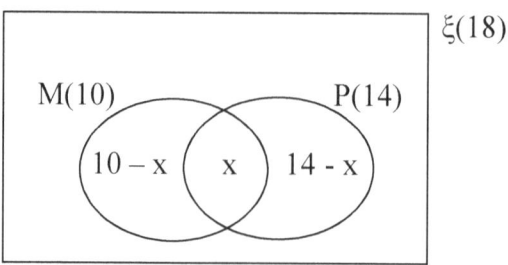

Figure 1.8

You can see that the number of girls in the class is

$$(10 - x) + x + (14 - x) = 18$$

Simplifying the expression on the left we have

$$24 - x = 18$$

Solving the equation gives

$$x = 6$$

Therefore the number of girls in the class who study both subjects is 6

This result can also be obtained, using the cardinal number formula for the union of sets. Substituting the given information in

$$n(M \cup P) = n(M) + n(P) - n(M \cap P)$$

we have

$$18 = 10 + 14 - x$$

$$x = 6$$

Try this 21

In a group of 30 men, 20 like music, 15 like reading and 3 do not like either. How many like both music and reading?

Three Sets Problems

Problems involving three sets can be solved, as illustrated in the examples below

Examples

In a class of 35 students, students may study at least one of the following subjects: Mathematics, Physics and Biology 15 students study Mathematics, 24 study Physics and 20 study Biology . 8 students study Mathematics and Physics, 10 study Physics and Biology, 6 students study Mathematics and Biology but not Physics and 3 students study all three subjects. How many students study only Mathematics or only Physics or only Biology?

Begin by constructing a Venn diagram with three overlapping circles, as shown in Figure 1.9. The circle M represents students who study mathematics, P students who study physics and B students who study Biology.

Always work from the centre of the diagram outwards. First find the intersection of all three sets i.e. $M \cap P \cap B$. Since 3 students study all three subjects, we place 3 in this region. Next we complete the region $M \cap P \cap B'$. Notice that the region $M \cap P$ consists of the region $M \cap P \cap B$ and $M \cap P \cap B'$. We have already placed 3 in region $M \cap P \cap B$, so we place $(8 - 3)$ i.e. 5 in the region $M \cap P \cap B'$. Similarly, we place 7 in $M' \cap P \cap B$. Finally, we place 6 in $M \cap P' \cap B$. The information is illustrated Figure 1.9.

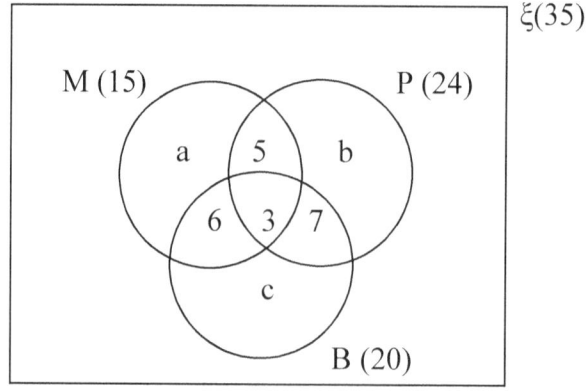

Figure 1.9

We represent the number of students who study only Mathematics, only Physics and only Biology by a, b and c respectively. From the Venn diagram the total number of students who study mathematics is $(a + 6 + 3 + 5)$. Since 15 students study Mathematics we have

$$a + 6 + 3 + 5 = 15$$

Solving this equation gives

$$a = 1$$

Similarly,

$$b + 7 + 3 + 5 = 24$$

$$b = 9$$

and

$$c + 7 + 3 + 6 = 20$$

$$c = 4$$

Hence the number of students who study only Mathematics or only Physics or only Biology is

$$1 + 9 + 4 = 14$$

Try this 22

In a class of students, 18 students offer Mathematics, 22 students offer Chemistry and 20 students offer Biology. 5 students offer all three subjects, 9 students offer Mathematics and Biology, 12 students offer Mathematics and Chemistry and 13 students offer Biology and Chemistry. Each student offers at least one of the subjects. How many students were in the class?

42 students like the following: Music, reading and sport. 27 students like music, 26 students like reading and 24 students like sport. 4 students like music only, 6 students like reading only and 2 students like sport only. 13 students like music and reading, 15 students like music and sport and 12 students like reading and sport. Find the number of students who like all three.

Let x represents the number of students who like all three items. The Venn diagram shown in Figure 1.10 is drawn using the procedure outlined in the preceding example.

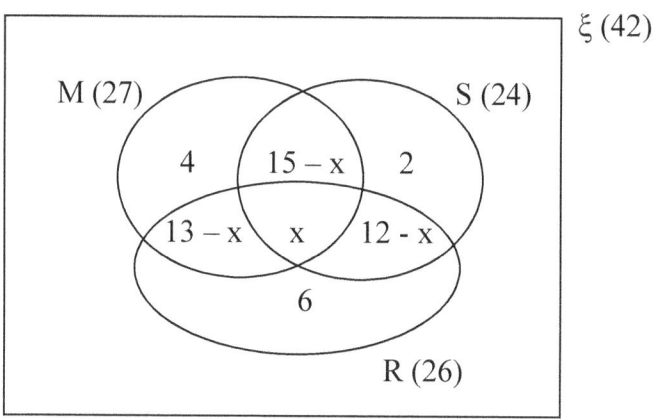

Figure 1.10

From the Venn diagram, we obtain the equation

$$6 + 13 - x + x + 12 - x + 4 + 15 - x + 2 = 42$$

$$52 - 2x = 42$$

$$x = 5$$

Hence, 5 students like music, reading and sport.

Alternatively, since the sum of all entries in set R is 26, we have

$$26 + 4 + 15 - x + 2 = 42$$

$$47 - x = 42$$

$$x = 5$$

45 students were asked to choose at least one of the following three programs: Science, Arts and Vocational. 18 of them chose Science, 21 chose Arts and 24 chose Vocational, 3 chose Science only, 4 chose Arts only and 10 chose Vocational only, 6 chose all three programs. Find the number of students who chose Vocational and Science programs only, and those who chose none of the three programs.

Let x represents the number of students who chose Vocational and Science only. The Venn diagram is shown in Figure 1.11

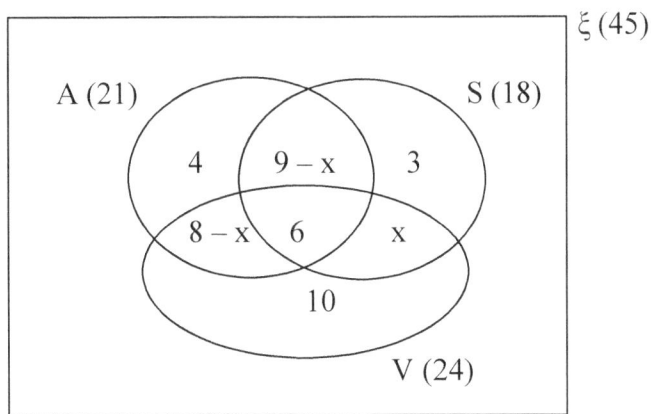

Figure 1.11

The sum of all the entries in set V is 24, so $24 - (10 + 6 + x)$ i.e. $8 - x$ students chose Arts and Vocational programs only. Similarly,

$9 - x$ students chose Arts and Science programs only. Since 21 students chose Arts program we have

$$4 + 8 - x + 6 + 9 - x = 21$$

$$27 - 2x = 21$$

$$x = 3$$

The total number of student who choose at least one program is

$$21 + 10 + 3 + 3 = 37$$

So, $45 - 37$ i.e. 8 chose none of the programs.

Exercise 1.4

1. In a class of 50 students, 40 of them study Chemistry and 30 study Physics. How many students study both Chemistry and Physics?

2. In a class of 20 boys, 16 of the boys play football, 12 play hockey. How many of the boys play both football and hockey?

3. In a class of 35 students, 19 study History and 16 of the students study Geography. 6 of them do not study any of the two subjects. How many of them study both subjects?

4. In a group of 50 men, 30 of them are fluent in English, 17 are fluent in French and 7 are neither fluent in English nor French. How many are fluent in both languages?

5. In an examination, every candidate took English language or French. 80% of the candidates took English language and 35% of the candidates took French. What percentage of the candidates took both subjects? If 600 candidates took the examination how many candidates took only French?

6. 60 students in a class study at least one of the following subjects: Mathematics, Physics and Biology. 28 students study Mathematics, 31 study Physics and 34 study Biology. 11 students study Mathematics and Physics, 14 study Physics and Biology, 8 students study Mathematics and Biology but not Physics and 5 students study all three subjects. How many students study only Mathematics or only Physics or only Biology?

7. A group of students were asked to state the type of music they like listening to; High life, Jazz and Pop. 14 like High life and

Jazz, 13 Highlife and Pop and 12 Jazz and Pop. 5 like all three type of music. 6 like Highlife only, 3 like Jazz only and 2 like Pop only. How many students were in the group?

8. There are 45 women in a market. 19 of them sell yam, 21 sell rice and 24 sell corn. 7 sell yam and rice, 9 sell rice and corn and 8 sell yam and corn. 7 sell only yam, 8 sell only rice and 10 sell only corn. 2 did not sell any of the three items. How many women sell all three items?

9. 40 students took a test in at least one of the following subjects: English, Mathematics and Science. 20 took English, 21 Science and 25 Mathematics. 11 took Science and Mathematics, 13 English and Mathematics and 8 English and Science. Find the number of students who took the test in all the three subjects.

10. A survey conducted at a refugee camp indicated that 50 refuges in the camp did not have at least one of the following documents : a passport, a health certificate and an identity card. 27 did not have a passport, 24 a health certificate and 20 an identity card. 10 did not have a passport only, 4 a health

certificate only, and 3 an identity card only. 6 did not have all three documents but 13 did not have a passport and health certificate. How many did not have exactly two of the documents? How many did not have any of the three documents?

11. A group of students study the following subjects: Mathematics, Physics and Economics. 16 of the students study Physics and 10 study Economics. 10 study Physics only, 4 study Mathematics only and 8 study Economics only. No student study both Physics and Economics. Find the number of students in the group. How many of them study Mathematics?

12. 50 Science students had a choice of the following subjects: Mathematics, Chemistry and Biology. 22 of the students chose Mathematics, 23 chose Chemistry and 20 chose Biology. 6 chose Biology and Chemistry. A student is allowed to choose either Biology or Mathematics but not both. Find the number that chose Mathematics and Chemistry. How many chose only Chemistry?

Review exercise 1

1. List the elements in the set

   (a) $A = \{x : x \text{ is a multiple of } 2 \text{ and } 6 < x \leq 18\}$

   (b) $B = \{x : x \in N \text{ and } 1 < x \leq 8\}$

   (c) $P = \{x : x \in Z \text{ and } -2 < x < 7\}$

   (d) $Q = \{x : x \in N \text{ and } x \text{ is a prime number less than } 10\}$

2. State whether the pair of sets are equal or equivalent

   (a) $\{x : x \in N \text{ and } x < 6\}$ $\{x : x \in Z \text{ and } 0 < x < 6\}$

   (b) $\{\text{prime factors of } 20\}$ $\{3, 4\}$

   (c) $\{x : x \in Z \text{ and } -1 < x < 5\}$ $\{0, 1, 3, 4, 5\}$

   (d) $\{x : x \in N \text{ and } 2 < x \leq 7\}$ $\{x : x \in N \text{ and } 3 \leq x < 8\}$

3. Determine whether each set is finite or infinite

   (a) $\{x : x \in N \text{ and } 0 < x \leq 50\}$ (b) $\{x : x \in R \text{ and } 2 < x < 3\}$

   (c) $\{\text{multiples of } 3\}$ (d) $\{x : x \in N \text{ and } 3x - 6 = 0\}$

4. Classify each statement as true or false

   (a) $\{\} \in \{1, 2, 3, 4\}$ (b) $\{\} \subset \{a, b, c\}$ (c) $4 \subset \{3, 4, 5\}$

   (d) $5 \notin \{3, 4, 6\}$ (e) $\{2, 3, 5\} \subset \{4, 7, 6\}$ (f) $\{1, 7, 8\} \supset \{1\}$

   (g) $\emptyset = \{\}$ (h) $\{0\} = \emptyset$ (i) $\{3, 5, 9\} \notin \{3, 9, 5\}$

(j) $\{6, 7, 10\} \subset \{7\}$

5. List all subsets of the given set. Identify which subsets are proper

and which are improper

(a) $B = \{a, b\}$ (b) $N = \{0\}$ (c) $S = \{1, 2, 3\}$

(d) $M = \{a, b, c, d\}$

6. State the number of subsets of the given set

(a) $A = \{a, b, c, d, e\}$ (b) $N = \{-3, -2, -1, 0, 1, 2, 3, 5\}$

7. Given $\xi = \{1, 2, 3, 4, 5, 6, 7, 8. 9\}$ $A = \{3, 4, 6, 8\}$ and

$B = \{3, 5, 7, 8\}$, find

(a) $A \cap B$ (b) $A \cup B$ (c) A' (d) B'

8. Given $\xi = \{1, 2, 3, 4, 5, 6, 7, 8, 9, 10\}$ $A = \{2, 3, 4, 5, 6\}$ and

$B = \{5, 6, 7, 8, 10\}$, find

(a) $A \cap B$ (b) $A \cup B$ (c) A' (d) B'

9. Given $\xi = \{1, 2, 3, 4, 5, 6, 7\}$, $A = \{1, 2, 3, 4, 5\}$ and $B = \{5, 6, 7\}$,

find

(a) $A \cap B$ (b) $A \cup B$ (c) B' (d) A'

(e) $A' \cup B$ (f) $A \cap B'$

10. For a set A, find each of the following

(a) $A \cup \emptyset$ (b) $A \cup A$ (c) $A \cap A$ (d) $A \cap \emptyset$

11. Fill in the blank

    (a) If $A \subset B$, then $A \cap B = $ ____

    (b) If $A \subset B$, then $A \cup B = $ ____

    (c) If $A \subset B$ and $B \subset A$, then $A = $ ____

12. Draw a Venn diagram with two overlapping sets A and B, and

    shade the region corresponding to the indicated set

    (a) $A \cap B$ (b) $A \cup B$ (c) A' (d) B'

    (e) $A' \cup B$ (f) $A \cup B'$ (g) $A' \cap B$ (h) $A \cap B'$

    (i) $A' \cup B'$ (j) $A' \cap B'$

13. Draw a Venn diagram with three overlapping sets A, B and C,

    shade the region corresponding to the indicated set

    (a) $A \cap B \cap C$ (b) $A \cup B \cup C$ (c) $(A \cup B)' \cap C$

    (d) $A \cap (B \cup C)$ (e) $B \cap (A \cup C')$ (f) $(A' \cup B) \cap C'$

14. Suppose $n(\xi) = 150$, $n(A) = 57$ and $n(B) = 73$. If

    $n(A \cap B) = 25$, find $n(A \cup B)$, and draw a Venn diagram

    illustrating the composition of ξ.

15. Suppose $n(\xi) = 120$, $n(A) = 36$ and $n(B) = 85$. If

$n(A \cup B) = 100$, find $n(A \cap B)$ and draw a Venn diagram

illustrating the composition of ξ.

16. Suppose $n(\xi) = 130, n(A) = 54$, $n(B) = 63$, and $n(C) = 58$.

If $n(A \cap B \cap C) = 12$, $n(A \cap B) = 27$, $n(A \cap B' \cap C') = 10$

and $n(A' \cap B' \cap C) = 13$,

(a) find $n(A \cap B' \cap C)$, $n(A' \cap B \cap C)$, $n(A' \cap B \cap C')$

(b) draw a Venn diagram illustrating the composition of ξ.

17. The result of a survey of a group of tourists who visited the

Central region in 2012, showed that 27 visited the Cape Coast

castle, 38 visited Elimina castle, and 16 visited both the Cape

Coast castle and Elimina castle. How many people visited

either the Cape coast castle or Elimina castle?

18. A survey of a group of people at a party indicated that 18 speak

Ga and 23 speak Twi. If 6 speak both Ga and Twi and 10 speak

neither, how many people were in the group?

19. A factory manufactured 170 cars in June. The factory manufactured 116 cars with automatic transmissions and 100 with power steering. How many cars were manufactured with both of these options?

20. A survey of 155 students who graduated from a university indicated that 85 took a course in mathematics, 96 in economics and 34 did not take a course in either. How many took a course in mathematics and economics?

21. Of 300 students who uses personal computers: 112 use IBM; 131 use Dell; 153 use Compaq; 35 use both IBM and Dell; 50 use both IBM and Compaq; 55 use Dell and Compaq; 15 use all three; and 29 use another computer brand. How many students use only IBM or only Dell or only Compaq?

22. A survey of 167 males who voted at a polling station indicated that 78 voted for the parliamentary candidate and 100 for the presidential candidate of the National Party. 25 voted for another party. How many voted for both the parliamentary and presidential candidates of the National Party?

Chapter Test 1

Take this test as you would take a test in class. After you are done, check your work against the answers in the back of the book.

1. List the elements in the set

   (a) $A = \{x : x \in Z \text{ and } -3 < x \leq 10\}$

   (b) $B = \{x : x \text{ is a factor of } 50\}$

   (c) $C = \{x : x \text{ is a prime factor of } 36\}$

2. State whether the pair of sets are equal or equivalent

   (a) $\{x : x + 2 = 0\}$ $\{-2\}$

   (b) $\{x : x \text{ is a factor of } 6\}$ $\{a, b, c, d\}$

   (c) $\{even\ prime\}$ $\{2\}$

   (d) $\{x : x \in N \text{ and } x < 5\}$ $\{3, 4, 7, 8\}$

3. Determine whether each set is finite or infinite

   (a) $\{2, 3, 5, 7, \cdots\}$

   (b) The set of multiples of 5 between 0 and 50

   (c) The set of real numbers between 1 and 2

4. Determine whether $A \subset B, B \subset A$ or $A = B$

   (a) $A = \{a, c, d, e\}$ $B = \{c, e\}$

(b) $A = \{2\}$ $B = \{prime\ numbers\}$

(c) $A = \{x: x \in N\ and\ x < 6\}$ $B = \{x: x \in N\ and\ 1 \leq x \leq 5\}$

5. (a) List all the subsets of $\{0, 1, 3\}$

(b) How many subsets has the set $\{a, b, c, d, e, f, g\}$

6. Given A = $\{1, 2, 4, 5, 6, 8\}$ and B = $\{2, 3, 4, 7, 8\}$, find

(a) $A \cup B$ (b) $A \cap B$

7. Given   A = $\{3, 4, 6, 7, 9, 10\}$,   B = $\{1, 3, 5, 9\}$ and

C = $\{1, 2, 4, 7, 10\}$, find

(a) $A \cap (B \cup C)$ (b) $A \cup (B \cap C)$

8. Given ξ = $\{1, 2, 3, 4, 5, 6, 7, 8, 9, 10\}$, A = $\{2, 4, 5, 9\}$ and

B = $\{1, 3, 4, 6, 8, 10\}$, find

(a) A' (b) B' (c) $A \cap B'$ (d) $(A' \cap B)'$

9. Given ξ = $\{x: x \in N\ and\ 1 \leq x \leq 12\}$, A = $\{multiples\ of\ 2\}$

and B = $\{factors\ of\ 12\}$, find

(a) $A' \cup B'$ (b) $(A \cup B)'$ (c) $A \cap (B \cup A)'$

10. Given ξ = $\{1, 2, 3, 4, 5, 6, 7\}$,   A = $\{1, 2, 3, 4\}$ and

B = $\{1, 3, 5, 6\}$,   draw a Venn diagram and place the

elements in the proper regions

11. Draw a Venn diagram illustrating the following sets.

ξ = {1, 2, 3, 4, 5, 6, 7, 8, 9, 10, 11, 12, 13, 14, 15},

A = {1, 2, 3, 4, 7, 9, 11},

B = {2, 3, 4, 5, 10, 12, 14} and

C = {1, 2, 4, 5, 8, 9}

12. In a group of 56 students, 16 own an iPad, 35 own a laptop and 7 own both iPad and laptop. How many students do not own either an iPad or a laptop?

13. In a science class of a school, 38 study mathematics, 47 study biology and 15 study both mathematics and biology. How many students study mathematics or biology?

14. A survey of 190 tourists who visited the country in June 2010 indicated that 108 visited the country in 2006, 95 visited in 2008 and 32 were visiting for the first time. How many tourists visited the country in both 2006 and 2008?

15. In a survey of 85 senior high school students, it was found that of the three weekly newspapers, Graphic, Guide and Times 37 read Graphic

34 read Guide

42 read Times

12 read Graphic alone

13 read Guide alone

18 read Times alone

10 read all three

(a) How many read Guide and Times alone

(b) How many read none of these three newspapers?

2

Logic

In this Chapter, we will study the principles and techniques use in distinguishing valid arguments from those that are not valid. One way to do this is to use Venn diagrams.

2.1 Illustrating Statements with Venn Diagrams

We can illustrate statements with Venn diagrams. The examples below illustrate the use of Venn diagrams to depict statements of the general form 'All A are B' (or 'If A, then B'), 'No A are B', 'Some 'A are B' and 'Not all A are B'

Example

Illustrate the statement below in a Venn diagram

'All my friends live in Accra'

Let ξ = {people}, F = {friends} and A = {people living in Accra}. All the members of set F are in set A, so F is a subset of set A, i.e. $F \subset A$. This relationship is shown in Figure 2.1. The inner circle represents F, and the outer circle represents A.

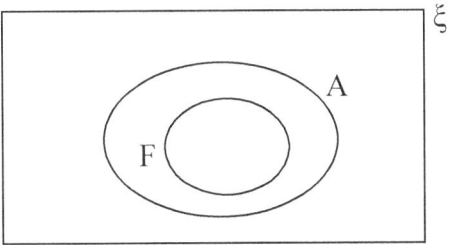

Figure 2.1

Try this 1

Illustrate the statement 'All boys play football' in a Venn diagram

Example

Illustrate the statement below in a Venn diagram

'Some policemen wear uniform'

Let ξ = {people}, P = {policemen} and W = {people who wear uniform}. There is at least one member of set P in set W. The Venn diagram illustrating the statement is shown in Figure 2.2. The shaded area represents policemen who wear uniform.

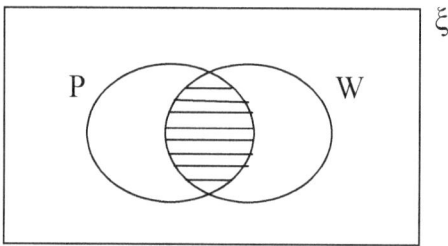

Figure 2.2

Try this 2

Illustrate the statement 'Some students are lazy' in a Venn diagram

Example

Illustrate the statement below in a Venn diagram

'No students are lazy'

Let ξ = {people}, S = {students} and L = {people who are lazy}. The set S and set L have no members in common, so the two sets are disjoint. The statement is illustrated as shown in Figure 2.3.

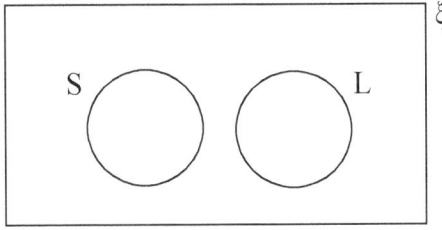

Figure 2.3

Try this 3

Illustrate the statement 'No animals are men' in a Venn diagram

Example

Illustrate the statement below in a Venn diagram

'Not all football players are strong'

Let ξ = {people}, F = {people who play football} and S = {people who are strong}. There is at least one member that is in set F that is not in set S. The statement is illustrated as shown in Figure 2.4. The shaded area represents football players who are not strong.

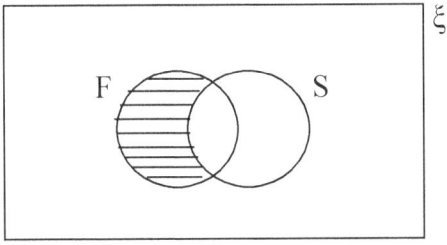

Figure 2.4

Try this 4

Illustrate the statement 'Some student are not clever' in a Venn diagram.

Exercise 2.1

Illustrate the following statements in Venn diagrams

1. Ama passed her exams

2. All my friends go to church on Sunday

3. None of the boys fail the test

4. Most girls like reading

5. Not all friends are loyal

6. Some students are intelligent

7. No boy is tall

 8. Some men are not rich

9. Some adults play football

10. All football players are not poor

2.2 Valid Arguments

One presents an argument when he makes a sequence of statements and then draws some conclusion from them. An argument consists of two components: the initial statements or premises, and the final statement or conclusion. An argument is said to be valid if its conclusion follows from the given set of premises. However, if the conclusion of an argument does not follow from the set of premises, the argument is said to be not valid. The validity of an argument can be shown by use of a Venn diagram as illustrated by the examples below.

Example

Draw a Venn diagram to verify the validity of the following arguments

Statement 1: All boys are intelligent

Statement 2: Efua is a boy

Conclusion: Efua is intelligent

Let ξ = {people}, B = {boys} and I = {people who are intelligent}. The statement 1 is of the form 'All B are I', and is represented by the Venn diagram shown in Figure 2.5.

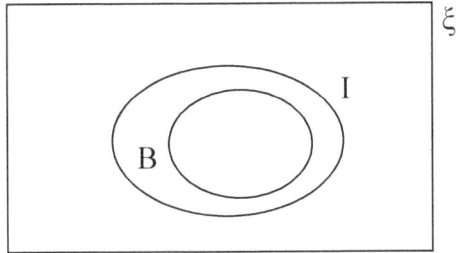

Figure 2.5

You can see from statement 2 that Efua is a member of set B. If we let x = Efua, the statement 'Efua is a boy' can then be represented by placing x within the circle labelled B, as shown in Figure 2.6.

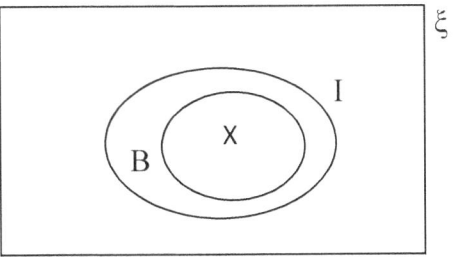

Figure 2.6

Since the set B is also in set I, then the element x is also in set I. Hence, the argument is valid. The argument is a valid argument even though the conclusion 'Efua is a boy' is obviously a false statement. Saying that an argument is valid merely means that the conclusion obtained logically follow from the given statements.

Try this 5

Draw a Venn diagram to verify the validity of the following argument.

Statement 1: All science students are clever

Statement 2: Ama is a science student

Conclusion: Ama is clever

Example

Draw a Venn diagram to determine the validity of the following argument

Statement 1: All workers are healthy

Statement 2: Kojo is healthy

Conclusion: Kojo is a worker

Let   ξ = {people}, W = {workers} and H = {people who are healthy}. The statement 1 is of the form 'All W are H'. The Venn diagram is shown in Figure 2.7. If we let x represent Kojo, statement 2 simply requires that we place x somewhere within the H circle; x could be placed in either of the two location shown in Figure 2.7 and Figure 2.8.

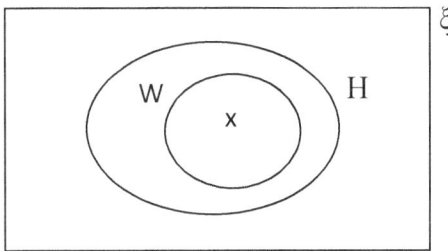

Figure 2.7

If x is placed as in Figure 2.7, the argument would appear to be valid, the Venn diagram supports the conclusion 'Kojo is a worker'. However, if x is placed as shown in Figure 2.8 the argument would not be valid. If there is at least one instance in which the conclusion does not follow from the given premises, then the argument is not valid.

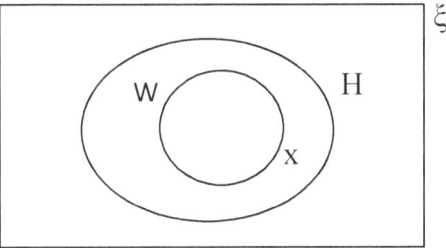

Figure 2.8

Saying that an argument is not valid does not mean that the conclusion is false. An argument that is not valid can have a true conclusion.

Try this 6

Draw a Venn diagram to determine the validity of the following argument

Statement 1: Students who studied hard pass their exams

Statement 2: Afua passed her exams

Conclusion: Afua studied hard

Exercise 2.2

In Exercises 1 – 8, use Venn diagrams to determine whether the following arguments are valid or not valid

1. All living things walk

   An orange tree is a living thing

   Therefore, an orange tree walks

2. All birds sing

   Mercy is a bird

   Therefore, Mercy sings

3. All doctors are healthy

   Ama is healthy

   Therefore, Ama is a doctor

4. All girls are bullies

   Kwesi is a girl

   Therefore, Kwesi is a bully

5. Some knowledgeable adults are teachers

   Mensah is a teacher

   Therefore, Mensah is knowledgeable

6. All students are intelligent

   Kwesi is not a student

   Therefore, Kwesi is not intelligent

7. Men do not eat grass

Afua is a man

Therefore, Afua does not eat grass

8. All boxers are men

Some women are boxers

Therefore, some women are men

In Exercises 9 – 12, draw a valid conclusion from the two statements

9. All teachers wear glasses

Sam is a teacher

10. All bicycle riders wear helmets

Kwesi is a bicycle rider

11. No public servant owns a gun

All policemen own guns

12. All students can read

Some adults are student

Review exercise 2

1. Illustrate the following statements in Venn diagrams

(a) All cats can fly

(b) Some dogs are brown

(c) All ants work hard

(d) No dogs read a book

(e) Most girls eat ice cream

(f) Not all students passed the mathematics test

In Exercises 2 – 8, draw a Venn diagram to verify whether the argument is valid or not valid

2. All priests are honest

Offei is honest

Therefore, Offei is a priest

3. All kind women are mothers

   Adisa is kind

   Therefore, Adisa is a mother

4. Adwoa is my friend

   All my friends are in the science class

   Therefore Adwoa is in the science class

5. All mathematicians are intelligent

   Kwame is not a mathematician

   Therefore, Kwame is not intelligent

6. All bullies are boys

   There is no bully who is not strong

   Mensah is not strong

   Therefore, Mensah is not a bully

7. All students work hard

   Some hardworking students pass their exams

   Esi is a hardworking student

   Therefore, Esi passes her exams

8. All mathematics students are clever

   Most mathematics students are in the science class

   Kofi is in the science class

   Therefore, Kofi is a mathematics student

In Exercises 9 – 12, draw a valid conclusion from the premises

9. All men are intelligent

   Esi is a man

10. Clever men do not steal

    Adu is a clever man

11. All students are intelligent

    All intelligent people are rich

12. No students wear uniform

    All policemen wear uniform

Chapter Test 2

Take this test as you would take a test in class. After you are done, check your work against the answers in the back of the book.

In questions 1 – 6, illustrate the statement in a Venn diagram

1. All cars have four doors

2. None of the boys like jazz music

3. Some students read every day

4. Not all the books in the library are new

5. Some of the students in the class are not clever

6. No dogs can fly

In questions 7 – 10, draw a Venn diagram to verify whether the argument is valid or not valid

7. All cars have engines

   The horse is a car

   Therefore, the horse has an engine

8. All my friends will attend the concert

   Kofi attended the concert

   Therefore, Kofi is my friend

9. All students love mathematics

   None of the workers love mathematics

   Efua loves mathematics

   Therefore, Efua is not a worker

10. All army officers wear uniform

    Some students wear uniform

    Kwame wears uniform

    Therefore, Kwame is a student

In questions 11 – 12, draw a valid conclusion from the premises

11. All homeless people are poor

     Mr Mensah is not poor

12. All good volleyball players are healthy

    Some doctors are good volleyball players

3

The Number System

3.1 The Real Number System

We carry out our mathematical calculations using a number system. The number system had evolved as a result of a process of successive expansion of the original system of natural numbers. Natural numbers came into existence when man first learnt counting. The set of natural numbers denoted by N is

$$\{1, 2, 3, 4, 5 \cdots\}$$

The three dots indicate that the pattern continues.

If we add two natural numbers, say 3 and 5, we obtain 8 which is also a natural number. However, subtracting 8 from 3 is not possible within the system of natural numbers. In order to do subtraction without restrictions the set of natural numbers was expanded to include the negative integers and the number zero, 0. The expanded set is called the set of integers, denoted by Z.

$$\{ \underbrace{\cdots -5, -4, -3, -2, -1}_{Negative\ integers}, 0, \underbrace{1, 2, 3, 4, 5 \cdots}_{Positive\ integers} \}$$

The subset $\{0, 1, 2, 3, \cdots\}$ of the set of integers is called the set of whole numbers, denoted by W.

If we divide 16 by 2 we get 8. 8 is an integer. However, dividing 8 by 16 is not possible within the system of integers. To make division always possible, except division by zero, the set of integers was expanded to include positive and negative fractions. The expanded set is called the set of rational numbers, denoted by Q.

Any rational number can be written as the ratio $\frac{p}{q}$ of two integers, where $q \neq 0$. Here are some examples of rational numbers

$$-\frac{3}{5}, \quad 2 = \frac{2}{1}, \quad \frac{1}{8} \quad and \quad \frac{5}{3}$$

The set of rational numbers is sufficient for performing the four basic operations of arithmetic. However, in order to be able to extract roots of all numbers, the system of rational numbers was expanded to include the set of irrational numbers. Square roots of numbers that are not perfect squares are called irrational numbers. The roots of these numbers are usually non-terminating and non-repeating decimals. Numbers like $\sqrt{2}$, π and $2.71828182\cdots$ are examples of irrational numbers.

The rational numbers and the irrational numbers together make up the set of real numbers. The set of real numbers is denoted by R. Note that all numbers are real numbers but not all real numbers are either rational numbers or integers. Figure 3.1 shows the relationships among various kinds of numbers.

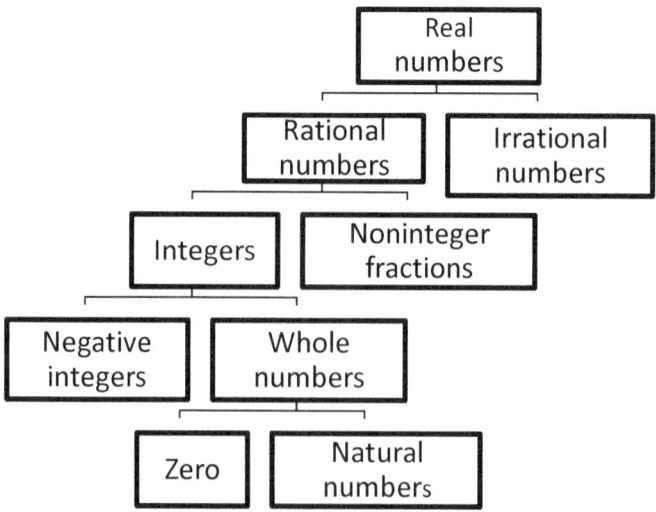

Figure 3.1

Example

Which of the numbers in the set $\{-9, -\sqrt{2}, -1, -\frac{1}{3}, 0, \frac{3}{4}, \sqrt{3}, \pi, 7\}$
are: (a) natural numbers (b) Integers, (c) rational numbers, and
(d) irrational numbers?

Solution

    (a) Natural numbers; $\{7\}$
    (b) Integers; $\{-9, -1, 0, 7\}$
    (c) Rational numbers; $\{-9, -1, -\frac{1}{3}, 0, \frac{3}{4}, 7\}$
    (d) Irrational numbers; $\{-\sqrt{2}, \sqrt{3}, \pi\}$

Try this 1

Which of the numbers in the set
$\{-\frac{9}{2}, -3, -2, -\sqrt{3}, -\frac{\pi}{2}, 0, \sqrt{7}, 5, \frac{17}{2}, 12\}$ are:

(a) natural numbers,   (b) integers,   (c) rational numbers, and
(d) irrational numbers?

The Real Number Line

Real numbers are represented geometrically by the number line.
The number line consists of a horizontal line with a point labelled
as 0, called the origin, and a convenient distance chosen as a unit
of length. Numbers to the left of the origin are negative and
numbers to the right of the origin are positive, as shown in Figure
3.2

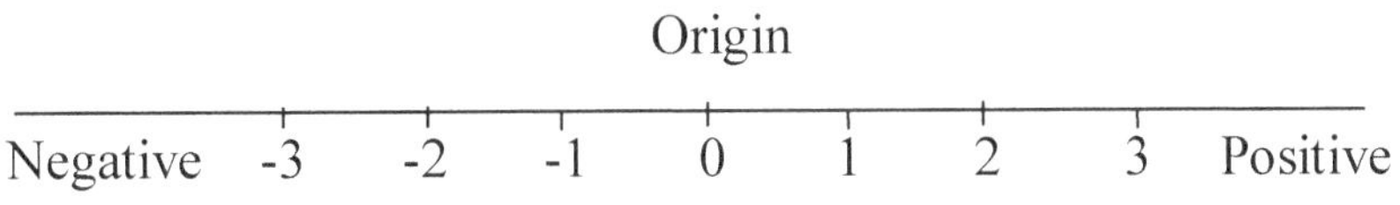

Origin

Negative   -3   -2   -1   0   1   2   3   Positive

Figure 3.2

Every real number is associated with exactly one point on the number line. The point that corresponds to a real number is called the graph of the number.

Representing Real Numbers on the Number Line

Recall that each point on the real number line represents a real number. Figure 3.3 shows the graphs of $-\frac{3}{2}$ and 2.

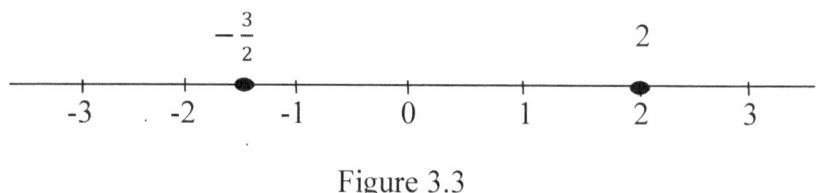

Figure 3.3

Try this 2

Represent the following real numbers on the real number line

(a) $-\frac{5}{2}$ (b) 3.7   (c) $\frac{8}{3}$ (d) 0.53

Ordering

You may have noticed that a number line has its numbers written in order. Smaller numbers are to the left, and lager numbers are to the right. For any two numbers, the one to the left is less than the one to the right.

To compare, say -6 and -1, you can draw a number line like the one shown in Figure 3.4.

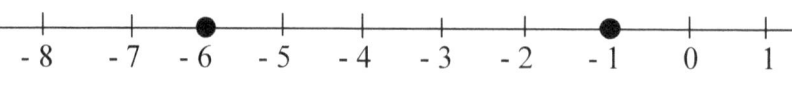

Figure 3.4

There are two ways of comparing these numbers. Notice that, -6 is to the left of -1. This means that -6 is less than -1. In symbol, we write $-6 < -1$. Now, -1 is to the right of -6. This means that -1 is greater than -6, written in symbol as $-1 > -6$.

Example

Place the correct symbol ($<$ or $>$) between each pair of numbers

(a) -4 -1 (b) -2 -7

(c) $\frac{2}{3}$ $\frac{1}{2}$ (d) $-\frac{1}{4}$ $-\frac{1}{2}$

(a) -4 lies to the left of -1 on the number line, so $-4 < -1$

(b) -2 lies to the right of -7 on the number line, so $-2 > -7$

(c) $\frac{2}{3}$ lies to the right of $\frac{1}{2}$ on the number line, so $\frac{2}{3} > \frac{1}{2}$

(d) $-\frac{1}{4}$ lies to the right of $-\frac{1}{2}$ on the number line, so $-\frac{1}{4} > -\frac{1}{2}$

Try this 3

Place the correct symbol ($<$ or $>$) between each pair of numbers

(a) 5 -3 (b) -10 -1

(c) $-\frac{1}{3}$ $-\frac{1}{2}$ (d) $\frac{3}{2}$ $\frac{5}{3}$

Exercise 3.1(a)

In Exercises $1 - 4$, which of the real numbers in the set are: (a) natural numbers, (b) integers, (c) rational numbers, and (d) irrational numbers?

1. $\{-12, -\sqrt{7}, -\frac{2}{3}, -\frac{1}{4}, 0, \frac{3}{8}, 1, \sqrt{3}, 4\pi, 8\}$

2. $\{-\frac{9}{2}, -\sqrt{6}, -\frac{\pi}{2}, -\frac{3}{8}, 0, \sqrt{13}, \frac{10}{3}, 8, 12\}$

3. $\{-3.6, -\sqrt{4}, -\frac{1}{2}, -0.\dot{3}, -0.3, \sqrt{5}, 3\pi, 26.4\}$

4. $\{-\sqrt{25}, -\sqrt{6}, -0.11, -\frac{5}{2}, 0, 0.75, 3, 20\}$

In Exercises 5 and 6, plot the real numbers on the real number line

5. (a) 5   (b) $\frac{3}{2}$ (c) $-\frac{7}{2}$ (d) - 4.2

6. (a) 7   (b) $\frac{4}{3}$ (c) - 7.65   (d) $-\frac{9}{2}$

In Exercises 7 – 9, place the correct symbol (< or >) between the pair of numbers

7. (a) 5   9   (b) 8   3   (c) -7 5   (d) 3   - 5

8. (a) - 8   - 2   (b) -1 - 12   (c) 0   - 3   (d) -32 - 7

9. (a) $\frac{3}{4}$ $\frac{1}{3}$ (b) 0 $-\frac{3}{4}$ (c) $-\frac{5}{2}$ $-\frac{3}{2}$ (d) $-\frac{2}{3}$ $-\frac{10}{3}$

Absolute Value

In signed numbers we are interested in two things, magnitude and direction. The signed numbers $+3$ and -3 are the same unit from 0 but opposite in direction as shown in Figure 3.5

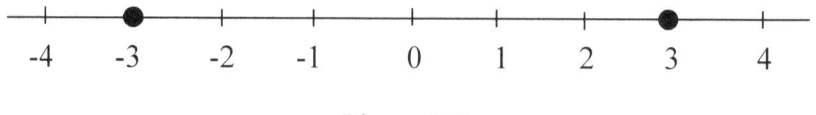

-4   -3   -2   -1   0   1   2   3   4

Figure 3.5

Notice that the opposite of -3 is 3, i.e. $-(-3) = 3$. Opposite numbers are also referred to as additive inverses because their sum is zero. For instance, $5 + (-5) = -5 + 5 = 0$. The number of units that a number is from zero on a number line is called the absolute value of the number. Because opposite numbers lie the same distance from 0 on the real number line, they have the same absolute value. Absolute value is denoted by double vertical bars $||$. So $|5| = 5$ and $|-5| = 5$.

Example

Find the absolute value of (a) $|-10|$ and (b) $|8|$

(a) $|-10| = 10$ Since -10 is 10 units from 0

(b) $|8| = 8$ Since 8 is 8 units from 0

Try this 4

Find the absolute value of (a) $|-31|$ and (b) $|6|$

Exercise 3.1(b)

In Exercises 1 – 10, find each absolute value

1. $|12|$ 2. $|85|$ 3. $|-12|$ 4. $|-15|$ 5. $-|-17|$

6. $|-18|$ 7. $-|16|$ 8. $-|-25|$ 9. $-|20|$ 10. $-|-31|$

In Exercises 11 – 16, place the correct symbol $<$, $>$ or $=$ between the pair of numbers.

11. $|-7|$ $|-2|$ 12. $|4|$ $|-9|$ 13. $|-8|$ $|8|$

14. $|25|$ $|-25|$ 15. $|-4|$ $|6|$ 16. $|-28|$ $|-17|$

Approximation

Rational numbers can be represented as either terminating or repeating decimals. For instance, the decimal representation of $\frac{1}{8} = 0.125$ is a terminating decimal, and the decimal representation of $\frac{7}{11} = 0.636363 \cdots = 0.\dot{6}\dot{3}$ is a repeating decimal. The dots over 63 indicate that the digits repeat.

The decimal representation of an irrational number neither terminates nor repeats. In calculations, involving non terminating decimals, the values found for the numbers are only approximations. For example, $\frac{2}{3}$ is approximating 0.667 to three decimal places, and we write $\frac{2}{3} \approx 0.667$.

Rounding to a number of decimal places

In any decimal representation of numbers, the accuracy of the numbers increases as the number of digits to the right of the decimal point increases. When a high degree of accuracy is not needed, the number can be rounded to a number of decimal places.

To round off, first identify the number in the place you are rounding to and the number that follows it. If the number to the right is 5 or more round up, and round down if the number to right is 4 or less. For example, to one decimal place 8.357 would round up to 8.4. Similarly, to two decimal places 2.4738 would round down to 2.47.

Example

Round the numbers to the number of decimal place(s) indicated in the bracket

(a) 12.963 (1)   (b) 73.483 (2)

(c) 6.5465 (3)   (d) 0.00573 (3)

(a) The first digit from the decimal point is 9. The number next to

    the 9 is 6, so 12.963 is written as 13.0 to 1 d.p.

(b) The second digit from the decimal point is 8. The number next

    to 8 is 3, so 73.483 is written as 73.48 to 2 d.p.

(c) The third digit from the decimal point is 6. The number next to

    6 is 5, so 6.5465 is written as 6.547 to 3 d.p.

(d) The third digit from the decimal point is 5. The number next to

    5 is 7, so 0.00573 is written as 0.006 to 3 d.p.

Try this 5

Round the numbers to the number of decimal place(s) indicated in the bracket

(a) 15.736 (1)   (b) 83.975 (2)   (c) 125.7896 (3)

Rounding to a number of significant figures

Suppose the length of a rod is stated as 7.06 m. This may be written in different ways, using other units. For example, 7060 mm = 706 cm = 7.06 m = 0.00706 km. Notice that the group of numbers 706 appears in each of the different ways of stating the measurement. The position of the decimal point and the number of zeros after or before 706 do not matter as far as the accuracy of the measurement is concerned, for they occur only as a result of changing the units in which the measurement is given. The digits 706 are called the significant figures.

Notice that the last zero of 7060 and the two zeros on the right of decimal point of 0.00706 are just placeholders, and are therefore not significant. However, the zero between the 7 and 6 is significant. In general, all non-zero digits are significant and all zeros between significant digits are significant.

To round to a number of significant figures, count digits from left to right and then round up if the next number is 5 or more, and round down if the next number is 4 or less.

Example

Round the numbers to the number of significant figure(s) indicated in the bracket

(a) 15.34 (3)   (b) 73.457 (4)

(c) 0.003954 (2)   (d) 57638 (2)

(a) Each digit is significant. The third digit is 3, and the next

   number is 4, so 15.34 is written as 15.3 to 3 s.f.

(b) Each digit is significant. The fourth digit is 5, and the next

   number is 7, so 73.457 is written as 73.46 to 4 s.f.

(c) The zeros at the beginning are not significant   The first

   significant figure is the first non zero digit from the   decimal

   point. The first significant number is 3 and the second is 9.

   The next number is 5 so 0.00395 written as 0.0040 to 2 s.f.

(d) Each digit is significant. The second digit is 7, and the next

   digit is 6, so 57638 is written as 58000 to 2 s.f. Notice that the three digits to the right are replaced with zeros.

Try this 6

Round the numbers to the number of significant figure(s) indicated in the bracket

(a) 32.547 (3)   (b) 641.96 (4)

(c) 0.000718 (2)   (d) 0.403 (2)

Exercise 3.1(c)

In Exercises, 1 – 4, round off the numbers to the number of decimal place(s) indicated in the brackets.

1. (a) 3.647 (1)   (b) 7.464 (1)   (c) 0.984 (1)

2. (a) 15.046 (1)   (b) 0.784 (2)   (c) 7.096 (2)

3. (a) 24.6953 (2)   (b) 0.80749 (3)   (c) 2.4865 (3)

4. (a) 12.30968 (3)   (b) 5.17539 (3)   (c) 6.41255 (3)

In Exercises, 5 – 9, round off the numbers to the number of significant figure(s) indicated in the brackets.

5. (a) 1.863 (2)   (b) 27.485 (2)   (c) 15.65 (3)

6. (a) 24.97 (3)   (b) 456.736 (4)   (c) 738 (1)

7. (a) 2714 (1)   (b) 0.0704 (2)   (c) 0.1953 (2)

8. (a) 0.5067 (3)   (b) 40.72 (3)   (c) 0.00897 (2)

9. (a) 7804 (2)   (b) 80.043 (3)   (c) 6496 (3)

Changing Recurring Decimals to Fractions

Recall that some fractions can be represented as repeating decimals. For instance, the decimal representation of $\dfrac{4}{11} = 0.363636\ldots$

Notice that the digits 3 and 6 repeats. Such decimals are called recurring decimals.

Recurring decimals are often represented by placing a dot above the number or numbers that repeat. The dots are placed on the first and the last recurring digits. So $0.363636\ldots$ is written as $0.\dot{3}\dot{6}$.

All recurring decimals can be expressed as fractions. To convert recurring decimals to fractions, follow the two steps below.

1. Form two equations which have the same recurring part after the decimal point.

2. Subtract one equation from the other to eliminate the recurring part of the decimal.

Examples below illustrate how these steps are applied.

Examples

Express $0.\dot{8}$ as fraction

$0.\dot{8}$ means $0.8888\ldots$

Let $x = 0.8888\ldots$

$10x = 8.8888\ldots$ Multiply both sides by 10

Subtracting we have

$\qquad 9x = 8$

$$x = \frac{8}{9} \qquad \text{Divide both side by 9}$$

This cannot be simplified any further so $0.\dot{8} = \dfrac{8}{9}$

Try this 7

Express $0.\dot{6}$ as a fraction

Express $0.\dot{6}\dot{3}$ as a fraction

$0.\dot{6}\dot{3}$ means $0.6363\ldots$

Let $x = 0.6363\ldots$

$100x = 63.6363\ldots \qquad \text{Multiply both sides by 100}$

Subtracting we have

$99x = 63$

$$x = \frac{63}{99} \qquad \text{Divide both sides by 99}$$

$$x = \frac{7}{11} \qquad \text{Simplify}$$

So in its lowest terms $0.\dot{6}\dot{3} = \dfrac{7}{11}$

Try this 8

Express $0.\dot{5}\dot{4}$ as a fraction

Express $0.5\dot{3}$ as a fraction

$0.5\dot{3}$ means $0.53333\ldots$

Let $x = 0.5333...$

To get two equations with the same recurring part after the decimal point, multiply by 10 and 100. This gives

$10x = 5.3333...$

$100x = 53.3333...$

Subtracting we have

$90x = 48$

$$x = \frac{48}{90}$$ Divide both sides by 90

$$x = \frac{8}{15}$$ Simplify

So in its lowest term $0.5\dot{3} = \dfrac{8}{15}$

Try this 9

Express $0.4\dot{6}$ as a fraction

Exercise 3.1(d)

In Exercises 1 – 12, express each decimal as a fraction

1. $0.\dot{3}$ 2. $0.\dot{4}$ 3. $0.\dot{7}$
4. $0.\dot{1}\dot{2}$ 5. $0.4\dot{5}$ 6. $0.\dot{5}\dot{5}$

7. $0.\dot{7}\dot{2}$ 8. $0.1\dot{0}\dot{8}$ 9. $0.\dot{1}8\dot{5}$
10. $0.4\dot{8}\dot{1}$ 11. $0.8\dot{6}\dot{4}$ 12. $0.\dot{8}\dot{1}$

In Exercises 13 – 20, express each decimal as a fraction

13. $0.1\dot{6}$ 14. $0.3\dot{8}$ 15. $0.5\dot{6}$ 16. $0.8\dot{3}$

17. $0.43\dot{6}$ 18. $0.58\dot{3}$ 19. $0.41\dot{6}$ 20. $0.18\dot{3}$

3.2 Operations with Real Numbers

The basic operations on real numbers are addition, subtraction, multiplication and division.

Addition of Integers

The result of adding two numbers is the sum.   The rules for adding integers are as follows.

1. Adding two integers with like sign

To add two integers with like signs, add their absolute values and attach the common sign to the result.

Example

Find (a) $32 + 48$ (b) $-16 + (-8)$

(a) $32 + 48 = +(32 + 48)$

$= 80$

(b) $-16 + (-8) = -(16 + 8)$

$= -24$

Try this 10

Find (a) $-9 + (-7)$ (b) $-23 + (-14)$

2. Adding two integers with unlike signs

To add two real numbers with unlike sign, subtract the smaller absolute value from the greater absolute value, and

1. if the positive number has the greater absolute value, the answer is positive

2. if the negative number has the greater absolute value, the answer is negative

Example

Find (a) $-20 + 8$ (b) $-7 + 12$

(a) $-20 + 8 = -(20 - 8)$

$$= -12$$

(b) $-7 + 12 = +(12 - 7)$

$$= 5$$

Try this 11

Find (a) $-18 + 7$ (b) $-9 + 12$

Exercise 3.2(a)

In Exercises 1 – 4, find the value of the following addition

1. (a) $7 + (-5)$ (b) $-9 + 12$ (c) $-15 + 27$

2. (a) $-13 + 20$ (b) $-6 + 14$ (c) $10 + (-6)$

3. (a) $-11 + 17$ (b) $-8 + 15$ (c) $-5 + 12$

4. (a) $21 + (-16)$ (b) $-32 + 45$ (c) $28 + (-19)$

In Exercises 5 – 8, find the value of the following addition

5. (a) $10 + (-13)$ (b) $-28 + 16$ (c) $-8 + 5$

6. (a) $13 + (-21)$ (b) $-14 + 9$ (c) $-11 + 6$

7. (a) $-17 + 12$ (b) $-15 + 7$ (c) $16 + (-31)$

8. (a) $-19 + 16$ (b) $35 + (-46)$ (c) $-28 + 19$

In Exercises 9 – 12, find the value of the following addition

9. (a) $-7 + (-9)$ (b) $-12 + (-16)$ (c) $-6 + (-8)$

10. (a) $-9 + (-11)$ (b) $-13 + (-5)$ (c) $-6 + (-3)$

11. (a) $-8 + (-15)$ (b) $-14 + (-21)$ (c) $-17 + (-23)$

12. (a) $-36 + (-15)$ (b) $-48 + (-12)$ (c) $-19 + (-37)$

Subtraction of Integers

Subtraction of two integers may be defined in terms of addition. For instance,

$$8 - 12 = 8 + (-12)$$

Example

Find (a) $7 - 21$ (b) $-17 - 5$

(a) $7 - 21 = 7 + (-21)$

$$= -(21 - 7)$$

$$= -14$$

(b) $-17 - 5 = -17 + (-5)$

$$= -(17 + 5)$$

$$= -22$$

Try this 12

Find (a) $11 - 17$ (b) $-18 + 25$

Exercise 3.2(b)

In Exercises 1 – 4, find the value of the following subtraction

1. (a) $8 - (-7)$ (b) $16 - (-9)$ (c) $14 - (-15)$

2. (a) $-9 - 5$ (b) $-6 - 4$ (c) $-13 - 12$

3. (a) $11 - (-17)$ (b) $-18 - 19$ (c) $-50 - 47$

4. (a) $16 - (-21)$ (b) $-63 - 37$ (c) $74 - (-12)$

In Exercises 5 – 8 find the value of the following subtraction

5. (a) $-4 - (-7)$ (b) $-5 - (-9)$ (c) $-16 - (-35)$

6. (a) $-6 - (-4)$ (b) $-8 - (-7)$ (c) $-21 - (-16)$

7. (a) $-13 - (-18)$ (b) $-35 - (-28)$ (c) $-11 - (-7)$

8. (a) $-10 - (-13)$ (b) $-42 - (-30)$ (c) $-51 - (-63)$

In Exercises 9 – 15, find the value of the following

9. (a) $8 - 15$ (b) $-9 + 13$ (c) $-10 - 6$

10. (a) $11 - 16$ (b) $37 - 45$ (c) $-12 - 18$

11. (a) $-23 + 15$ (b) $-14 + 21$ (c) $0 - 5$

12. (a) $0 - (-5)$ (b) $-17 + (-8)$ (c) $20 - 35$

13. (a) $-9 - (-6)$ (b) $-16 - (-26)$ (c) $-32 - 16$

14. (a) $-48 + 60$ (b) $100 - 125$ (c) $-167 + 135$

15. (a) $-96 - (-100)$ (b) $72 + (-98)$ (c) $64 - (-36)$

Multiplication of Integers

Multiplication of two integers can be considered as repeated addition. For instance, we can write 2×5 as $2 + 2 + 2 + 2 + 2$ or $5 + 5$. The multiplication, 2×5 may also be denoted as $2 \cdot 5$, $2(5)$ and $(2)(5)$. The result of multiplying two integers is their product, and each of the two numbers is a factor of the product.

To multiply integers, find the product of their absolute values and then use the rules below to determine the sign of the answer.

1. When you multiply two integers with the same signs, the result is always positive.

Example

Find the product (a) 6×4 (b) -8×-3

 (a) $6 \times 4 = 24$

(b) $-8 \times -5 = 40$

Try this 13

Find the product (a) 7×8 (b) -12×-5

2. When you multiply two integers with different signs, the result is always negative.

Example

Find the product (a) -8×3 (b) 7×-4

(a) $-8 \times 3 = -24$

(b) $7 \times -4 = -28$

Try this 14

Find the product (a) -9×6 (b) 15×-3

Note that

1. the product of zero and any other integer is zero

2. the product of an even number of negative factors is positive

3. the product of an odd number of negative factors is negative

Exercise 3.2(c)

In Exercises, 1 – 4, find the product

1. (a) -6×-7 (b) -8×-9 (c) -12×-8

2. (a) -5×-12 (b) -9×-4 (c) -12×-3

3. (a) -7×-8 (b) -14×-6 (c) -15×-6

4. (a) -5×-9 (b) -20×-7 (c) -15×-12

In Exercises, 5 – 8, find the product

5. (a) -5×6 (b) -7×10 (c) -12×8

6. (a) 4×-11 (b) 9×-4 (c) 13×-2

7. (a) -7×8 (b) 14×-5 (c) -12×4

8. (a) 5×-9 (b) 20×-6 (c) -15×11

Division of Integers

To divide, say 8 by 2 is the same as to multiply 8 by the reciprocal of 2, i.e. $8 \div 2 = 8 \times \frac{1}{2}$

The result of dividing two integers is the quotient of the numbers. The number 8 is the dividend and the number 2 is the divisor. The sign rules for division are the same as those for multiplication

1. The quotient of two integers with the same signs is positive

Example

Find (a) $16 \div 8$ (b) $-24 \div -6$

(a) $16 \div 8 = 2$

(b) $-24 \div -6 = 4$

Try this 15

Find (a) $21 \div 7$ (b) $-32 \div -8$

2. The quotient of two integers with different signs is negative

Example

Find (a) $-16 \div 8$ (b) $36 \div -9$

(a) $-16 \div 8 = -2$

(b) $36 \div -9 = -4$

Try this 16

Find (a) $-18 \div 3$ (b) $12 \div -4$

Note that zero divided by a number is zero, and a number divided by zero is undefined.

Exercise 3.2(d)

In Exercises 1 – 4, find the quotient

1. (a) $-12 \div -6$ (b) $-20 \div -5$ (c) $-16 \div -8$

2. (a) $-45 \div -15$ (b) $-63 \div -9$ (c) $-56 \div -8$

3. (a) $-24 \div -8$ (b) $-40 \div -8$ (c) $-36 \div -9$

4. (a) $-72 \div -36$ (b) $-90 \div -10$ (c) $-48 \div -16$

In Exercises 5 – 8, find the quotient

5. (a) $-15 \div 3$ (b) $-21 \div 7$ (c) $-30 \div 6$

6. (a) $45 \div -9$ (b) $72 \div -12$ (c) $90 \div -15$

7. (a) $-20 \div 4$ (b) $35 \div -7$ (c) $-80 \div 10$

8. (a) $48 \div -12$ (b) $77 \div -11$ (c) $-18 \div 9$

Properties of Integers

1. Commutative Property

We can add two numbers, say -8 and 6, to obtain a unique number, denoted by $-8 + 6$ i.e. -2. Also, $6 + (-8) = -2$. Notice that changing the order of the two numbers does not change the result, i.e. $-8 + 6 = 6 + (-8)$. This example illustrates the commutative property of addition of integers. Similarly, we can show that multiplication is commutative. For instance, $-3(4) = 4(-3) = -12$. The commutative property tells us that we can add or multiply in any order.

2. Associative Property

We can add three integers by grouping in any manner. For instance, we can add 4, 6 and 8 as follows: First, we add 6 and 8 together then add that sum to 4, i.e. $(6 + 8) + 4 = 14 + 4 = 18$.

We can also add 8 and 4, together and then add that sum to 6. i.e. $(8 + 4) + 6 = 12 + 6 = 18$. So $(6 + 8) + 4 = 6 + (8 + 4)$. This is possible because of the associative property.

We can also group integers in a product in any way we want and still get the same answer.

For instance, $(-4 \times 5)(3) = (-4)(5 \times 3) = -60$.
The associative property says that integers can be grouped in any manner to find a sum or a product.

3. Distributive Property

The distributive property involves addition and multiplication together. We can illustrate this property with an application.

Suppose that we want to find the total of the two areas shown in Figure 3.6.

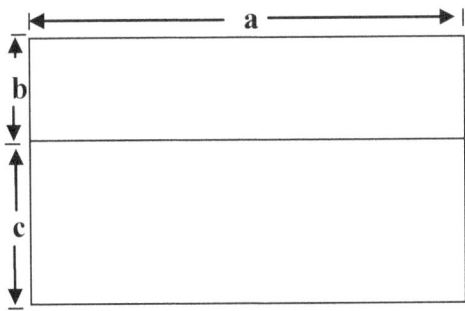

Figure 3.6

We may find the total area by multiplying the length by the overall width, which is found by adding the two widths.

$a \cdot (b + c)$

We can find the total area as a sum of the two areas

$a \cdot b + a \cdot c$

So $a \cdot (b + c) = a \cdot b + a \cdot c$

Notice that the answer is the same in both cases. The distributive property tells us that we can add first and then multiply, or multiply first and then add. We say that multiplication distributes over all the terms inside the brackets.

We may use the distributive property to find the product 79×6 quickly. We express 79 as $80 - 1$ and then multiply 80 by 6 and 1 by 6. The difference of the two products, $480 - 6$, is 474.

Example

Use the distributive property to evaluate 17×32

$$17 \times 32 = 17(30 + 2)$$
$$= 510 + 34$$
$$= 544$$

Try this 17

Use the distributive property to evaluate 15×98

Exercise 3.2(e)

1. Use the distributive property to evaluate

(a) $5 \times 7 + 5 \times 13$ (b) $36 \times 4 - 11 \times 4$ (c) $64 \times 12 + 36 \times 12$

(d) $8 \times 48 - 8 \times 28$ (e) $99 \times 25 + 25$ (f) $9 \times 148 - 9 \times 28$

2. Use the distributive property to evaluate

(a) 7×99 (b) 48×7 (c) 197×6

(d) 999×17 (e) 13×57 (f) 298×4

Order of Operations

How should $8 + 3 \times 10$ be computed? If we multiply 3 by 10 and then add 8, the result is 38. However if we add 8 and 3 first and then multiply the result by 10, we get 110. These results differ, and to avoid confusion, the following rules exist for determining the order in which the operations should be performed.

1. First do operations that occur within brackets

2. Next simplify all powers

3. Then do multiplications and divisions in the order in which they occur from left to right

4. Finally, do additions and subtractions in the order in which they occur from left to right

Examples

Find the value of $12 + (3 \times -6) - 9$

Working out the bracket first, we obtain

$12 + (-18) - 9$

We then do the addition first, to obtain

$-6 - 9$

Finally, subtracting we obtain -15

Try this 18

Find the value of $15 - (-4 \times 7) + 12$

Simplify $21 \div 7 \times 2 + 12$

$21 \div 7 \times 2 + 12 = 3 \times 2 + 12$

$$= 6 + 12$$

$$= 18$$

Try this 19

Simplify $16 \div 8 \times 2 - 7$

Simplify $9 \times 4 \div 12 - 8$

$9 \times 4 \div 12 - 8 = 36 \div 12 - 8$

$$= 3 - 8$$

$$= -5$$

Try this 20

Simplify $5 \times 6 \div 15 + 7$

Exercise 3.2(f)

In Exercises 1 – 12, find the value of each of the following

1. $18 - 25 + 3$ 2. $8 + 3 \times 4$ 3. $12 + 20 \times -3$

4. $15 - 18 \div -3$ 5. $8 \div (10 - 14)$ 6. $9 \times 4 \div 12$

7. $(10 - 4) \div 2 \times 3$ 8. $15 + 6 \div 3 - 18$

9. $8 + 5 \times 3 - 3 + 8 \div 4$ 10. $20 \div (18 - 25 + 3)$

11. $15 \times 2 \div 6 + (12 - 20)$ 12. $(4 \times 3 - 16) \div (3 - 14 \div 2)$

Application of Integers

The set of integers has practical applications. When we state that the temperature of a refrigerator is $-5°C$, it means that the temperature is $5°C$ below zero. The temperature of $-5°C$ is considered cold. If it was $-25°C$ outside a few minute ago, and the temperature drops 2 degrees, then the temperature is now $-27°C$

Distances above sea level are indicated as positive and distances below sea level are indicated as negative. For example, a town 90 m below sea level is represented as -90 m. The distance between two towns, one 64 m above sea level and the other 86 m below sea level is 150 m.

Finally, banks frequently use positive and negative to indicate credits and debits respectively. For example, if Kwesi deposits GH¢100 and withdraws GH¢ 120, his balance will be - GH¢ 20.

Other quantities usually considered positive are profit, assets, dates A.D. and price increases. Losses, liabilities, date B.C., and price decreases are negative.

Examples

The highest elevation in a country is 1250 m above sea level. The lowest elevation is 1530 m below sea level. Find the difference between these two elevations?

The distance $= 1250 - (-1530)$

$$= 2780$$

The difference between the two elevations is 2780 m.

Try this 21

Town A is 270 m above sea level and Town B is 380 m below sea level. Find how high Town A is above Town B.

The temperature at noon was $-8° C$. If the temperature dropped $7° C$, find the temperature now.

Change in temperature $= -8 - 7$

$$= -15$$

The temperature is $15° C$ below zero.

Try this 22

The temperature at 1.00 p m was $-12° C$. If the temperature dropped $8° C$, find the temperature now.

Exercise 3.2(g)

1. A girl has GH¢ 125 in her personal account. She writes a cheque for GH¢ 130, what is her new balance?

2. A man was born in 30 BC and died in 52 AD. How old was he when he died?

3. A shop made a profit of GH¢ 875 in June. In July, the shop made a loss of GH¢ 900, what is the profit or loss at the end of July?

4. A man deposits GH¢180 into his personal bank account. He bought 3 shirts at GH¢ 40 each and 2 pairs of jeans which cost GH¢65 each. If he issued a cheque to cover the total cost, what is the balance in his account?

5. A boy standing at a point 75 m above sea level descends to a point 80 m. What is his new position?

6. A submarine was situated 850 m below sea level. If it ascends 350 m, what is its new position?

7. A submarine was situated 480 m below sea level. If it descends 320 m, what is its new position?

8. The temperature was 28^0 C at noon and - 32^0 C at 4 p.m. What is the difference between the two temperature?

9. The temperature at 8 am was -10° C, and was rising at a uniform rate at 4° C per hour. What will be the temperature at 10 am?

10. The temperature is -8° C at 6 o'clock in the evening. If the Temperature drops 2° C every hour, what is the temperature at 1.00 am?

11. The melting point of mercury is -39° C. The freezing point of alcohol is -114° C. How much warmer is the melting point of

mercury than the freezing point of alcohol?

12. Roman civilization began in 509 BC and ended in 476 AD. How long did Roman civilization last?

3.3 Operations with Fractions

Adding Common Fractions

Fractions such as $\frac{3}{5}$ and $\frac{7}{4}$ are called common fractions. The top number in each fraction is called the numerator and the bottom number is called the denominator.

A fraction that has its numerator less than its denominator is called a proper fraction, while a fraction that has its numerator greater than its denominator is called an improper fraction. For example, $\frac{4}{5}$ is a proper fraction and $\frac{9}{4}$ is an improper fraction. Fractions such as $2\frac{1}{4}$ are called mixed numbers

Adding or subtracting fractions with the same denominator

To add or subtract fractions with the same denominators add or subtract their numerators and place the result over the common denominator.

Example

Find $\frac{5}{12} + \frac{2}{12}$

$$\frac{5}{12} + \frac{2}{12} = \frac{7}{12}$$

Try this 23

Work out:

$$\frac{7}{15} - \frac{4}{15}$$

Example 3.3(a)

In Exercises 1 – 5, find the value of:

1. $\frac{3}{5} + \frac{1}{5}$ 2. $\frac{4}{7} + \frac{2}{7}$ 3. $\frac{5}{14} + \frac{3}{14}$ 4. $\frac{5}{12} + \frac{3}{12}$ 5. $\frac{7}{15} + \frac{11}{15}$

In Exercises 6 – 10, find the value of:

6. $\frac{5}{8} - \frac{3}{8}$ 7. $\frac{7}{9} - \frac{5}{9}$ 8. $\frac{13}{24} - \frac{5}{24}$ 9. $\frac{19}{30} - \frac{7}{30}$ 10. $\frac{13}{25} - \frac{8}{25}$

Adding or subtracting fractions with different denominators

Fractions with different denominators can only be added or subtracted if the fractions are first changed to equivalent fractions with a common denominator. The fraction is then added or subtracted as described above. It is customary to use the lowest common multiple (LCM) of the denominators as the common denominator.

Examples

Find $\frac{3}{8} + \frac{5}{12}$

The lowest common multiple of 8 and 12 is 24 and will become the common denominator. First we convert $\frac{3}{8}$ and $\frac{5}{12}$ to equivalent fractions with 24 as the denominator.

We change $\frac{3}{8}$ to $\frac{9}{24}$ by multiplying both the numerator and the denominator by 3

We change $\frac{5}{12}$ to $\frac{10}{24}$ by multiplying both the numerator and the denominator by 2. Therefore $\frac{3}{8} + \frac{5}{12}$ is the same as $\frac{9}{24} + \frac{10}{24}$

Now

$$\frac{3}{8} + \frac{5}{12} = \frac{9}{24} + \frac{10}{24} = \frac{19}{24}$$

In practice, we write the common denominator only once:

$$\frac{3}{8} + \frac{5}{12} = \frac{9 + 10}{24} = \frac{19}{24}$$

Try this 24

Find $\frac{1}{4} + \frac{2}{5}$

Find $\frac{5}{8} - \frac{2}{5}$

The LCM of 8 and 5 is 40, so we change $\frac{5}{8}$ and $\frac{2}{5}$ to equivalent fractions with 40 as the denominator.

$$\frac{5}{8} - \frac{2}{5} = \frac{25 - 16}{40} = \frac{9}{40}$$

Try this 25

Find $\frac{3}{4} - \frac{2}{7}$

Example

Find $\dfrac{7}{15} + \dfrac{5}{6}$

The LCM of 15 and 6 is 30. Then

$$\frac{7}{15} + \frac{5}{6} = \frac{14 + 25}{30} = \frac{39}{30} = \frac{13}{10}$$

$\dfrac{13}{10}$ is an improper fraction, so we change it to a mixed number

$$\frac{13}{10} = 1\frac{3}{10}$$

Therefore $\dfrac{7}{15} + \dfrac{5}{6} = 1\dfrac{3}{10}$

Try this 26

Find $\dfrac{5}{9} + \dfrac{3}{5}$

Exercise 3.2(b)

In Exercises 1 – 5, find the value of:

1. $\dfrac{2}{3} + \dfrac{3}{5}$ 2. $\dfrac{3}{4} + \dfrac{5}{6}$ 3. $\dfrac{2}{9} + \dfrac{4}{15}$ 4. $\dfrac{5}{12} + \dfrac{7}{20}$ 5. $\dfrac{5}{6} + \dfrac{7}{12} + \dfrac{1}{4}$

In Exercises 6 – 10, find the value of:

6. $\dfrac{2}{3} - \dfrac{3}{5}$ 7. $\dfrac{4}{7} - \dfrac{3}{8}$ 8. $\dfrac{5}{6} - \dfrac{3}{8}$ 9. $\dfrac{7}{9} - \dfrac{5}{12}$ 10. $\dfrac{5}{6} - \dfrac{3}{8} - \dfrac{5}{12}$

Adding or subtracting mixed numbers

To add or subtract mixed numbers, first add or subtract the whole numbers and then add or subtract the fractions. Finally add or subtract the two answers.

Example

Work out $4\frac{1}{3} + 2\frac{3}{4}$

First add the whole numbers

$4 + 2 = 6$

Next add the fraction

$$\frac{1}{3} + \frac{3}{4} = \frac{4+9}{12} = \frac{13}{12} = 1\frac{1}{12}$$

Therefore $4\frac{1}{3} + 2\frac{3}{4} = 6 + 1\frac{1}{12} = 7\frac{1}{12}$

In practice, we may present the solution as

$$4\frac{1}{3} + 2\frac{3}{4} = 6\frac{4+9}{12} = 6\frac{13}{12} = 7\frac{1}{12}$$

This result can also be obtained as follows:

You may first change each mixed number to equivalent improper fractions and then add or subtract as described above.

$$4\frac{1}{3} + 2\frac{3}{4} = \frac{13}{3} + \frac{11}{4} = \frac{52+33}{12} = \frac{85}{12} = 7\frac{1}{12}$$

Try this 27

Find $5\frac{4}{5} - 2\frac{2}{3}$

Exercise 3.3(c)

In Exercises 1 – 5, find the value of:

1. $2\frac{1}{2} + 3\frac{3}{4}$

2. $5\frac{1}{3} + 2\frac{1}{2}$

3. $3\frac{2}{15} + 4\frac{5}{9}$

4. $6\frac{7}{8} + 5\frac{3}{10}$

5. $1\frac{5}{12} + 3\frac{7}{20}$

In Exercises 6 – 10, find the value of:

6. $3\frac{1}{2} - 2\frac{1}{8}$

7. $5\frac{5}{8} - 3\frac{1}{4}$

8. $7\frac{1}{9} - 6\frac{1}{2}$

9. $8\frac{1}{8} - 5\frac{3}{5}$

10. $3\frac{1}{6} - 1\frac{2}{3}$

Multiplying Common Fractions

To multiply two fractions multiply the numerator together to give the numerator of the product and then multiply the denominators together to give the denominator of the product. Reduce the answer to the lowest term if possible.

Example

Find $\frac{8}{20} \times \frac{6}{9}$

$$\frac{8}{20} \times \frac{6}{9} = \frac{48}{180} = \frac{4}{15}$$

You may first reduce the fraction to the lowest terms and then multiply

$$\frac{8}{20} \times \frac{6}{9} = \frac{2}{5} \times \frac{2}{3} = \frac{4}{15}$$

Try this 28

Find $\frac{6}{7} \times \frac{2}{3}$

Multiplying Mixed Numbers

Mixed numbers must be changed to improper fractions before multiplying. Then multiply as described above.

Example

Find $2\frac{2}{3} \times 5\frac{3}{4}$

$$2\frac{2}{3} \times 5\frac{3}{4} = \frac{8}{3} \times \frac{23}{4} = \frac{184}{12} = \frac{46}{3} = 15\frac{1}{3}$$

Try this 29

Find $1\frac{3}{4} \times 2\frac{2}{5}$

Exercise 3.3(d)

In Exercises 1 – 5, find the value of:

1. $2 \times \frac{5}{8}$ 2. $20 \times \frac{3}{16}$ 3. $\frac{9}{10} \times \frac{5}{6}$ 4. $\frac{15}{24} \times \frac{8}{9}$ 5. $\frac{18}{35} \times \frac{7}{12}$

In Exercises 6 – 10, find the value of:

6. $6 \times 3\frac{2}{3}$ 7. $1\frac{3}{4} \times 8$ 8. $2\frac{1}{4} \times 3\frac{1}{3}$ 9. $3\frac{2}{9} \times 1\frac{2}{7}$ 10. $4\frac{1}{5} \times 7\frac{1}{2}$

Dividing Fractions

To divide fractions multiply the first fraction by the reciprocal of the second fraction. Find the reciprocal by inverting the fraction.

Example

Find $\dfrac{5}{16} \div \dfrac{3}{8}$

$$\frac{5}{16} \div \frac{3}{8} = \frac{5}{16} \times \frac{8}{3} = \frac{5}{2} \times \frac{1}{3} = \frac{5}{6}$$

Try this 30

Find $\dfrac{5}{8} \div \dfrac{3}{4}$

Dividing Mixed Numbers

Mixed numbers must be changed to equivalent improper fractions before dividing.

Example

Find $4\dfrac{1}{2} \div 1\dfrac{2}{7}$

$$4\frac{1}{2} \div 1\frac{2}{7} = \frac{9}{2} \div \frac{9}{7} = \frac{9}{2} \times \frac{7}{9} = \frac{7}{2} = 3\frac{1}{2}$$

Try this 31

Find $3\dfrac{2}{3} \div 2\dfrac{1}{3}$

Exercise 3.3(e)

In Exercises 1 – 5, find the value of:

1. $8 \div \frac{4}{9}$ 2. $\frac{7}{10} \div 14$ 3. $\frac{5}{6} \div \frac{2}{3}$ 4. $\frac{7}{8} \div \frac{3}{4}$ 5. $\frac{3}{10} \div \frac{9}{25}$

In Exercises 6 – 10, find the value of:

6. $3\frac{1}{8} \div 5$ 7. $18 \div 2\frac{1}{4}$ 8. $4\frac{1}{2} \div 1\frac{1}{5}$ 9. $2\frac{3}{4} \div 1\frac{3}{8}$ 10. $6\frac{5}{12} \div 1\frac{5}{6}$

Order of Operations Involving Fractions

The order of operations in fractions is the same as in integers.

Example

Simplify $1\frac{2}{3} \times \left(\frac{3}{4} - \frac{2}{5} \right)$

First work out the operation in the bracket

$$\frac{3}{4} - \frac{2}{5} = \frac{15 - 8}{20} = \frac{7}{20}$$

Next work out the multiplication

$$1\frac{2}{3} \times \left(\frac{3}{4} - \frac{2}{3} \right) = 1\frac{2}{3} \times \frac{7}{20} = \frac{5}{3} \times \frac{7}{20} = \frac{7}{12}$$

Try this 32

Simplify $\left(4\frac{2}{3} + 1\frac{3}{4} \right) \div 1\frac{5}{6}$

Exercise 3.3(f)

Simplify the following fractions

1. $1\frac{1}{5} \times \left(3\frac{1}{4} - 2\frac{1}{5}\right)$ 2. $1\frac{5}{8} - \left(\frac{7}{8} + \frac{1}{4}\right)$ 3. $3\frac{1}{2} - 1\frac{3}{5} \div 2\frac{2}{3}$

4. $\left(2\frac{7}{9} \div \frac{5}{21}\right) \times 1\frac{1}{7}$ 5. $4\frac{2}{5} \times \frac{5}{6} \div \left(1 + \frac{5}{6}\right)$ 6. $2\frac{3}{4} + 3\frac{2}{5} - 1\frac{2}{3}$

7. $5\frac{1}{2} - \left(3\frac{1}{4} - 1\frac{7}{8}\right)$ 8. $1\frac{1}{2} - \frac{3}{5} \div \frac{2}{3}$ 9. $\left(2\frac{2}{3} - 1\frac{3}{5}\right) \div \left(2\frac{1}{5} - 1\frac{2}{3}\right)$

10. $\left(2\frac{1}{2} + 1\frac{3}{4}\right) \div 4\frac{1}{4} - \frac{3}{5}$ 11. $\left(1\frac{1}{4} \times 7\right) \div \left(12\frac{7}{12} - 11\frac{1}{3}\right)$

12. $4\frac{1}{2} \times 4\frac{2}{3} \div \left(5\frac{1}{2} + 3\frac{1}{4}\right)$ 13. $1 + 2 \times 1\frac{1}{2} + 3 \times 1\frac{7}{9}$

14. $\frac{1}{4} + \frac{1}{6} \div \frac{1}{7} - \frac{3}{5} \times 1\frac{2}{3}$ 15. $\frac{1}{4} + \frac{1}{3} \div \left(\frac{2}{3} - \frac{4}{7}\right) - \frac{3}{4}$

Problems Involving Fractions

Example

A man spent $\frac{1}{4}$ of his income on rent, $\frac{3}{5}$ on food and the rest on transportation. If he spent GH¢ 450 on transportation, how much was his income?

The man spent $\frac{1}{4} + \frac{3}{5} = \frac{5+12}{20} = \frac{17}{20}$ on rent and food

Hence, he spent $1 - \frac{17}{20} = \frac{3}{20}$ on transportation

$\frac{3}{20}$ of his income is GH¢ 450

Therefore his income is $450 \times \frac{20}{3} = 3,000$

The man's income was GH¢ 3,000.00

Try this 33

Kofi, Kojo and Ama shared a sum of money. Kofi had $\frac{1}{3}$, Kojo had $\frac{1}{5}$ and Ama had the remainder. If the amount share by them is GH¢ 900, how much did Ama get?

Exercise 3.3(g)

1. A father gave GH¢ 3,600 to his son. The son spent $\frac{1}{4}$ of this amount and saved $\frac{1}{3}$ of it. If he used the rest to buy a computer, how much did the computer cost?

2. A man paid $\frac{7}{12}$ of the price of a television set as down payment. He paid $\frac{7}{10}$ of the remainder the following month. If he had GH¢ 120 more to pay find the price of the television set?

3. A businessman paid $\frac{3}{8}$ of his annual income as tax. He gave $\frac{1}{5}$ of his net income to charity. If he had GH¢ 16,000 left, what was his income?

4. Mr. Ofori spent $\frac{3}{4}$ of his April salary, $\frac{2}{3}$ of his May salary and $\frac{3}{5}$ of his June salary. He saved the rest of his salary. If his monthly salary is GH¢ 3,600, how much did he save in the 3 months?

5. $\frac{1}{4}$ of candidates who took a test in mathematics and science failed both subject. $\frac{1}{2}$ of the candidates passed both subjects and the rest

passed in one subject. $\frac{1}{3}$ of those who passed one subject failed in mathematics. How many candidates took the test if 10 of those who passed in one subject passed mathematics?

6. In a school $\frac{5}{12}$ of the students are in the first year, $\frac{1}{3}$ are in the second year and the rest are in the third year. How many of the students are in the third year if there are 1500 students in the school?

7. In a school easy competition, the first prize winner receives $\frac{4}{9}$ of the prize, the second prize winner receives $\frac{1}{3}$ of the prize and the third prize winner receives the remainder. If the third prize winner received GH¢ 1000, what was the total prize money?

8. A man goes $\frac{3}{5}$ of a journey by train and $\frac{1}{3}$ by bus. He walked the rest of the journey. If the total distance he travelled is 4.5 km, how long did he walk?

9. Kwame, Esi and Afua each won a prize. Their total prize money was GH¢ 300. Kwame won $\frac{7}{12}$ of the prize, Esi won $\frac{3}{10}$ of the prize, Afua won the rest of the prize. Calculate the amount Afua received.

10. Kojo buys a bicycle for GH¢ 240. During the first year it losses $\frac{9}{40}$ of its value. During the second year it losses $\frac{7}{9}$ of its value at the end of the first year. Calculate its value at the end of the

second year.

3.3 Binary Operation

In addition, we add two real numbers to obtain a single real number called the sum. For instance $3 + 5 = 8$. An operation such as addition that is applied to two elements of a set to produce a unique element of the set is called a binary operation. Multiplication is also a binary operation.

Binary operations are often written using notation such as $a * b$, $a \oplus b$ and $a \circ b$. A binary operation may be defined by a rule that combines two elements of a set to obtain another element of the set. For instance, an operation $*$ on real numbers may be defined as $m * n = 2m - n$. Then, $3 * -2 = 2(3) - (-2) = 8$. Notice that 3 and -2 are substituted for m and n respectively.

Example

Given $x \circ y = x + y - xy$, find $2 \circ -3$

$$2 \circ -3 = 2 + (-3) - (2)(-3)$$

$$= -1 + 6$$

$$= 5$$

Try this 34

Given $a * b = a^2 + 3b$, find $-3 * -4$

Properties of Binary Operation

Commutative Property

Recall that the commutative property states that changing the order of an operation does not change the answer. A binary operation $*$ on a set S is said to be commutative if for all $a, b \in S$

$$a * b = b * a$$

Associative Property

Recall that the associative property allows us to add or multiply three or more numbers by grouping in any manner. A binary operation $*$ on a set S is said to be associative if for all elements a, b and $c \in S$, we have

$$(a * b) * c = a * (b * c)$$

Identity Element

The identity element is the element of a set of numbers that when combined with another number in a particular operation leaves that number unchanged. For example, 0 is the identity element under addition for real numbers, since for any real number $a, a + 0 = 0 + a = a$. Similarly, 1 is the identity element under multiplication for real numbers, since for any real number $a, a \times 1 = 1 \times a = a$. Generally, an element e is an identity element of a set S, if

$$a * e = e * a = a$$

Inverse Element

The inverse of an element is the element of a set of numbers that when combined with another number in a particular operation gives the identity element.

Every real number a has an additive inverse given by $-a$, since $a + (-a) = -a + a = 0$. Also, every non zero real number a has a multiplicative inverse given by $\frac{1}{a}$, since $a \times \frac{1}{a} = \frac{1}{a} \times a = 1$.

Generally, if e is an identity element of a set S, then a^{-1} is an inverse of a if

$$a * a^{-1} = a^{-1} * a = e$$

A binary operation on a finite set is often displayed in a table that demonstrates how the operation is performed. It is easy to check whether an operation defined by a table is commutative. If the table is symmetric about the diagonal, then the operation is commutative.

Example

Operation $*$ is defined on the set $\{2, 3, 4, 5\}$ as shown in the table below

*	2	3	4	5
2	5	4	3	2
3	4	2	5	3
4	3	5	2	4
5	2	3	4	5

(a) Is this operation commutative?

(b) Name the identity element

(c) For each element having an inverse, name the element and its

    inverse

(d) Show that $(2 * 3) * 4 = 2 * (3 * 4)$

(a) Yes, it is commutative. The table is symmetric with respect to

    the diagonal line.

(b) The identity element is 5 because all of the values in its row or

    column are the same as the row or column headings

(c) What element, when paired with a specific value will return the

identity element 5? The inverse of 2 is 2; the inverse of 3 is 4;

the inverse of 4 is 3; and the inverse of 5 is 5.

(d) $(2 * 3) * 4 = 2 * (3 * 4)$

$$4 * 4 = 2 * 5$$

$$2 = 2$$

Therefore $(2 * 3) * 4 = 2 * (3 * 4)$

Try this 35

Operation ∘ is defined on a set $\{a, b, c, d\}$ as shown in the table below

*	a	b	c	d
a	d	c	a	b
b	c	a	b	d
c	a	b	c	c
d	b	c	d	a

(a) Is this operation commutative?

(b) Name the identity element

(c) For each element having an inverse, name the element and its

inverse

(d) Show that $d * (c * b) = (d * c) * b$

Exercise 3.4

In Exercises 1 – 10, use the given rule to evaluate each binary operation.

1. $x * y = 3x - 4y$; $-2 * 3$ 2. $a * b = a - b + ab$; $3 * -2$

3. $x \oplus y = xy - y$; $-3 \oplus 2$ 4. $a * b = a^2 - b$, $-4 * -3$

5. $x \circ y = 3x + 2y - 5xy$; $2 \circ 1$ 6. $a * b = a^2 + 2b - 5$; $-3 * -2$

7. $x \otimes y = 2x - y$; $(2 \otimes 3) \otimes 5$ 8. $a * b = a^3 - 3b$; $-2 * (3 * 4)$

9. $x * y = x + y - 2xy$; $(4 * 3) * 2$
10. $a * b = 2a + b - ab$; $(-5 * 3) * 2$

11. Operation $*$ is defined on a set $\{2, 3, 4, 5\}$ as shown in the table

    below

*	2	3	4	5
2	5	2	3	4
3	2	3	4	5
4	3	4	5	2
5	4	5	2	3

    (a) Is this operation commutative?

    (b) Name the identity element if it exists.

    (c) For each element having an inverse name the element and

       its inverse.

    (d) Show that $(2 * 3) * 4 = 2 * (3 * 4)$

12. Operation $*$ is defined on a set $\{1, 2, 3, 4\}$ as shown in the table

    below

*	1	2	3	4
1	1	3	1	2
2	3	1	2	4
3	1	2	4	3
4	2	4	3	1

(a) Is this operation commutative?

(b) Name the identity element if it exists

(c) For each element having an inverse, name the element and its inverse

(d) Show that $4 * (3 * 2) = (4 * 3) * 2$

Review Exercise 3

In Exercises 1 and 2, which of the real numbers in the set are:
(a) natural numbers   (b) integers,   (c) rational numbers and
(d) irrational numbers?

1. $\{- 3.5, \ -\sqrt{9}, -2.3, - 1, - 0.75, 0, \sqrt{3}, 2\pi, \frac{19}{2}, 12\}$

2. $\{-\frac{9}{2}, -\sqrt{5}, -\frac{3}{4}, 0, \sqrt{36}, \frac{15}{2}, 9, 23\}$

3. Plot the real numbers on the real number line

    (a) -4 (b) $\frac{1}{2}$ (c) -5.2 (d) $\frac{5}{2}$

4. Place the correct symbol $<$ or $>$ between the pair of numbers

    (a) -7 -3 (b) 5.23 5.0 (c) -12 0 (d) $\frac{3}{4}$ $\frac{1}{2}$

5. Find each absolute value

    (a) $|-8|$ (b) $|7.3|$ (c) $-|-32|$ (d) $-|15|$

6. Place the correct symbol $<$, $>$ or $=$ between the pair of numbers

    (a) $|-8|$ $|-12|$ (b) $|-13|$ $|7|$ (c) $|-5|$ $|5|$

7. Round off the numbers to the number of decimal place(s) indicated in the bracket

(a) 7.326 (2)   (b) 0.964 (1)   (c) 0.8997 (3)   (d) 13.743 (2)

8. Round off the numbers to the number of significant figure(s) indicated in the bracket

(a) 0.00734 (1)   (b) 64.975 (3)   (c) 0.0608 (2)

(d) 1399.7547 (4)

9. Express each decimal as a fraction

(a) $0.\dot{6}$ (b) $0.\dot{7}\dot{2}$ (c) $0.3\dot{6}$ (d) $0.27\dot{9}$

10. Find the value of:

(a) $2 - (-9)$ (b) $-5 - (-7)$ (c) $-8 - (-8)$

(d) $12 - 16$

(e) $15 - 3 - (-12)$ (f) $25 - 36 + 7$ (g) $-42 + 50 - 16$

(h) $12 - (-18) - 40$

11. Find the value of:

(a) 8×-3 (b) -9×4 (c) -6×-13 (d) -15×8

(e) $24 \div -3$ (f) $-32 \div -4$ (g) $-57 \div 19$ (h) $-52 \div -13$

12. Use the distributive property to evaluate

(a) $6 \times 21 + 6 \times 29$ (b) $83 \times 64 - 83 \times 14$

(c) 197×7 (d) 6×248

13. Find the value of:

    (a) $5 + 3 \times 7$ (b) $19 - 5 \times 3 + 3$

    (c) $32 - 8 \div 4 - 2$ (d) $18 - 6 \div 3 \times 2 + 7$

    (e) $28 \div 4 + 3 \times 5$ (f) $8 + (15 \div 3) - 7 \times 3$

    (g) $12 - (14 \div 7 \times 4) + 3$ (h) $16 \div 8 \times 3 - (17 - 23)$

14. A boy has GH¢ 320 in her personal account. He wrote two cheques for GH¢ 135 and GH¢ 270 in June and July respectively, what is his new balance?

15. A girl standing at a point 12 m above sea level descends to a point 18 m. What is her new position?

16. The temperature at noon on a December day was $-7°\,C$. If the temperature drops $3°\,C$ every hour, what is the temperature at 2.00 pm?

17. Find the value of:

    (a) $\frac{7}{8} + \frac{3}{8}$ (b) $\frac{9}{14} - \frac{3}{14}$ (c) $\frac{2}{5} + \frac{8}{15}$ (d) $\frac{7}{9} - \frac{5}{12}$

    (e) $2\frac{5}{6} - 1\frac{7}{12}$ (f) $4\frac{2}{3} + 3\frac{1}{4}$ (g) $5\frac{2}{3} - 2\frac{3}{4}$ (h) $2\frac{5}{6} + 3\frac{7}{10}$

18. Find the value of:

    (a) $9 \times \frac{2}{3}$ (b) $\frac{7}{15} \times \frac{25}{21}$ (c) $2\frac{1}{8} \times 12$ (d) $1\frac{3}{5} \times 4\frac{3}{4}$

    (e) $6 \div \frac{2}{3}$ (f) $\frac{5}{8} \div \frac{3}{4}$ (g) $4\frac{1}{2} \div 3$ (h) $3\frac{5}{8} \div 1\frac{3}{4}$

19. Find the value of:

(a) $\frac{2}{3} \times \left(\frac{3}{4} + 1\frac{1}{2}\right)$

(b) $\left(3\frac{1}{4} + 2\frac{2}{3}\right) \div 7\frac{8}{9}$

(c) $1\frac{4}{5} - \frac{3}{7} \times \left(2\frac{1}{3} + \frac{3}{5}\right)$

(d) $\left(\frac{5}{6} - \frac{2}{3}\right) \div \frac{1}{2}$

(e) $1\frac{2}{3} \div 2\frac{1}{4} - \frac{4}{9}$

(f) $2\frac{1}{4} - 3\frac{1}{4} \div 1\frac{3}{4} \times \frac{7}{12}$

20. Use the given rules to find the value of each operation

(a) $x * y = x + y - xy$ $-3 * -2$

(b) $a * b = 2a + 3b + 4ab$ $2 * -3$

(c) $x * y = 3x^2 + 2y^2 + xy$ $-1 * 3$

(d) $a * b = a - b + ab$ $(3 * 2) * 4$

21. Operation $*$ is defined on a set $\{3, 4, 5, 6\}$ as shown in the table below

*	3	4	5	6
3	5	3	5	4
4	3	4	5	6
5	4	5	5	5
6	4	6	5	3

(a) State the identity element for the operation $*$

(b) Which of the elements has no inverse?

(c) Evaluate $(4 * 3) * (6 * 5)$

Chapter Test 3

Take this test as you would take a test in class. After you are done, check your work against the answers in the back of the book.

1. Which numbers in the set $\{- 6.7, -\sqrt{25}, -\frac{9}{2}, -\pi, - 1.2, 0, 2,$

   $\sqrt{7}, \frac{7}{2}, 5\}$ are:

   (a) natural numbers   (b) integers   (c) rational numbers

   (d) irrational numbers

2. Plot the numbers on the number line

   (a) $- 1.75$ (b) 3.2 (c) $\frac{13}{6}$

3. Evaluate

   (a) $|-9|$ (b) $|8 - 17|$ (c) $-|6| + |-3|$

4. Find the value of:

   (a) $-18 - (-15)$ (b) 6×-9 (c) $-45 \div 15$

5. The temperature at noon on a June day was $-9°\,C$. If the temperature rises by $2°\,C$ every hour, what is the temperature at 3.00 pm?

6. A man standing at a point 100 m above sea level descends to a point 160 m, what is his new position?

7. Identify the property that is illustrated by each statement

   (a) $(-7 \times 5) \times 4 = -7 \times (5 \times 4)$

   (b) $6(9 - 4) = 6 \times 9 - 6 \times 4$

   (c) $-8 + (-6) = -6 + (-8)$

8. Round off the numbers to the number of place(s) indicated

(a) 19.962 (1 d.p.)   (b) 0.0007083 (2 s.f.)   (c) 0.8994 (2 d.p.)

9. Express each decimal as a fraction

(a) $0.1\dot{8}$ (b) $3.7\dot{9}$

10. Find the value of:

(a) $10 - 18 \div 6 \times 3 + 9$ (b) $32 \div (19 - 6 \times 3 + 7) - 12$

11. Find the value of:

(a) $\left(4\frac{3}{4} - 1\frac{5}{6}\right) \div 1\frac{1}{24}$ (b) $1\frac{1}{2} - \frac{2}{3} \div 1\frac{2}{3} + 2\frac{1}{3} \times 1\frac{2}{7}$

12. Given the binary operation $x * y = 2x + 3y - 4xy$, evaluate

$2 * 3$

13. Operation $*$ is defined on the set $\{-1, 0, 1, 2\}$ as shown in the

table below

*	-1	0	1	2
-1	2	-1	0	1
0	-1	0	1	2
1	0	1	2	-1
2	1	2	-1	0

(a) Is this operation commutative?

(b) Name the identity element if it exists

(c) For each element having an inverse name the element and

its inverse

(d) Evaluate $(-1 * 0) * 1$

4

Algebraic Expressions

In Chapter 3, you learned how to do calculations with numbers by using the four basic operations of addition, subtraction, multiplication, and division. This chapter introduces some basic concepts in algebra. In algebra, you will use numbers as well as letters. The letters called variables, represent numbers. By convention, letters towards the end of the alphabet e.g. x, y and z are used to represent variables. Letters at the beginning of the alphabet e.g. a, b and c are typically used to represent constants.

4.1 Algebraic Expressions

An expression containing a number, a variable, the product of a number and variable(s), and one or more arithmetic operations is called an algebraic expression. Examples are:

$$3x, \quad 5x + 2, \quad 2xy - 7xz, \quad x^2 + 6x - 3$$

Terms in an Algebraic Expression

The terms of an algebraic expression are those parts that are separated by addition. A term in an algebraic expression may be a number, a variable or a product of numbers and variables. For example, $3x^2 - 4x + 2$ has three terms. The first term is $3x^2$, the second term is $-4x$ and the third term is 2. $3x^2$ and $-4x$ are called the variable terms and 2 is called the constant term. Generally, a term that contains a variable is called a variable term, and a term that contains only a number is called the constant term.

Coefficient of a Variable Term

The number in front of the variable or the numerical factor of a variable term is called the coefficient. Consider the algebraic expression $5x^2 - 3xy^2 - 2$. The coefficient of the term $5x^2$ is 5, and the coefficient of $-3xy^2$ is -3. If no coefficient is indicated, then it is understood to be 1. For example, x^2y has a numerical coefficient of 1.

Like Terms

Two or more terms that have exactly the same variable factors are called like terms. The coefficients do not have to be the same. For example, for the expression $3x + 2y - 7x + 4y^2$, $3x$ and $-7x$ are like terms. However, $2y$ and $4y^2$ are not like terms because their variable factors y and y^2 are different.

Simplifying Algebraic Expressions

You can only combine like terms. To combine like terms, we use the distribution property in the reverse form, $ac + bc = (a + b) \cdot c$

The distributive property allows you to add the coefficients of the like terms, as illustrated in the examples below:

Example

Simplify:

(a) $3a + 5a$ (b) $-6x + 4x$ (c) $7x^2y - 12x^2y$

(a) $3a + 5a = (3 + 5)a$ using the distributive property

$\qquad = 8a$ Adding the numbers

You can omit the distributive property step when combining like terms. Just add the coefficients.

(b) $-6x + 4x = -2x$ Adding the coefficient

(c) $7x^2y - 12x^2y = -5x^2y$ Subtracting the coefficient

Try this 1

Simplify:

(a) $4x + 7x$ (b) $-6x + 4x$ (c) $-6xy - 9xy$

Exercise 4.1(a)

Simplify:

1. (a) $4x + 2x$ (b) $7y + y$ (c) $6a + 3a$

    (d) $8b + 5b$ (e) $9x^2 + 3x^2$

2. (a) $8x - 3x$ (b) $5y - y$ (c) $2a - 3a$

    (d) $4b - 7b$ (e) $10x^2 - 6x^2$

3. (a) $-2x + 7x$ (b) $-10a + 6a$ (c) $8y + (-5y)$

    (d) $-x + 3x$ (e) $-3y^2 + 2y^2$

4. (a) $-2x - 7x$ (b) $-y - 3y$ (c) $2x - (-3x)$

    (d) $4ab - 9ab$ (e) $-5xy^2 - 2xy^2$

Example

Simplify:

(a) $3a + 5b + 2a + 3b$ (b) $6p - 3q - 4p + 8q$

(a) Using the commutative property we have

$$3a + 5b + 2a + 3b = 3a + 2a + 5b + 3b$$

Adding coefficients of like terms we have

$$= 5a + 8b$$

This cannot be simplified any further.

You can omit the first step

(b) $6p + 3q - 4p - 8q = 2p - 5q$ Combining like terms

Try this 2

Simplify

(a) $5x + 8y + 4x - 3y$ (b) $4a + 7b - 5a - 3b$

Exercise 4.1(b)

Simplify

1. (a) $8x + 2y + 5y$ (b) $a + 4a + 3b$

    (c) $5x + 3y + 2x + y$ (d) $-6a + 4b + 10a + b$

    (e) $5x^2 + 2y + 3x^2 + 3y$

2. (a) $-a - 3a - 2a$ (b) $4x - 3y - 2x - 5y$

    (c) $7x - 4y - 2x - 5y$ (d) $6a - 3b - 4a - 5b$

    (e) $8y^2 - 6y - 5y^2 - 10y$

3. (a) $7x^2 + 5x + 3x^2 + 8x$ (b) $6ab^2 + 5a^2b + 2ab^2 + 2a^2b$

    (c) $6xy + 5yz + 3xy + 2yz$ (d) $-5x^3 + 2x^2 + 7x^3 + x^2$

    (e) $-8x^2y + 3xy^2 + 5x^2y + 2xy^2$

4. (a) $3x^2 + 5x - x^2 - 2x$ (b) $5ab^2 + 4a^2b - 2ab^2 - 2a^2b$

(c) $-3x^3 - 2x^2 + 5x^3 - 3x^2$ (d) $4x^2y - 3xy^2 - 2x^2y - xy^2$

Multiplication of Algebraic Expressions

The product of 3 and y is written as $3 \cdot y$, $(3)(y)$ or $3y$. $3y$ is the shortest and the most common way of writing the product. Note that the centred dot means multiplication.

Consider the product $2a \cdot 3b$

$$2a \cdot 3b = 2 \cdot a \cdot 3 \cdot b \qquad \text{Factorise each term}$$

$$= 2 \cdot 3 \cdot a \cdot b \qquad \text{Commutative property}$$

$$= 6ab \qquad \text{Simplest form}$$

The product of a number of algebraic expressions is the product of their numerical coefficients multiplied by the product of their letters. The product can be written in any order but it is most convenient to write the number first and the letters in alphabetical order.

Example

Simplify

(a) $4x \cdot 2y$ (b) $-3a \cdot 5b$ (c) $-2p \cdot -3q$

(a) $4x \cdot 2y = 8xy$

(b) $-3a \cdot 5b = -15ab$

(c) $-2p \cdot -3q = 6pq$

Try this 3

Simplify

(a) $2a \cdot 7b$ (b) $5x \cdot -2y$ (c) $-4p \cdot -3q$

Exercise 4.1(c)

Simplify the following

1. (a) $3a \cdot 4b$ (b) $5x \cdot 2y$ (c) $2p \cdot 6q$

   (d) $8u \cdot 7v$ (e) $s \cdot 5t$

2. (a) $-2r \cdot 3s$ (b) $4p \cdot -2q$ (c) $5a \cdot -3b$

   (d) $-6x \cdot 4y$ (e) $7p \cdot -3q$

3. (a) $-a \cdot -4b$ (b) $-5x \cdot -2y$ (c) $-6p \cdot -7q$

   (d) $-8r \cdot -3s$ (e) $-9x \cdot -2y$

Simplifying Powers

For any natural number n, a^n means $\underbrace{a \cdot a \cdot a \cdots a}_{n \, factors}$

The number n is called the index (or exponent) and a is called the base. The index indicates the number of times to repeat the base as a factor.

For example, $x^5 = x \cdot x \cdot x \cdot x \cdot x$

Now look at the product $x^3 \cdot x^4$

$x^3 \cdot x^4 = (x \cdot x \cdot x)(x \cdot x \cdot x \cdot x)$

$\qquad = x \cdot x \cdot x \cdot x \cdot x \cdot x \cdot x$

$\qquad = x^7$

So $x^3 \cdot x^4 = x^{3+4} = x^7$

Notice that the product has the same base and an index equal to the sum of the original indices. Now the product

$$2x^3 \cdot 3x^5 = (2 \cdot 3)(x^3 \cdot x^5)$$

$$= 6x^8$$

In the expression $2x^5$, the power 5 applies only to the letter x. However, in the expression $(2x)^5$, the power 5 applies to both 2 and x. That is $(2x)^5 = 2^5 \cdot x^5 = 32x^5$.

Example

Simplify:

(a) $3xy \cdot 2x$ (b) $-4p^2q \cdot -5q^2$ (c) $(-4x^2y)^2$

(a) $3xy \cdot 2x = 6x^2y$

(b) $-4p^2q \cdot -5q^2 = 20p^2q^3$

(c) $(-4x^2y)^2 = 16x^4y^2$

Try this 4

Simplify:

(a) $5xy^2 \cdot 3x^2$ (b) $-7a^3b \cdot 2ab^2$ (c) $(-3x^2y^3)^2$

Exercise 4.1(d)

Simplify:

1. (a) $2(-y)^2$ (b) $(2y)^2$ (c) $(3x^2y)^2$

   (d) $(-2xy^2)^3$ (e) $4(-x)^3y^2$

2. (a) $3a^2b \cdot (ab)^3$ (b) $(-2a)^3 \cdot 3b^2$ (c) $-4c \cdot (-3cd)^2$

(d) $5b^2 \cdot (-b)^3$ (e) $-2a^2b^3 \cdot -3ab^2$

3. (a) $2a \cdot 3a$ (b) $2ab \cdot 3a$ (c) $5ab \cdot 6a^3b$

(d) $4x^3y^2 \cdot 2x^2y$ (e) $a^3b \cdot 2bc$

4. (a) $-3ab \cdot 2a$ (b) $-4xy \cdot -2x$ (c) $5x^2y \cdot -3y^2$

(d) $-6x^2y^3 \cdot -xy^2$ (e) $-2a^3b \cdot 3b^3c^2$

5. (a) $3xy^2 \cdot 2xy^5$ (b) $3p^3q \cdot 2pq^2$ (c) $5a^2b^3 \cdot 4a^3b^4$

(d) $5xy^2 \cdot 3x^2y$ (e) $6a^2b \cdot 7ab^3$

6. (a) $-3x^2y^5 \cdot 2x^3y$ (b) $-2a^2b \cdot -5a^5b^2$ (c) $7b^2c^3 \cdot -b^3c^4$

(d) $-8x^3y^5 \cdot -4x^3y$ (e) $7a^2b^2c^2 \cdot 3a^3b^2c^3$

Division of Algebraic Expressions

Consider the division $a^5 \div a^3$

You can write the division in the form of a fraction, and divide as follows:

$$\frac{a^5}{a^3} = \frac{a \cdot a \cdot a \cdot a \cdot a}{a \cdot a \cdot a} = a \cdot a = a^2$$

The quotient has the same base and an index equal to the index of the numerator minus the index of the denominator.

Example

Simplify

(a) $6a^3b \div 3a^2b$ (b) $15x^3y^2 \div -5xy$ (c) $-20a^4b^3 \div -4a^2b^2$

(a) $6a^3b \div 3a^2b = \dfrac{6a^3b}{3a^2b} = 2a$

(b) $-15x^3y^2 \div -5xy = \dfrac{-15x^3y^2}{-5xy} = 3x^2y$

(c) $-20a^4b^3 \div 4a^2b^2 = \dfrac{-20a^4b^3}{4a^2b^2} = -5a^2b$

Try this 5

Simplify

(a) $8a^2b \div 4ab$ (b) $-6x^4y^3 \div -3x^2y$ (c) $-14p^3q^2 \div 7p^2q$

Exercise 4.1(e)

Simplify the following

1. (a) $\dfrac{4ab}{a}$ (b) $\dfrac{5xyz}{xy}$ (c) $\dfrac{15ab}{5a}$

   (d) $\dfrac{27xyz}{9xz}$ (e) $\dfrac{20x^3}{4x^2}$

2. (a) $\dfrac{16xy^2}{8xy}$ (b) $\dfrac{28x^3y^2}{7x^2y}$ (c) $\dfrac{24x^5y^4}{8x^3y^3}$

   (d) $\dfrac{7a^3b^2}{21a^2b^3}$ (e) $\dfrac{15x^3yz}{18x^2y^2}$

3. (a) $\dfrac{-12x^3y^2}{4xy}$ (b) $\dfrac{-10xy^2}{-5x^2y}$ (c) $\dfrac{18xy}{-12xy^2}$

   (d) $\dfrac{-35y^4z^3}{7y^3z^4}$ (e) $\dfrac{-32a^5b^2z^3}{-8a^3b^3z^2}$

4. (a) $\dfrac{18x^2y}{(-3x)(-2x)}$ (b) $\dfrac{(-2ab)^2}{4ab}$ (c) $\dfrac{(-3b)^2 \times 2a^2b}{27ab^2}$

   (d) $\dfrac{(-3a)(-2ab)}{(-ab)^2}$ (e) $\dfrac{8x(-y)^2}{2(-x)^2y}$

Evaluating Algebraic Expressions

We can find the value of an algebraic expression if numerical values are assigned to the letters. Finding the value of an expression is called evaluating the expression. To evaluate an algebraic expression, replace each variable with the given value and then simplify the numerical expression that results.

Example

Find the value of $3a + 2b$ when $a = -2$ and $b = 4$.

Replace a with -2, and b with 4, and use rules for signed numbers.

$3a + 2b = 3(-2) + 2(4) = -6 + 8 = 2$

Try this 6

Find the value of $2a - 3b$ when $a = 3$ and $b = -1$

Exercise 4.1(f)

In Exercises 1 – 10, evaluate each of the expressions if $a = -3$,

$b = 2$, $c = 6$ and $d = 5$

1. $7ab$ 2. $4cd$ 3. $3a^2c$ 4. $5b^2d$

5. $a + 3c$ 6. $4c - 3d$ 7. $4b + d$ 8. $-2a + d$

9. $3d - 2c$ 10. $3ab + 2cd$

11. Find the value of $(-x)^2 - 5x$ when $x = -3$

12. Find the value of $2x^2 - 6x$ when $x = 2$

13. Find the value of $a(2b - c)$ when $a = 2, b = 3$ and $c = -4$

14. Evaluate $\dfrac{3a^2-4a}{a^2-a}$ for $a = -2$

15. Evaluate $\dfrac{2a^3+c}{b+3a}$ when $a = -2, b = 3$ and $c = 4$

Writing Algebraic Expressions for Statements

Sometimes we find it useful to represent real-life situations with algebraic expressions. To do this, you must be able to translate words and phrases into symbols. In the following table, we list some verbal expressions that indicate the four basic operations of arithmetic and show their translation into algebraic expression.

Operations	Verbal expressions	Algebraic expression
Addition	The sum of a and b	$a + b$
	4 plus x	$4 + x$ or $x + 4$
	7 more than m	$m + 7$
	y increased by 3	$y + 3$
Subtraction	a minus 3	$a - 3$
	The difference of x and y	$x - y$
	5 less than b	$b - 5$
	4 fewer than y	$y - 4$
	t decreased by 2	$t - 2$
Multiplication	The product of a and b	ab
	7 times x	$7x$
	Twice y	$2y$
Division	The quotient of m and n	m/n
	x divided by 3	$x/3$
	One-third of a	$a/3$

Note that each of the algebraic expressions represents one number, just as the sum of 6 and 3 is 9.

Try this 7

Write an algebraic expression for each of the following statement

(a) 6 times x (b) y more than 5   (c) 7 fewer than a

(d) x less than 8   (e) Three times m (f) The quotient of 3 and n

(g) One-half of x (h) 9 increased by y

Exercise 4.1(g)

Write an algebraic expression for the following statements

1. The sum of x and y 2. x increased by 3

3. 2 more than a 4. p increased by 5

5. a minus c 6. 7 less than s

7. z fewer than 7   8. p times q

9. The product of 6 and b 10. Eight less than twice x

11. One more than three times m

12. The product of 2 and the sum of a and b

13. Twice the sum of x and y

14. Three times the difference of x and y

15. The difference of a and b divided by 3

16. The sum of 2 times x and y divided by 5

17. 2 more than three times x

18. Three times the difference of x and 4

19. The product of 3 less than x and 2

20. 5 less than twice a divided by 3

4.2 Expansion

Recall the distributive property $a(b + c) = ab + ac$. We can use this property to remove brackets as illustrated below

Now $3(2a + b) = 3 \cdot 2a + 3 \cdot b$

$$= 6a + 3b$$

Notice that each term inside the bracket is multiplied by the expression in front of the bracket. The process of removing brackets is called expansion.

Examples

Expand $-2(3x - 5y)$

$-2(3x - 5y) = -2 \cdot 3x - (-2) \cdot 5y$

$$= -6x + 10y$$

Notice that a negative sign preceding a bracket changes the sign of all terms inside the bracket.

Try this 8

Expand

(a) $-3(2x + y)$ (B) $-2(x - 3y)$

Expand (a) $5xy(x - 2y)$ (b) $-2x^2y(x - y)$

(a) $5xy(x - 2y) = 5xy \cdot x - 5xy \cdot 2y$

$$= 5x^2y - 10xy^2$$

With practice you will be able to leave out the first step in these expansions.

(b) $-2x^2y(x - 3y^2) = -2x^3y + 6x^2y^3$

Try this 9

Expand

(a) $3p(2p - q)$ (b) $-2ab(a + 3b)$

Exercise 4.2(a)

Expand the following:

1. (a) $3(x + 2y)$ (b) $4(3a + 5b)$ (c) $8(c - d)$

   (d) $2(4r - 3s)$ (e) $7(2p - 3q)$ (f) $5(2x + y)$

2. (a) $-6(a - b)$ (b) $-3(2x - 3y)$ (c) $-5(c - 3d)$

   (d) $-4(2a + b)$ (e) $-9(q + 2p)$ (f) $-7(3x + 2y)$

3. (a) $8(a + 2)$ (b) $10(3 - b)$ (c) $4(13 + x)$

   (d) $-2(12 - 7a)$ (e) $-12(p + 3)$ (f) $-7(3a - 4)$

4. (a) $4(3a + 2b + c)$ (b) $-5(x - 2y - 3z)$

   (c) $3(a - b + 2)$ (d) $2(6 + 3a - 2b)$

   (e) $7(2x + y - 3)$ (f) $-4(2 - 3b + c)$

5. (a) $3x(2x + 4y)$ (b) $-4y(3 - 2y)$

   (c) $5a(a - 2b)$ (d) $2r(3r - 4s)$

   (e) $-x^2y(2y - 3)$ (f) $-3p^2q^2(p + q)$

6. (a) $5x(3x + 2xy - y)$ (b) $2ab(6a - 7b + c)$

(c) $-2x(1 - 2xy - 4y^2)$ (d) $-3a^2(1 + 2a - 3a^2)$

(e) $4r^3(2 - 3s + 4rt)$ (f) $-8p(p - 2pq + 3q)$

The distributive property often allows us simplify expressions with brackets. This is illustrated by the example below.

Examples

Simplify

$3(x - 2y) + 5(2x + 3y)$

Expanding this expression we have

$3(x - 2y) + 5(2x + 3y) = 3x - 6y + 10x + 15y$

$$= 13x + 9y$$

Try this 10

Simplify

(a) $5x - 2(x + 6y)$ (b) $-3(x + 2y) + 8y$

Simplify $4(2a + b) - 3(2a + 3b)$

Expanding this expression we have

$4(2a + b) - 3(2a + 3b) = 8a + 4b - 6a - 9b$

$$= 2a - 5b$$

Try this 11

Simplify $3(x - 2y) - (x - 3y)$

Exercise 4.2(b)

Simplify the following

1. (a) $6(a + 1) + 2(2a + 3)$ (b) $5(3x + 4) + 2x$

(c) $4(x + 2) + 5(x + 1)$ (d) $3(2a + 1) + 2(2 + 3a)$

(e) $2(3p - 2q) + 3(p + 2q)$ (f) $7(2x - 1) + 4(3x + 1)$

(g) $2m + 5(1 - m)$ (h) $2(y - 1) + 3(y - 2)$

2. (a) $6(r + 2) - 3(2 + r)$ (b) $4(2x + 3) - 3(6x + 1)$

(c) $2(a - 3) - 3(a - 4)$ (d) $2a - 4(a - 2b)$

(e) $3(4p - q) - 2(p + 2q)$ (f) $5(r + 3) - 3(3r + 1)$

(g) $8p - 3(2p + 3q)$ (h) $6(2x + 1) - 4(2x - 1)$

3. (a) $2x(x - y) + 3x(x - 2y)$ (b) $-7y(3y - 2) + 10y^2$

(c) $s(2r + s) - s(r - s)$ (d) $2xy(y - x) + 3xy(x - y)$

(e) $p^2(1 - q) + p(2p + 3pq)$ (f) $r^2(2 - r) - r^3(1 - 2r)$

(g) $3p^2 - 2p(p - 3)$ (h) $-2e(e + 3f) + e(3e - 5f)$

4.3 Factorisation

You can use the distributive property in reverse $ab + ac = a(b + c)$ to write the sum of the terms of an algebraic expression as a product of factors. The process of writing sums of the terms of an algebraic expression as product of factors is called factorisation.

The first step in factorisation is to determine the greatest common factor of a set of terms. The greatest common factor is the product of the largest integer factor of the coefficients and the largest common variable power of each term.

Consider the terms of the expression $8a^2 + 6a$. The largest common factor of 8 and 6 is 2, and the variable factor with the largest common power of a^2 and a is a. So, $2a$ is the greatest common factor. Write

$$8a^2 + 6a = 2a(4a + 3)$$

Examples

Factorise $9x^2y + 12xy^2$

The largest common factor of 9 and 12 is 3, x is the largest common power of x^2 and x, and y is the largest common power of y^2 and y. So, $3xy$ is the greatest common factor. Write

$$9x^2y + 12xy^2 = 3xy(3x + 4y)$$

Try this 12

Factorise:

(a) $3a^2 + 15a$ (b) $7a^2b - 14ab^2$

Factorise $-10p^3q^2 + 15p^2q$

$$-10p^3q^2 + 15p^2q = -5p^2q(2pq - 3)$$

The greatest common factor is usually considered to have a positive coefficient. However, it is sometimes convenient to factorise a negative number out of the expression.

Try this 13

Factorise:

(a) $-6a^3b + 8ab^2$ (b) $-15xy^2 - 6x^2y$

Exercise 4.3(a)

1. State the greatest common factor in each case

(a) $5a + 15b$ (b) $9x - 6y$ (c) $8y + 20z$

(d) $4ac + 8ab$ (e) $10p - 8pq$ (f) $7xy - 14y^2$

(g) $5pq + 10q^2$ (h) $7xy^2 + 5x^2y$ (i) $6a^3 - 2a^2$

(j) $3r^2s + 15rs^3$ (k) $12y + 8y^2$ (l) $10r^2s - 5r^3$

2. State the second factor

(a) $6a + 12 = 6(\quad)$ (b) $8 - 4y = 4(\quad)$

(c) $3c + 6d = 3(\quad)$ (d) $10r - 5s = 5(\quad)$

(e) $7x^2 + 14x^3 = 7x^2(\quad)$ (f) $2p^3 - 4p^2 = 2p^2(\quad)$

(g) $a^2b + ab^2 = ab(\quad)$ (h) $a^2x^3 - ax^2 = ax^2(\quad)$

(i) $9x^3y - 18x^2y^2 = 9x^2y(\quad)$ (j) $-6ab - 4ac = -2a(\quad)$

(k) $-3p^2 + 4pq = -p(\quad)$ (l) $-8p^2q - 4pq^2 = -4pq(\quad)$

3. Factorise:

(a) $3a + 12$ (b) $4y - 8$

(c) $5p + 10q$ (d) $7x - 21y$

(e) $8xy - 12yz$ (f) $6p^2 - 15pr$

(g) $6p^2q - 18pq^2$ (h) $7xy^2 + 4x^2y$

4. Factorise:

(a) $-4x^2y - 6xy^2$ (b) $-10a^2 + 15ab$

(c) $-12a^2b + 8ab^2$ (d) $-5xy^2 - 10x^2y$

(e) $-12p^3q^2 + 8p^2q$ (f) $-2p^2 - 3pq$

(g) $-35xy^2 - 14x^2y$ (h) $-7a^3b^2 + 21a^2b^3$

Factorisation by Grouping

If an algebraic expression can be split into groups of terms and the groups share a common factor, then the original expression can be factorised. This method is called factorisation by grouping.

Examples

Factorise $2x^2 + 6x + 3xy + 9y$

We group the first two terms and the last two terms, and look for a common factor for the first two terms, and then a common factor for the second two terms.

Grouping the terms we have

$$2x^2 + 6x + 3xy + 9y = (2x^2 + 6x) + (3xy + 9y)$$

$$= 2x(x + 3) + 3y(x + 3)$$

$$= (x + 3)(2x + 3y)$$

Try this 14

Factorise $3x^2 + 12x + 2xy + 8y$

Factorise $xy - yz + y^2 - xz$

We have to rearrange the expression since grouping the original expression does not give a common factor.

$$xy - yz + y^2 - xz = xy + y^2 - xz - yz$$
$$= y(x + y) - z(x + y)$$
$$= (x + y)(y - z)$$

Try this 15

Factorise $5x^2 - 2xy - 15x + 6y$

Exercise 4.3(b)

Factorise the following

1. $ab + ac + bd + cd$ 2. $pr + ps - qr - qs$

3. $5ax + 5ay + 2bx + 2by$ 4. $3ac + 6bc + 4ad + 8bd$

5. $2xy + 4xz - 3y^2 - 6yz$ 6. $6xy - 9xz - 8wy + 12wz$

7. $3ax - ay + 3bx - by$ 8. $10ay - 6b - 2by + 30a$

9. $4pr + 10qs - 5qr - 8ps$ 10. $4au + 6bv + 4bu + 6av$

11. $7x^2 + 6yz - 14xy - 3xz$ 12. $10ax + 8bx + 10ay + 8by$

4.4 Multiplying Binomial

Again we can use the distributive property to multiply the two binomial $(a + 2)$ and $(a + 3)$ as follows:

Using the distributive property

$$(a + 2)(a + 3) = (a + 2) \cdot a + (a + 2) \cdot 3$$
$$= a^2 + 2a + 3a + 6$$
$$= a^2 + 3a + 2a + 6$$
$$= a^2 + 5a + 6$$

You can obtain the third step by multiplying each term of the second bracket by the first term and then by the second term in the first bracket.

Examples

Find the $(x - 3)(x + 2)$

$$(x - 3)(x + 2) = x^2 + 2x - 3x - 6$$
$$= x^2 - x - 6$$

Try this 16

Find the product $(x + 5)(x - 3)$

Find the product $(y - 3)(y - 4)$

$$(y - 3)(y - 4) = y^2 - 7y + 12$$

Try this 17

Find the product $(x - 3)(x - 2)$

With practice you can write the product directly, and you should try to do so.

Exercise 4.4(a)

Find the product

1. $(x + 4)(x + 5)$ 2. $(x + 7)(x + 6)$

3. $(x + 1)(x + 9)$ 4. $(x + 12)(x + 5)$

5. $(x + 2)(x + 10)$ 6. $(x + 3)(x + 15)$

7. $(2x + 3)(x + 4)$ 8. $(3x + 2)(2x + 1)$

9. $(5x + 4)(2x + 7)$ 10. $(6x + 5)(3x + 4)$

11. $(x - 3)(x + 5)$ 12. $(x - 7)(x + 10)$

13. $(x - 4)(x + 8)$ 14. $(2x - 5)(x + 3)$

15. $(x - 8)(x + 9)$ 16. $(3x - 10)(x + 3)$

17. $(7x - 15)(x + 2)$ 18. $(x - 9)(x + 5)$

19. $(x - 10)(x + 7)$ 20. $(x - 12)(x + 6)$

21. $(x + 9)(x - 4)$ 22. $(x + 8)(x - 3)$

23. $(x + 7)(x - 2)$ 24. $(x + 13)(x - 5)$

25. $(x + 5)(x - 8)$ 26. $(x + 4)(x - 11)$

27. $(x + 6)(x - 9)$ 28. $(2x + 3)(3x - 2)$

29. $(4x + 3)(2x - 7)$ 30. $(x + 8)(5x - 2)$

31. $(x - 2)(x - 5)$ 32. $(x - 8)(x - 4)$

33. $(x - 7)(x - 3)$ 34. $(x - 3)(x - 10)$

35. $(x - 6)(x - 3)$ 36. $(x - 12)(x - 10)$

37. $(3x - 4)(5x - 3)$ 38. $(x - 6)(5x - 1)$

39. $(3x - 5)(2x - 7)$ 40. $(4x - 3)(2x - 5)$

Squaring Binomials

Consider the square of a binomial, such as $(x + y)^2$. This can be expressed as $(x + y)(x + y)$. Since this is the product of two binomials, we have

$$(x + y)^2 = (x + y)(x + y)$$

$$= x^2 + 2xy + y^2$$

Similarly, $(x - y)^2 = x^2 - 2xy + y^2$

The expression obtained when a binomial is squared is called a perfect square trinomial.

You may have noticed that

1. The first term is the square of the first term of the binomial

2. The middle term is twice the product of the two terms of the

   Binomial. The middle term can be positive or negative

3. The last term is the square of the last term of the binomial

Example

Find:

(a) $(x + 5)^2$ (b) $(2x - 3)^2$

(a) $(x + 5)^2 = (x)^2 + 2(x)(5) + (5)^2$

$$= x^2 + 10x + 25$$

(b) $(2x - 3)^2 = (2x)^2 - 2(2x)(3) + (-3)^2$

$$= 4x^2 - 12x + 9$$

Again we have shown all the steps. With practice you can write just the square.

Try this 18

Find (a) $(x - 7)^2$ (b) $(3x + 2)^2$

Exercise 4.4(b)

Expand

1. $(x + 3)^2$ 2. $(x + 7)^2$

3. $(x + 6)^2$ 4. $(x + 12)^2$

5. $(2x + 3)^2$ 6. $(5x + 4)^2$

7. $(x - 5)^2$ 8. $(x - 2)^2$

9. $(x - 10)^2$ 10. $(x - 3)^2$

11. $(6x - 1)^2$ 12. $(7x - 2)^2$

Difference of Squares

Consider the product of the sum and difference of the same two terms, such as

$(x + y)(x - y)$.

Now $(x + y)(x - y) = (x)^2 - (x)(y) + (y)(x) - (y)^2$

$$= x^2 - xy + yx - y^2$$

$$= x^2 - y^2$$

This is called the difference of two squares

Notice that the product of two binomials that differ only in the sign between the terms is the difference of their squares.

Example

Multiply:

(a) $(x - 3)(x + 3)$ (b) $(2x - 3)(2x + 3)$

(a) $(x - 3)(x + 3) = (x)^2 - (3)^2$

$$= x^2 - 9$$

(b) $(2x - 3)(2x + 3) = (2x)^2 - (3)^2$

$$= 4x^2 - 9$$

Try this 19

Multiply:

(a) $(x + 6)(x - 6)$ (B) $(x - 2y)(x + 2y)$

Exercise 4.4(c)

Expand:

1. $(x + 4)(x - 4)$ 2. $(x + 5)(x - 5)$

3. $(x - 3)(x + 3)$ 4. $(x - 8)(x + 8)$

5. $(x - 7)(x + 7)$ 6. $(2x + 3)(2x - 3)$

7. $(4x + 5)(4x - 5)$ 8. $(3x - 9)(3x + 9)$

9. $(7x - 4)(7x + 4)$ 10. $(6x + 13)(6x - 13)$

4.5 Factorising Expressions of the form $ax^2 + bx + c$

You may have noticed in section 4.4 that the product of two binomials often results in an expression with three terms. In this section, we will learn how to reverse the multiplication in section 4.4 and look for two binomial factors whose product is an expression of the form $ax^2 + bx + c$, where $a \neq 0$.

Factorising an Expression of the Type $x^2 + bx + c$

We will consider two cases:

1. When Constant Terms are Positive

Examples

Factorise $x^2 + 7x + 10$

To factorise an expression such as $x^2 + 7x + 10$, you need to find two factors of 10 whose sum is 7. Those numbers are 2 and 5. The x^2 resulted from x times x,which suggest that the first term of each binomial factor is x. So

$$x^2 + 7x + 10 = (x + 2)(x + 5)$$

If you have trouble, finding the factors list all the distinct pair of factors of the constant term and then choose the pair from the list whose sum is the coefficient of the term containing x.

Try this 20

Factorise $x^2 + 11x + 18$

Factorise $x^2 - 5x + 6$

Since the constant term is positive and the coefficient of the middle term is negative, we look for two factors of 6 which are both negative, and their sum must be -5. Recall that the product of two negative numbers is always positive, whereas the sum of two negative numbers is always negative. The numbers we need are -3 and -2, so

$$x^2 - 5x + 6 = (x - 2)(x - 3)$$

Try this 21

Factorise $x^2 - 6x + 8$

Exercise 4.5(a)

Factorise:

1. $x^2 + 9x + 20$ 2. $x^2 + 10x + 21$

3. $x^2 + 14x + 48$ 4. $x^2 + 12x + 20$

5. $x^2 + 14x + 45$ 6. $x^2 + 15x + 36$

7. $x^2 + 12x + 32$ 8. $x^2 + 9xy + 18y^2$

9. $x^2 + 5xy + 6y^2$ 10. $x^2 + 7xy + 12y^2$

11. $x^2 + 7xy + 10y^2$ 12. $x^2 + 15xy + 56y^2$

13. $x^2 - 9x + 14$ 14. $x^2 - 8x + 15$

15. $x^2 - 13x + 36$ 16. $x^2 - 11x + 30$

17. $x^2 - 17x + 70$ 18. $x^2 - 13x + 40$

19. $x^2 - 12x + 27$ 20. $x^2 - 10xy + 16y^2$

21. $x^2 - 7xy + 6y^2$ 22. $x^2 - 11xy + 28y^2$

23. $x^2 - 8xy + 15y^2$ 24. $x^2 - 15xy + 54y^2$

2. When constant terms are negative

Examples

Factorise $x^2 + 7x - 18$

The constant term $- 18$, must be expressed as the product of a negative number and a positive number. Since the sum of these two numbers must be positive 7, the positive number must have the greater absolute value. The two numbers are $- 2$ and 9. So

$$x^2 + 7x - 18 = (x - 2)(x + 9)$$

Try this 22

Factorise $x^2 + 2x - 24$

Factorise $x^2 - 2x - 15$

Since the sum of the two numbers must be $- 2$, the negative number must have the greater absolute value. The numbers are $- 5$ and 3. So

$$x^2 - 2x - 15 = (x - 5)(x + 3)$$

Try this 23

Factorise $x^2 - x - 12$

You may have noticed that if the constant term is positive, the two factors have the same sign and it is the same as the sign of the middle term. If the constant term is negative, the two factors have opposite signs. If the middle term is positive, then the larger absolute value is positive and if the middle term is negative the larger absolute value is negative.

Exercise 4.5(b)

Factorise

1. $x^2 + x - 20$ 2. $x^2 + 4x - 21$

3. $x^2 + 2x - 48$ 4. $x^2 + 5x - 50$

5. $x^2 + 3x - 70$ 6. $x^2 + 3x - 40$

7. $x^2 + 6x - 27$ 8. $x^2 + 2xy - 15y^2$

9. $x^2 + xy - 6y^2$ 10. $x^2 + 4xy - 32y^2$

11. $x^2 + 7xy - 30y^2$ 12. $x^2 + 5xy - 36y^2$

13. $x^2 - 5x - 14$ 14. $x^2 - 7x - 30$

15. $x^2 - 4x - 96$ 16. $x^2 - x - 56$

17. $x^2 - 2x - 24$ 18. $x^2 - 3x - 54$

19. $x^2 - 6x - 27$ 20. $x^2 - 5xy - 24y^2$

21. $x^2 - 2xy - 35y^2$ 22. $x^2 - xy - 20y^2$

23. $x^2 - 6xy - 16y^2$ 24. $x^2 - xy - 42y^2$

Factorising Expression of the Type $ax^2 + bx + c$, where $a > 1$

One method of factorising expressions of the type $ax^2 + bx + c$ is known as the grouping method. We use two examples to illustrate this method.

Examples

Factorise $3x^2 - 10x - 8$

Begin by multiplying the coefficient of the term in x^2, and the constant term i.e. $3(-8) = -24$. Next look for two factors of the product -24 whose sum is the coefficient of the middle term -10. Now the two factors whose sum is -10 are -12 and 2. Express the middle term as the sum $-12x + 2x$, using the two factors, and then factorise by grouping.

$$3x^2 - 10x - 8 = 3x^2 - 12x + 2x - 8$$

$$= 3x(x - 4) + 2(x - 4)$$

$$= (x - 4)(3x + 2)$$

Try this 24

Factorise $2x^2 + 7x - 15$

Factorise $6x^2 - 19x + 10$

The product of the coefficient of the term in x^2 and the constant term is $6(10) = 60$. The two factors of 60 whose sum is -19 are -4 and -15. Rewrite the middle term as $-4x - 15x$. So

$$6x^2 - 19x + 10 = 6x^2 - 4x - 15x + 10$$

$$= 2x(3x - 2) - 5(3x - 2)$$

$$= (3x - 2)(2x - 5)$$

Try this 25

Factorise $3x^2 - 10x + 8$

Exercise 4.5(c)

Factorise:

1. $2x^2 + 9x + 4$

2. $3x^2 + 11x + 6$

3. $5x^2 + 26x - 24$

4. $6x^2 - 5x - 6$

5. $10x^2 - 9x - 9$

6. $21x^2 - 2x - 8$

7. $24x^2 - 10x - 21$

8. $6x^2 + 17x - 45$

9. $2x^2 + 11x + 15$

10. $15x^2 - 19x - 10$

11. $12x^2 - 29x + 15$

12. $8x^2 + 18x - 35$

13. $6x^2 - x - 15$

14. $5x^2 + 2x - 3$

15. $3x^2 + 10x + 8$

Factorising Perfect Squares

You may recall that squares of binomials result in perfect square trinomials.

$$(x + y)^2 = x^2 + 2xy + y^2$$

$$(x - y)^2 = x^2 - 2xy + y^2$$

A perfect square trinomial can be factorised as follows: find the square root of the first and last terms and add if the middle term is positive or subtract if the middle term is negative.

Examples

Factorise $x^2 + 16x + 64$

The expression is a perfect square, so we have

$$x^2 + 16x + 64 = (x + 4)^2$$

Try this 26

Factorise $x^2 + 8x + 16$

Factorise $x^2 - 10x + 25$

$$x^2 - 10x + 25 = (x - 5)^2$$

Try this 27

Factorise $x^2 - 18x + 81$

Exercise 4.5(d)

Factorise the following

1. $x^2 + 6x + 9$ 2. $x^2 + 10x + 25$

3. $x^2 + 14x + 49$ 4. $x^2 + 12x + 36$

5. $x^2 + 18x + 81$ 6. $x^2 + 24x + 144$

7. $x^2 + 30x + 225$ 8. $4x^2 + 16xy + 16y^2$

9. $9x^2 + 30xy + 25y^2$ 10. $16x^2 + 24xy + 9y^2$

11. $25x^2 + 40xy + 16y^2$ 12. $4x^2 + 20xy + 25y^2$

13. $x^2 - 4x + 4$ 14. $x^2 - 12x + 36$

15. $x^2 - 16x + 64$ 16. $x^2 - 20x + 100$

17. $x^2 - 22x + 121$ 18. $x^2 - 26x + 169$

19. $x^2 - 12xy + 36y^2$ 20. $x^2 - 10xy + 25y^2$

21. $9x^2 - 12xy + 4y^2$ 22. $4x^2 - 20xy + 25y^2$

23. $16x^2 - 24xy + 9y^2$ 24. $49x^2 - 126xy + 81y^2$

Factorising the Difference of two Squares

Recall the expression for the product of a sum and difference of two terms

$$(a + b)(a - b) = a^2 - b^2$$

This also means that a binomial of the form $a^2 - b^2$, called difference of two squares, has its factors as $a + b$ and $a - b$. To factorise a difference of two squares, we reverse the expression.

1. Find the square root of each term

2. Write two binomials that are the sum and difference of those

   square roots.

$$a^2 - b^2 = (a + b)(a - b)$$

Example

Factorise $x^2 - 81$

Since $x^2 - 81$ is a difference of squares, we have

$$x^2 - 81 = (x)^2 - (9)^2$$

$$= (x - 9)(x + 9)$$

Try this 28

Factorise $x^2 - 25$

Factorise $25x^2 - 36y^4$

$$25x^2 - 36y^4 = (5x)^2 - (6y^2)^2$$

$$= (5x - 6y^2)(5x + 6y^2)$$

Try this 29

Factorise $16x^6 - 25y^4$

Factorise $18x^2 - 32$

Notice that 2 is a common factor. If all the terms contain a common factor, it should be taken out first.

$$18x^2 - 32 = 2(9x^2 - 16)$$

$$= 2(3x - 4)(3x + 4)$$

Try this 30

Factorise $8x^2 - 50$

Exercise 4.5(e)

Factorise the following

1. $x^2 - 16$ 2. $x^2 - 64$ 3. $x^2 - 144$

4. $9y^2 - 64$ 5. $4x^2 - 25$ 6. $169 - x^2$

7. $25x^2 - 16y^2$ 8. $27x^2 - 12y^2$ 9. $8x^2 - 72y^2$

10. $5x^3 - 45x$ 11. $3x^4 - 48$ 12. $1 - 49x^2$

13. $20x^3 - 45xy^2$ 14. $32x^2y - 18y^3$ 15. $50x^3 - 8xy^2$

Using the Difference of Two Squares

We can sometimes use the identity for the difference of two squares to simplify a calculation.

Example

Find $8.4^2 - 1.6^2$

$$8.4^2 - 1.6^2 = (8.4 - 1.6)(8.4 + 1.6)$$

$$= 6.8 \times 10.0$$

$$= 68$$

Try this 31

Find $17.3^2 - 2.7^2$

Find 83×77

$$83 \times 77 = (80 + 3)(80 - 3)$$

$$= 80^2 - 3^2$$

$$= 6400 - 9$$

$$= 6391$$

Try this 32

Find 53×47

Exercise 4.5(f)

1. $12.8^2 - 7.2^2$ 2. $8.4^2 - 1.6^2$ 3. $86.5^2 - 13.5^2$

4. $73^2 - 27^2$ 5. $836^2 - 164^2$ 6. 88×92

7. 204×196 8. 75×85 9. 57×63

10. 1010×990

4.6 Algebraic Fractions

An algebraic expression of the form $\frac{a}{b}$, $b \neq 0$ is known as algebraic fraction. Some examples of algebraic fractions are:

$$\frac{ax}{by}, \qquad \frac{2x}{3x+4}, \qquad \frac{x^2 + 5x + 6}{x^2 - 9}$$

Equivalent Fractions

Consider the algebraic fraction $\frac{3}{x}$. If we multiply both the numerator and denominator by xy, we obtain $\frac{3xy}{x^2y}$. If $x = 3$ and $y = 2$, then $\frac{3}{x} = \frac{3}{3} = 1$ and $\frac{3xy}{x^2y} = \frac{3(3)(2)}{(3^2)(2)} = 1$. The two fractions look different but they have the same value. The two fractions are said to be equivalent. The value of a fraction does not change when the numerator and the denominator is multiplied or divided by the same number.

Simplifying Fractions

An algebraic fraction is said to be in simplified form if its numerator and denominator have no common factors. To simplify algebraic fractions, you will use exactly the same procedure used in simplifying common fractions. Factorise both the numerator and the denominator, and then divide both the numerator and the denominator by the common factor.

Examples

Simplify $\frac{9x^2y}{12xy^2}$

Look for a factor common to both the numerator and the denominator. Then divide the common factor out of both the numerator and the denominator.

$$\frac{9x^2y}{12xy^2} = \frac{3xy \cdot 3x}{3xy \cdot 4y}$$

Dividing the numerator and denominator by $3xy$, we have

$$\frac{9x^2y}{12xy^2} = \frac{3x}{4y}$$

The simplified form is $\frac{3x}{4y}$

Try this 33

Simplify:

(a) $\frac{18x^2y}{15xy^4}$ (b) $\frac{16y^3z}{32y^2z^3}$

Simplify $\frac{6xy}{3x^2-15x}$

Factorise both the numerator and denominator and then divide out the common factor.

$$\frac{6xy}{3x^2 - 15x} = \frac{3x \cdot 2y}{3x(x - 5)} = \frac{2y}{x - 5}$$

The simplified form is $\frac{2y}{x-5}$

Try this 34

Simplify:

(a) $\frac{4xy-12x}{8xy}$ (b) $\frac{5x^2y}{10x^3+15x^2}$

Simplify $\frac{x^2-2x-15}{x^2-25}$

Factorise both the numerator and denominator and then divide out the common factor.

$$\frac{x^2 - 2x - 15}{x^2 - 25} = \frac{(x - 5)(x + 3)}{(x - 5)(x + 5)} = \frac{x + 3}{x + 5}$$

The simplified form is $\frac{x+3}{x+5}$

Try this 35

Simplify:

(a) $\frac{4x^2-12x}{x^2+5x-24}$ (b) $\frac{x^2+x-6}{3xy+9y}$

Exercise 4.6(a)

Simplify the following

1. $\dfrac{20a^2b}{30ab^2}$

2. $\dfrac{10x^2y^3}{12x^3y^2}$

3. $\dfrac{12xy}{9x^2y}$

4. $\dfrac{3a+12b}{a^2+4ab}$

5. $\dfrac{4x^2-1}{2x^2+x}$

6. $\dfrac{2xy+8x}{2xy^2-32x}$

7. $\dfrac{x^3-16x}{x^2-2x-8}$

8. $\dfrac{x^2+2x-15}{3x^2-9x}$

9. $\dfrac{x^2-x-6}{x^2-9}$

10. $\dfrac{x^2-3x-28}{x^2-x-42}$

11. $\dfrac{2x^2-9x+4}{12+x-x^2}$

12. $\dfrac{x^2+4x-21}{x^2+15x+56}$

13. $\dfrac{x^2-4y^2}{x^2+4xy+4y^2}$

14. $\dfrac{x^2-4x+4}{4-x^2}$

15. $\dfrac{x+3y}{9y^2-x^2}$

Multiplication of Algebra Fractions

The rule for multiplying algebraic fractions is the same as the rule for multiplying numerical fractions. The numerator and the denominator of the product are obtained by multiplying the numerators and the denominators of the fractions. The new fraction is then written in simplified form. Sometimes, it is usually most convenient to divide out the common factor(s) before actually performing the multiplication.

Example

Multiply $\dfrac{12xy^2}{15x^3}\cdot\dfrac{20xy}{8y^2}$

$$\dfrac{12xy^2}{15x^3}\cdot\dfrac{20xy}{8y^2}=\dfrac{240x^2y^3}{120x^3y^2}=\dfrac{2y}{x}$$

You can simplify the fraction first before you multiply.

$$\frac{12xy^2}{15x^3} \cdot \frac{20xy}{8y^2} = \frac{4y^2}{5x^2} \cdot \frac{5x}{2y} = \frac{2y}{x}$$

Multiply: $\frac{x^2-16}{x^2+4x} \cdot \frac{2x^2-6x}{x^2-7x+12}$

Factorise the numerator and denominator of each fraction. Then divide out the common factors.

$$\frac{x^2-16}{x^2+4x} \cdot \frac{2x^2-6x}{x^2-7x+12} = \frac{(x-4)(x+4)}{x(x+4)} \cdot \frac{2x(x-3)}{(x-3)(x-4)}$$

$$= 2$$

Multiply: $\frac{x^2+5x+6}{x^2-x-6} \cdot \frac{x^2+3x-10}{x^2+8x+15}$

Factorise the numerator and denominator of each fraction. Then divide out the common factors.

$$\frac{x^2+5x+6}{x^2-x-6} \cdot \frac{x^2+3x-10}{x^2+8x+15} = \frac{(x+2)(x+3)}{(x+2)(x-3)} \cdot \frac{(x-2)(x+5)}{(x+3)(x+5)}$$

$$= \frac{x-2}{x-3}$$

Try this 36

Multiply:

(a) $\frac{3x^2}{5y^2} \cdot \frac{10y^5}{15x^3}$ 　　　(b) $\frac{x^2-81}{x^2+9x} \cdot \frac{5x^2}{x^2-7x-18}$

Exercise 4.6(b)

Multiply the following

1. $\frac{6x}{8y} \cdot \frac{20y^2}{18x^2}$ 　　　2. $\frac{5x^3}{3x} \cdot \frac{9}{20x}$ 　　　3. $\frac{8xy^5}{5x^3y^2} \cdot \frac{15y^2}{16xy^3}$

4. $\dfrac{2x^2y}{3x^3y^2} \cdot \dfrac{6xy^2}{8x}$ 5. $\dfrac{7xy}{3xz} \cdot \dfrac{15x^2z^2}{21x^2y^2}$ 6. $\dfrac{x^2-y^2}{4x} \cdot \dfrac{4}{x+y}$

7. $\dfrac{3x^2y}{12x^2y} \cdot \dfrac{8x^2}{xy+2y^2}$ 8. $\dfrac{4x+12y}{8xy} \cdot \dfrac{3x^2y}{x^2+3xy}$ 9. $\dfrac{a^2+b^2}{x+2y} \cdot \dfrac{x^2-4y^2}{a^3+ab^2}$

10. $\dfrac{a^2-b^2}{a^2+ab} \cdot \dfrac{a+b}{(a-b)^2}$ 11. $\dfrac{x^2+3x}{x^2-9} \cdot \dfrac{x^2-3x}{x^2}$ 12. $\dfrac{x^2-9}{x^2+4x+3} \cdot \dfrac{2x+6}{2x-6}$

Dividing Algebraic Fractions

To divide, we invert the divisor and multiply using the steps for multiplying algebraic fractions. The two examples below illustrate this approach.

Examples

Divide the following

(a) $\dfrac{8x^2}{10y^3} \div \dfrac{12x^2}{25y^2}$ (b) $\dfrac{x^2-x-12}{x^2-5x} \div \dfrac{x^2-16}{x^2-x-20}$

First invert the divisor and then multiply as described above. Do not simplify before it is written as a multiplication.

(a) $\dfrac{8x^2}{10y^3} \div \dfrac{12x^2}{25y^2} = \dfrac{8x^2}{10y^3} \cdot \dfrac{25y^2}{12x^2}$

$= \dfrac{5}{3y}$

(b) First invert the divisor. Then factorise the numerator and denominator of each fraction and multiply as described above.

$\dfrac{x^2-x-12}{x^2-5x} \div \dfrac{x^2-16}{x^2-x-20} = \dfrac{x^2-x-12}{x^2-5x} \cdot \dfrac{x^2-x-20}{x^2-16}$

$= \dfrac{(x+3)(x-4)}{x(x-5)} \cdot \dfrac{(x-5)(x+4)}{(x-4)(x+4)}$

$= \dfrac{x+3}{x}$

Try this 37

Divide the following

(a) $\dfrac{3x^2y}{8xy^3} \div \dfrac{9x^3}{4y^4}$

(b) $\dfrac{x^2-x-6}{2x-6} \div \dfrac{x^2-4}{4x^2}$

Exercise 4.6(c)

Divide the following

1. $\dfrac{4xy}{5z} \div \dfrac{8y^2}{15z^2}$

2. $\dfrac{4x^2y^2}{9x^3} \div \dfrac{8y^2}{27xy}$

3. $\dfrac{8x^3y}{27xy^3} \div \dfrac{16x^3y}{45y}$

4. $\dfrac{6x^2y^2}{14ab^2} \div \dfrac{12xy^2}{7b^2}$

5. $\dfrac{3x^3}{2y^4} \div \dfrac{6x^2}{4y^3}$

6. $\dfrac{12xy}{x^2-y^2} \div \dfrac{3y}{x-y}$

7. $\dfrac{x}{x-2} \div \dfrac{x^2}{x^2-4}$

8. $\dfrac{3}{x^2-16} \div \dfrac{6}{2x+8}$

9. $\dfrac{4x-12}{5x+15} \div \dfrac{8x^2}{x^2+3x}$

10. $\dfrac{a^2+2ab}{a^2} \div \dfrac{ab+2b^2}{b^2}$

11. $\dfrac{x^2-xy}{9x^2y} \div \dfrac{y-x}{6x}$

12. $\dfrac{a+b}{a-b} \div \dfrac{a^2+2ab+b^2}{a-b}$

Addition and Subtraction of Algebraic Fractions

Addition and subtraction of algebraic fractions follow the rules for adding and subtracting numerical fractions.

Finding the Least Common Denominator (LCD)

The least common denominator (LCD) is the lowest common multiple (LCM) of the denominators of the fractions. The lowest common multiple of the denominators is the smallest product that contains every factor from every term.

To find the LCD factorise each denominator completely, and then find the product of the highest power of each factor that appears in any denominator. When the denominators have no common factors, their LCM is simple their product.

Examples

Find the LCM of the following:

(a) $3a, 2b$ and c (b) $4x^2, 6x^2y$ and $3y^3$

(c) $3x + 6, 4x + 12$ and $x^2 + 5x + 6$

(a) Since, $3a, 2b$ and c have no common factor, the LCD is the product $3a \cdot 2b \cdot c = 6abc$

(b) First factorise each expression, writing any numerical factor as prime factors.

$4x^2 = 2 \cdot 2 \cdot x \cdot x$

$6x^2y = 2 \cdot 3 \cdot x \cdot x \cdot y$

$3y^3 = 3 \cdot y \cdot y \cdot y$

The LCM is the product of the highest power of each distinct factor. Hence, the LCM is the product $2 \cdot 2 \cdot 3 \cdot x \cdot x \cdot y \cdot y \cdot y$ or $12x^2y^3$

(c) First factorise each expression completely, writing any numerical factor as prime factors

$3x + 6 = 3(x + 2)$

$4x + 12 = 2 \cdot 2 \cdot (x + 3)$

$x^2 + 5x + 6 = (x + 2)(x + 3)$

The LCM is $2 \cdot 2 \cdot 3 \cdot (x + 2) \cdot (x + 3) = 12(x + 2)(x + 3)$

Try this 38

Find the LCM of the following

(a) $2x, \ 3y, \ 5z$ (b) $4x^2y, \ 6xy^3, \ 10x^3y^2$

(c) $2x^2 + 6x, \ x^2 + x - 6$

Exercise 4-6(d)

Find the LCM of the following

1. $3x, \ 4x^2, \ 12x$ 2. $2x^2, \ 3xy, 6y^2$

3. $a - b, \ a + b$ 4. $3x + 3, \ x^2 - 1$

5. $2x + y, \ 2x - y, \ 4x^2 - y^2$ 6. $x - 2, \ x^2 + x - 6$

7. $x + 3, \ x^2 + 3x, \ x^2 + 5x + 6$

8. $x^2 - 5x + 6, \ x^2 - 7x + 12, \ x^2 - 6x + 8$

9. $x + 2, x + 1, \ x^2 + 3x + 2$

10. $x - y, x + y, \ x^3 - xy^2$

Adding and Subtracting Fractions

Example

Work out:

(a) $\dfrac{3x}{x^2+2x} + \dfrac{6}{x^2+2x}$ (b) $\dfrac{6x}{x^2-4} - \dfrac{3}{x-2}$ (c) $\dfrac{3}{x^2+x-6} + \dfrac{2}{x^2+5x+6}$

(a) Since the denominator is the same, add the numerators. Then place the sum over the common denominator and simplify the result if possibly.

$$\frac{3x}{x^2+2x} + \frac{6}{x^2+2x} = \frac{3x}{x(x+2)} + \frac{6}{x(x+2)}$$

$$= \frac{3x+6}{x(x+2)}$$

$$= \frac{3}{x}$$

(b) First find the LCM of the denominators. The LCM is $(x-2)(x+2)$. Next convert each fraction into an equivalent fraction with the common denominator $(x-2)(x+2)$ and add as described above. You can write the common denominator only once.

$$\frac{6x}{x^2-4} - \frac{3}{x-2} = \frac{6x}{(x-2)(x+2)} - \frac{3(x+2)}{(x-2)(x+2)}$$

$$= \frac{3x-6}{(x-2)(x+2)}$$

$$= \frac{3(x-2)}{(x-2)(x+2)}$$

$$= \frac{3}{x+2}$$

(c) First factorise the expression in the denominator and then find the LCM. The LCM is $(x-2)(x+2)(x+3)$. Hence

$$\frac{3}{x^2+x-6} + \frac{2}{x^2+5x+6} = \frac{3}{(x-2)(x+3)} + \frac{2}{(x+2)(x+3)}$$

$$= \frac{3(x+2)+2(x-2)}{(x-2)(x+2)(x+3)}$$

$$= \frac{5x+2}{(x-2)(x+2)(x+3)}$$

Try this 39

Work out:

(a) $\frac{2x+y}{3x} + \frac{2y+x}{3x}$ (b) $\frac{5}{3x-2} - \frac{3}{2x+1}$ (c) $\frac{3}{x^2+5x+6} + \frac{2}{x^2+x-2}$

Exercise 4.6(e)

Work out:

1. $\frac{x}{2} + \frac{x}{2}$ 2. $\frac{3y}{4} - \frac{y}{2}$ 3. $\frac{5}{x} + \frac{3}{2x}$

4. $\frac{5}{2y} - \frac{7}{3y}$ 5. $\frac{x+y}{4} + \frac{x-y}{3}$ 6. $\frac{2x-3}{4} - \frac{3x-2}{6}$

7. $\frac{4x-3y}{4} - \frac{3x+5y}{3}$ 8. $\frac{2x+3}{2} + \frac{4-2x}{4}$ 9. $\frac{7}{2x-3} + \frac{3}{2x+3}$

10. $\frac{12}{x^2-9} - \frac{2}{x-3}$ 11. $\frac{5}{x-y} - \frac{2x}{x^2-xy}$ 12. $\frac{3}{x-y} - \frac{2}{x+y}$

13. $\frac{x}{x^2-x-30} - \frac{1}{x+5}$ 14. $\frac{3}{x^2+4x+3} - \frac{1}{x^2-9}$ 15. $\frac{2}{y^2+y-6} + \frac{3y}{y^2-2y-15}$

16. $\frac{5}{x^2-3x-10} - \frac{4}{x^2+2x-35}$ 17. $\frac{x+2}{x^2-x-12} - \frac{x}{x^2+6x+9}$

18. $\frac{x-1}{x^2+4x+4} + \frac{1-x}{x^2-x-6}$ 19. $\frac{5}{x^2-4} - \frac{7}{x+2}$ 20. $\frac{1}{x-y} - \frac{3}{x+y} + \frac{3x-y}{x^2-y^2}$

Review exercise 4

1. Multiply:

(a) $5x^3 \cdot 3xy^2$ (b) $2x^2y \cdot 3xy^3$ (c) $5y^2 \cdot 3y^3$

(d) $7x^5 \cdot 4x^3$ (e) $-2x^2 \cdot 7x^3$ (f) $4x^5 \cdot -3x$

(g) $4x^3y^2 \cdot 8x^2y$ (h) $-3x^4y^2 \cdot -7x^2y^3$ (i) $-3x^5y^2 \cdot 2x^4y$

(j) $(2x^2y^3)^2$ (k) $(-5ab^2)^3$ (m) $(-3x^2y^3)^4$

2. Simplify the following:

(a) $\frac{15x^6}{5x^2}$ (b) $\frac{-16x^7}{4x^5}$ (c) $\frac{-45x^5}{-15x^3}$

(d) $\frac{40m^3n^2}{-8m^2n}$ (e) $\frac{-72x^4y^6}{9x^3y^5}$ (f) $\frac{35x^5y^2}{7x^2y}$

(g) $\frac{-42m^7n^5}{-6m^4n^3}$ (h) $\frac{36p^5q^3}{6p^3q^2}$ (i) $\frac{-24m^4n^2}{8m^2n^2}$

(j) $\frac{-81x^7y^6}{-27x^4y^5}$

3. Evaluate each of the expressions if $w = -2, x = 5, y = -4$ and $z = 6$

(a) $8x + 2y$ (b) $3w^2 + 4y$ (c) $4y - 2x$

(d) $4(2w - z)$ (e) $y(3w - x)$ (f) $w(x + 3y)$

4. Write an algebraic expression for the following statements

(a) The sum of a and b (b) x increased by 2 (c) 3 more than x

(d) x minus y (e) 6 less than a (f) 8 times x

(g) The product of 5 and the sum of a and b

(h) Twice the difference of x and y

5. Simplify the following

(a) $3x + 4x$ (b) $5a - 7a$ (c) $9y - 5y + 4y$

(d) $8y + 7y - y$ (e) $3x - 2y + 5x + 3y$ (f) $-2a + 4b - 7a - b$

6. Simplify the following

(a) $x(5x + 2)$ (b) $3x(5 - 2x)$ (c) $5y(2y - 1)$

(d) $-5t(3 - 2t)$ (e) $-6x(3y - 2)$ (f) $-y(3y + 2x)$

(g) $2x(3y + x)$ (h) $-(5x - 3y)$ (i) $4x(x + 2y - 3)$

(j) $3x(x - 2y + 1)$ (k) $-2p(p - 3q + 4)$

7. Simplify the following

(a) $2x - (5x - 3)$ (b) $5a - (4a - 3)$

(c) $7y - (2y + 5)$ (d) $3(3y - 1) - 2(y - 5)$

(e) $-3(y^2 - 2) + y^2(y + 3)$ (f) $x(x^2 + 3) - 3(x + 4)$

(g) $3x - 2(4x - 5)$ (h) $7 - 3(x + 2)$

(i) $8y^2 + y - 2(y + 3y^2)$ (k) $4x - 9y - 3(2x - y)$

8. Multiply:

(a) $(x + 6)(x + 3)$ (b) $(x + 5)(x + 2)$ (c) $(x + 5)(x - 2)$

(d) $(x + 6)(x - 2)$ (e) $(a - 6)(a - 7)$ (f) $(y - 4)(y - 8)$

(g) $(x - 3)(x + 3)$ (h) $(x + 6)(x - 6)$ (i) $(3x - 2)(3x + 2)$

(j) $(x + 3)^2$ (k) $(2x - 3)^2$ (m) $(3x + 2)^2$

9. Factorise:

(a) $2x^2 + 6x$ (b) $10t^2 - 5t$ (c) $5x^3 + 10x^2$

(d) $12y^2 - 6y$ (e) $27pq + 18p^2q$ (f) $12x^2 + 8x$

(g) $8xy - 24xy^2$ (h) $3x^2y + 6xy^2$ (i) $7x^2y - 21xy^3$

(j) $36x^3y^2 - 8x^2y^3$

10. Factorise by grouping:

(a) $x^2 + 2x + xy + 2y$ (b) $2x^2 - 4x + xz - 2z$

(c) $x^2 + y + x + xy$ (d) $a^2 - 3y - 3a + ay$

(e) $x^3 + 3x^2 - 4x - 12$ (f) $2x^3 + 12x^2 - 5x - 30$

11. Factorise

(a) $x^2 + 6x + 5$ (b) $x^2 + 7x + 10$ (c) $y^2 + 11y + 28$

(d) $y^2 + 4y - 45$ (e) $x^2 + 7x - 60$ (f) $x^2 - 2x - 15$

(g) $a^2 - 4a - 12$ (h) $x^2 - 8x + 15$ (i) $a^2 - 7a + 10$

(j) $y^2 - 11y + 10$ (k) $x^2 + x - 42$ (l) $a^2 - 2a - 35$

12. Factorise

(a) $2x^2 + 7x - 4$ (b) $6x^2 - 23x + 7$ (c) $7x^2 + 15x + 2$

(d) $9a^2 - 6a - 8$ (e) $4a^2 - 4a - 15$ (f) $3x^2 - 5x - 2$

13. Factorise:

(a) $x^2 - 18x + 81$ (b) $x^2 + 14x + 49$ (c) $16x^2 - 24x + 9$

(d) $4x^2 + 12x + 9$ (e) $64x^2 + 16x + 1$ (f) $9x^2 - 30x + 25$

14. Factorise:

(a) $x^2 - 36$ (b) $x^2 - 9$ (c) $9y^2 - 4$

(d) $16a^2 - 9$ (e) $16a^2 - 9b^2$ (f) $18x^2 - 8y^2$

(g) $49a^2 - 16b^2$ (h) $75xy^6 - 3x^5$

15. Simplify the following:

(a) $\dfrac{4x^5}{6x^2}$ (b) $\dfrac{10x^2y}{15x^3}$ (c) $\dfrac{12a^2b^5}{16ab^2}$

(d) $\dfrac{18x^5y^4}{24x^3y^6}$

(e) $\dfrac{-15x^3y^4}{-20x^2y^5}$

(f) $\dfrac{-4x^3y}{6xy^2}$

(g) $\dfrac{14x^3y^2}{-21x^2y^5}$

(h) $\dfrac{-8a^2b^4}{-16a^4b^2}$

(i) $\dfrac{x^2+3x+2}{3x+6}$

(j) $\dfrac{5y^2+15y}{y^2-2y-15}$

(k) $\dfrac{x^2-16}{x^2-x-20}$

(l) $\dfrac{x^2-4x+4}{x-2}$

(m) $\dfrac{x^2+x-6}{x^2-3x+2}$

(n) $\dfrac{x^2+xy-6y^2}{4y^2-x^2}$

(o) $\dfrac{ab-2b+4a-8}{2b+6-ab-3a}$

16. Multiply:

(a) $\dfrac{3x^4y^2}{10x^2y^4} \cdot \dfrac{5x^2y^3}{9x^2y^3}$

(b) $\dfrac{8x^2y^6}{5x^4y^3} \cdot \dfrac{15y^3}{16x^2y^4}$

(c) $\dfrac{-4a^2b^3}{15a^4} \cdot \dfrac{25a^2b^2}{-16b^4}$

(d) $\dfrac{7x^2y^3}{-12x^3y^2} \cdot \dfrac{-24x^4y^6}{21x^3y^8}$

(e) $\dfrac{x^2-5x}{3x^2} \cdot \dfrac{10x}{5x-25}$

(f) $\dfrac{x^2+3x-10}{5x} \cdot \dfrac{15x^2}{3x+15}$

(g) $\dfrac{x^2+8x}{4x} \cdot \dfrac{12x}{x^2-64}$

(h) $\dfrac{x^2-81}{x^2+9x} \cdot \dfrac{5x^2}{x^2-7x-18}$

(i) $\dfrac{2x-6}{x^2+2x} \cdot \dfrac{x^2+4x+4}{3-x}$

(j) $\dfrac{3x-15}{x^2+3x} \cdot \dfrac{x^2+x-6}{5-x}$

17. Divide:

(a) $\dfrac{4x^3y^3}{9x^4} \div \dfrac{8y^3}{27x^2y^2}$

(b) $\dfrac{8x^3y^2}{27x^2y^4} \div \dfrac{16x^3y^2}{45y^2}$

(c) $\dfrac{3x-6}{8} \div \dfrac{5x-10}{6}$

(d) $\dfrac{x^2+2x}{4x} \div \dfrac{6x+12}{8}$

(e) $\dfrac{4a+12}{5a-15} \div \dfrac{8a^2}{a^2-3a}$

(f) $\dfrac{6p+18}{9p} \div \dfrac{3p+9}{p^2-2p}$

(g) $\dfrac{16x}{4x^2-16} \div \dfrac{4x+12}{x^2+x-6}$

(h) $\dfrac{x^2-9}{2x^2-6x} \div \dfrac{2x^2+5x-3}{4x^2-1}$

(i) $\dfrac{x^2-2x-8}{9x^2} \div \dfrac{x^2-16}{3x+12}$

(j) $\dfrac{x^2-9y^2}{4x^2+12xy} \div \dfrac{x^2-xy-6y^2}{12xy}$

18. Work out:

(a) $\dfrac{3}{7xy} + \dfrac{4}{7xy}$

(b) $\dfrac{8}{3ab} - \dfrac{2}{3ab}$

(c) $\dfrac{7x}{x+3} + \dfrac{21}{x+3}$

(d) $\dfrac{6x-y}{4y} - \dfrac{2x+3y}{4y}$ (e) $\dfrac{2}{x+1} - \dfrac{3}{x+3}$ (f) $\dfrac{3}{x^2-9} + \dfrac{1}{2x+6}$

(g) $\dfrac{2}{y^2-y-6} + \dfrac{3y}{y^2+2y-15}$ (h) $\dfrac{2a}{a^2+a-12} - \dfrac{3}{a^2+2a-8}$

(i) $\dfrac{6x}{x^2-9} - \dfrac{5x}{x^2-x-6}$ (j) $\dfrac{4y}{y^2-6y+5} + \dfrac{3y}{y^2-1}$

Chapter Test 4

Take this test as you would take a test in class. After you are done, check your work against the answers in the back of the book.

1. Simplify

(a) $3xy^2 \cdot 2x^2y^3$ (b) $(-2x^2y^3)^2$ (c) $\dfrac{-48x^3y^5}{-16x^2y^3}$

2. Expand:

(a) $3x(2y - x)$ (b) $-2p(3p - 4q)$ (c) $xy(y - x)$

3. Simplify:

(a) $5p - 2q - (p + q)$ (b) $5x - 2(3 + 2x)$

(c) $6(y + 1) - 2(2y - 3)$

4. Simplify:

(a) $4y + 3y - 2y$ (b) $3a - 2b - 4a + 5b$

(c) $9k - 4t - 3k + 5t$

5. Write algebraic expression for each statement

(a) 7 more than x (b) 8 less than a (c) 12 times b

(d) Three times the quotient of a and b

6. Evaluate each of the expression if $a = -2, b = 7$ and $c = -8$

(a) $a - 3b - 2c$ (b) $-3a + b + c$ (c) $a^2b + 4c$

7. Multiply:

(a) $(x + 7)(x - 8)$ (b) $(2x - 3)(2x + 3)$ (c) $(3x - 2)^2$

8. Factorise:

(a) $2x^2 - 6x$ (b) $14x^2y + 7x^3$ (c) $-18ab^2 + 12a^2b$

9. Factorise:

(a) $x^2 - 2xy - 3x + 6y$ (b) $x^3 - 5x^2 + 3x - 15$

(c) $2pr + ps - 6qr - 3qs$

10. Factorise:

(a) $x^2 + 8x - 20$ (b) $x^2 - xy - 42y^2$ (c) $x^2 + 7xy + 10y^2$

11. Factorise:

(a) $3x^2 + 4x - 7$ (b) $7x^2 + 10x + 3$

12. Factorise:

(a) $x^2 + 18x + 81$ (b) $4x^2 - 12xy + 9y^2$ (c) $81x^2 - 100y^6$

13. Use the difference of two squares to evaluate

(a) $5.8^2 - 4.2^2$ (b) $73.6^2 - 26.4^2$ (c) $973^2 - 27^2$

14. Simplify:

(a) $\dfrac{6a^3b}{8a^2b^2}$ (b) $\dfrac{4a^2+8a}{6a+12}$ (c) $\dfrac{x^2-x-6}{x^2-7x+12}$

15. Simplify:

(a) $\dfrac{5x}{3y^2} \cdot \dfrac{9y^3}{10x^2}$ (b) $\dfrac{x-1}{x^2} \div \dfrac{x^2-1}{x^3}$ (c) $\dfrac{12}{x^2-9} - \dfrac{2}{x-3}$

5

Linear Equations and Inequalities

5.1 Linear Equations

An equation is a statement that two algebraic expressions on either side of an equal sign are equal. Some examples are $x = 3$, $2x + 3 = 7$, $3x + 15 = 3(x + 5)$, and $x^2 - x - 6 = 0$. The expression on the left is called the left hand side and the expression on the right is called the right hand side.

An equation may be either true or false. However, some equations are true for all values of the variable. Such equations are called identities, and their solution set is the set of all real numbers. For example, the equation $3x + 15 = 3(x + 5)$ is an identity.

Here, we will consider equations in one variable, where the index of the variable is 1. These equations are called linear equations.

Solution of an Equation

A solution for an equation is any value for the variable that makes the equation a true statement. For instance, 2 is a solution of $3x + 4 = 10$ because $3(2) + 4 = 10$ is a true statement.

Equivalent Equations

Equations that have the same solution are called equivalent equations. The following are all equivalent equations:

$$3x - 2 = 4, \ x + 5 = 7 \text{ and } x = 2$$

You can transform an equation to an equivalent equation by:

1. adding the same number to both sides of the equation

2. subtracting the same number from both sides of the equation

3. multiplying both sides of the equation by the same number

   (except 0)

4. dividing both sides of the equation by the same number (except

   0)

5. simplifying both sides of the equation

Solving Linear Equations

We say that a linear equation is solved when it is transformed to an equivalent equation of the form $x = a$, where a is some number. That is, solving an equation is a process of writing a set of equivalent equations until you isolate the variable on one side.

Example

Solve:

(a) $2x + 5 = 11$ (b) $3x - 2 = 10$

(a) $2x + 5 = 11$

     $2x + 5 - 5 = 11 - 5$ Subtract 5 from both sides

         $2x = 6$ Simplify both sides

         $x = 3$ Divide both sides by 2

(b) $3x - 2 = 10$

     $3x - 2 + 2 = 10 + 2$ Add 2 to both sides

         $3x = 12$ Simplify both sides

         $x = 4$ Divide both sides by 3

Try this 1

Solve:

(a) $3x + 2 = 5$ (b) $2x - 3 = 7$

Exercise 5.1(a)

Solve the following equations

1. $5x + 3 = 18$ 2. $3x + 2 = 8$ 3. $1 + 2x = 7$

4. $4x - 7 = 5$ 5. $6x - 5 = 7$ 6. $3x - 2 = 4$

7. $15 = 3 - 2x$ 8. $5x + 2 = 17$ 9. $1 - 3x = 7$

10. $7x - 2 = 12$ 11. $4x + 1 = 2$ 12. $2x + 5 = 8$

You may have noticed that when any term is removed from either side of an equation by addition or subtraction, it appears on the other side but with the opposite sign. We work two examples to illustrate this.

Example

Solve:

(a) $4x - 3 = 9$ (b) $5x + 2 = 12$

(a) $4x - 3 = 9$

$$4x = 9 + 3 \qquad \text{Add 3 to both sides}$$

$$4x = 12 \qquad \text{Simplify the right side}$$

$$x = 3 \qquad \text{Divide both sides by 4}$$

(b) $5x + 2 = 12$

$$5x = 12 - 2 \qquad \text{Subtract 2 from both sides}$$

$$5x = 10 \qquad \text{Simplify the right side}$$

$$x = 2 \qquad \text{Divide both sides by 5}$$

Try this 2

Solve:

(a) $5x + 3 = 18$ (b) $2x - 5 = 7$

Exercise 5.1(b)

Solve the following equations

1. $6x - 15 = 9$ 2. $9x - 2 = 7$ 3. $4x + 5 = 9$

4. $3x + 4 = 1$ 5. $8 + 5x = -2$ 6. $9 - 2x = 5$

7. $12 = 3x + 6$ 8. $18 = 8 - 5x$ 9. $4x + 3 = -9$

10. $7x - 4 = 17$ 11. $9x + 2 = 11$ 12. $3 - 7x = 24$

Equations with variables on both sides

Sometimes you will be asked to solve an equation with variable terms on both sides. In such cases, rewrite the equation with all like terms on one side.

Examples

Solve $5x - 3 = 3x + 7$

We must first bring all terms with the variable to one side of the equation and all constant terms to the other side. This can be done by adding 3 to both sides, to get all constant terms on the right, and then subtract $3x$ from both sides, to get all variable terms on the left. You can do both steps in one step.

$$5x - 3 = 3x + 7$$

$$5x - 3 + 3 = 3x + 7 + 3 \qquad \text{Add 3 to both sides}$$

$$5x = 3x + 10$$

$$5x - 3x = 3x - 3x + 10 \qquad \text{Subtract } 3x \text{ from both sides}$$

$$2x = 10$$

$$x = 5 \qquad \text{Divide by 2}$$

Solve $4x + 5 = 16 - 7x$

$$4x + 5 = 16 - 7x$$

$$4x + 7x = 16 - 5 \qquad \text{Add } 7x \text{ and subtract 5 in one step}$$

$$11x = 11$$

$$x = 1 \qquad \text{Divide by 11}$$

Try this 3

Solve:

(a) $5x + 12 = 2x + 3$ \qquad (b) $1 - 2x = 3x - 9$

Exercise 5.1(c)

Solving the following equations

1. $3x = 15 - 2x$ \qquad 2. $5x - 16 = x$

3. $9x = 2x - 21$ \qquad 4. $-6y = 12 - 2y$

5. $9x - 7 = 4x + 8$ \qquad 6. $7x - 4 = 3x + 12$

7. $8 - 5x = 7x - 16$ \qquad 8. $3x - 7 = 11 - 6x$

9. $8x - 3 - 2x = -3x + 9$ \qquad 10. $4x + 5 - 3x = 5x - 3$

11. $4x - 8 = 7x + 7$ 12. $7x + 1 + 5x = 6x + 25$

Equations with Brackets

To solve equations containing brackets, you should first multiply out the brackets.

Examples

Solve $3(2x - 1) = 9$

$3(2x - 1) = 9$

$6x - 3 = 9$ Remove bracket

$6x = 12$

$x = 2$ Divide both sides by 6

Try this 4

Solve $2(3x + 4) = 14$

You must be careful if a negative sign precedes a bracket. The sign of each term inside the bracket must be changed.

Solve $4 - 3(x - 1) = 1$

$4 - 3(x - 1) = 1$

$4 - 3x + 3 = 1$ Remove bracket

$-3x = -6$

$x = 2$ Divide by -3

Try this 5

Solve $1 - 2(x + 3) = 5$

Exercise 5.1(d)

Solve the following equations

1. $2(x - 4) = 6$ 2. $5(x + 3) = 20$

3. $4(x - 2) = 5(x - 3)$ 4. $3(2x + 5) = 2(x + 9)$

5. $3(x + 7) = 7x - 3$ 6. $8 - (5x - 6) = 2x$

7. $1 - 2(x + 3) = 7$ 8. $3 - 4(x - 2) = 2(4 - 3x)$

9. $4(2x - 7) - 3(x - 5) = 2$ 10. $5(x - 2) - 6 = 8 + 3(x + 4)$

Equations with Fractional Coefficients

Equations containing one or more fractions are generally easier to solve if you first clear all the fractions by multiplying all terms on each side of the equation by the least common denominator (LCD).

Example 7

Solve:

(a) $\frac{1}{3}x + 2 = \frac{1}{2}x$ (b) $\frac{1}{4}(x + 1) = \frac{1}{3}(x + 2)$ (c) $\frac{x+3}{2} - \frac{x-5}{4} = \frac{5}{8}$

(a) First, find the LCM of the numbers in the denominator. The

     denominators are 2 and 3. So the LCD is 6.

$$\frac{1}{3}x + 2 = \frac{1}{2}x$$

$$2x + 12 = 3x \qquad \text{Multiply each term by 6}$$

$$12 = x$$

It is usually easier to isolate the variable term on the side that will result in a positive coefficient.

(b) The denominators are 3 and 4. The LCD is 12.

$$\frac{1}{4}(x+1) = \frac{1}{3}(x+2)$$

$$3(x+1) = 4(x+2) \qquad \text{Multiply both sides by 12}$$

$$3x+3 = 4x+8 \qquad \text{Remove brackets}$$

$$-5 = x$$

(c) The denominators are 2, 4 and 8. The LCD is 8.

$$\frac{x+3}{2} - \frac{x-5}{4} = \frac{5}{8}$$

$$4(x+3) - 2(x-5) = 5 \qquad \text{Multiply each term by 8}$$

$$4x+12 - 2x+10 = 5 \qquad \text{Remove brackets}$$

$$2x = -17$$

$$x = -6\frac{1}{2}$$

Any attempt on the part of the beginner to omit the line with the brackets is likely to cause trouble, particularly where negative signs are involved. You must be careful to multiply every term on both sides by the LCD.

Try this 6

Solve:

(a) $\frac{1}{3}x - 3 = \frac{1}{6}x + 2$ (b) $\frac{2}{5}(x-1) = \frac{3}{10}(x+2)$

Exercise 5.1(e)

Solving the following equations

1. $\frac{2}{3}x + 5 = 17$

2. $\frac{3}{4}x - 5 = 4$

3. $\frac{1}{2}x + 3 = \frac{5}{6}x$

4. $\frac{1}{4}(x - 2) = \frac{1}{3}(x + 1)$

5. $\frac{2x-3}{5} = \frac{2+x}{4}$

6. $3 - \frac{1}{2}(x + 8) = \frac{3}{4}x$

7. $\frac{1}{5}(2x + 3) - \frac{3}{2} = \frac{1}{2}(x + 1)$

8. $\frac{x+6}{3} - \frac{x+7}{4} = \frac{x+3}{6}$

9. $\frac{x+6}{5} = \frac{x-4}{10} + \frac{1}{2}$

10. $\frac{2}{3x} + 1 = 3 - \frac{1}{2x}$

11. $5 - \frac{6}{x} = \frac{x+6}{2x}$

12. $\frac{4}{x+5} = \frac{7}{x+8}$

5.2 Solving Word Problems

You can use equations to model and solve problems. In Chapter 4, you may have acquired the skills to translate English phrases to algebraic expressions. If this is the case, then you are ready to use this skills and your equation solving skills to solve word problems.

Translating Verbal Statements into Algebraic Statements

To solve problems requires the ability to translate conditions stated in words into algebraic statements. The following exercises provide a practice.

Exercise 5.2(a)

Express the condition stated in words as algebraic expression

1. 3 more than twice x

2. 6 less than 3 times y

3. The product of twice x and y

4. The sum of a and b, divided by 5

5. A class consists of x students. Five new students joined the class. How many student are now in the class

6. A boy spent GH¢x of his pocket money. If his pocket money is GH¢12, how much did he have left?

7. Kofi is x years old. How old was he 3 years ago?

8. The sum of the ages of two boys is 32 and one of them is x years old. How old is the other boy?

9. If the perimeter of a square is x cm, what is the length of a side?

10. Ama is x years old. How old will she be in 5 years time?

11. The smallest of three consecutive odd numbers is x. What is the odd number next to x?

12. Mr Mensah is 5 years more than three times his son's present age. If his son is x years old, how old is Mr Mensah?

13. Ama had GH¢15. She bought 9 pencils at GH¢x per pencil. How much has she left?

14. A boy has GH¢x and his sister has twice as much in her savings account. If his sister saved a further GH¢50, how much has she got in her saving account?

Translating Verbal Statements into Algebraic Equations

Most problems in algebra are solved by first translating a condition stated in words into an equation. We start with examples to illustrate the process. Here, we are only interested in the setting up of the equation.

Examples

The sum of three consecutive even numbers is 12

First, assign a variable to the unknown and then translate the problem into an equation. Let x be the smallest even number. Then the next larger even number is $x + 2$ and the largest even number is $x + 4$

Therefore $x + (x + 2) + (x + 4) = 12$

3 less than twice a number is 11

Let x represent the number. Then twice the number is $2x$

Therefore $2x - 3 = 11$

The sum of the present ages of a man and his son is 50 years. In 5 years time the man will be twice as old as the son

Let the son's present age be x years

Then the father's age is $(50 - x)$ years

In five years the son will be $(x + 5)$ years old and the father will be $(50 - x) + 5$ years old.

Therefore $(50 - x) + 5 = 2(x + 5)$

The examples given above suggest the following steps:

1. Read the problem carefully, and make sure you completely

    understand what the problem is asking you to do

2. Represent one of the unknown quantities with a variable and try

    to relate all the other unknown quantities to this variable

3. Form an equation that will relate know quantities to the unknown

    quantities

Forming the correct equation is very important. We hope with practice, you will be able to form the correct equation of any given problem. It is important that you state exactly what the variable represents and in what units it is measured if any.

Exercise 5.2(b)

Express the condition stated in words as an algebraic equation

1. One number is thrice another and their sum is 12

2. Twice a number exceeds the same number by 5

3. The sum of three consecutive odd numbers is 4 less than 19

4. If 3 is added to 5 times a number the result is 9 less than twice

    the number

5. The sum of the ages of Obeng and Mensah is 30 years. Three

    years ago Mensah was twice as old as Obeng

6. A boy bought two pens and three pencils. The price of a pen is 5

    Gp more than the price of a pencil. Together they cost GH¢ 8.50

7. A rectangle is twice as long as it is wide and its perimeter is 72 cm

8. One number is four times another number. Their sum is 25

9. Three more than twice a number is nine less than five times the number

10. When 1 is added to a number, and the result is multiplied by 3, you obtained the number you started with

Using Linear Equations to Solve Problems

In this section we combine our ability to translate problems into equations with our skills in solving equations. The two examples below illustrate how we solve word problems.

Examples

5 more than three times a number is 17. Find the number

Represent the number by x, then three times the number is $3x$

5 more than three times the number is $3x + 5$

Thus, $3x + 5 = 17$

Next solve the equation

$3x + 5 = 17$

$3x = 12$

$x = 4$

The number is 4

Try this 7

3 less than twice a number is 7, find the number.

A boy bought 12 pencils from two shops. He bought a certain number of pencils at 50 Gp each at one shop and the rest at 20 Gp each from another shop. How many pencils did he buy from each shop if the total cost of pencils is GH¢ 3.60?

Represent the number of pencils bought at 50 Gp each by x

Then $(12 - x)$ pencils were bought at 20 Gp each. Hence,

$50x + 20(12 - x) = 360$

$50x + 240 - 20x = 360$

$$30x = 120$$

$$x = 4$$

He bought 4 pencils at 50 Gp each and 8 pencils at 20 Gp each

Try this 8

Kofi has 14 coins. Some are 50 Gp coins and some are 20 Gp coins. The total value of the coins is GH¢ 4.60, how many of each coin does Kofi have?

Five hundred tickets were sold for a concert. Adult tickets were GH¢ 4 while student tickets were GH¢ 3. If the total sales were GH¢1650, how many of each type of tickets were sold?

Let x be the number of adult tickets sold

Then $(500 - x)$ student tickets were sold

The total value of adult tickets sold is $4x$ and the total value of student tickets sold is $3(500 - x)$

So $4x + 3(500 - x) = 1650$

$$4x + 1500 - 3x = 1650$$

$$x = 150$$

150 adult tickets and 350 student tickets were sold

Try this 9

At a certain theatre, an adult's ticket cost GH¢ 9 and a child's ticket cost GH¢ 5. One night 2500 people were at the theatre. If the total ticket sales were GH¢ 15,700 that night, how many adults were at the theatre?

Exercise 5.2(c)

Solve the following problems

1. One number is three times a second number. The sum of the two numbers is 12. Find the number.

2. 3 is added to five times a number and the result is multiplied by 2. The final result is 36. What is the number?

3. One number is 5 more than a second number. If two times the smaller plus three times the larger is 80, find the two numbers.

4. A number is added to 4 and the result divided by 3. The answer is the same as you get by subtracting the number from 6 and dividing by 2. What is the number?

5. One number is three times a second number and the difference between the first number and the second number is 18. Find the number.

6. A class consists of 40 students. If there are 12 more boys than girls in the class, how many boys are in the class?

7. The sum of three consecutive even numbers is 6 more than 24. Find the numbers.

8. The sum of three consecutive integers is 99. Find the three integers.

9. The sum of three consecutive odd numbers is 4 less than 91. Find the three numbers.

10. The length of a rectangle is 5 cm more than its width. Find the width if the perimeter is 34 cm.

11. The length of a rectangle is 2 cm less than three times the width. If the perimeter is 60 cm, find the dimensions of the rectangle.

12. Kwame builds a fence around a rectangular garden. The perimeter of the garden is 84 cm. The width is 8 cm less than the length. Find the dimensions of the garden.

13. Ama bought 35 Gp stamps and 15 Gp stamps at the post office. If she purchased 15 stamps at a cost of GH₵ 3.65, how many of

each kind did she buy?

14. Kojo has 5 Gp and 10 Gp coins. There were 55 coins in all. If the total value of the coins is GH¢ 3.75, how many of 5 Gp coins does he have?

15. Tickets for a concert were GH¢ 5 for children and GH¢ 10 for adult. Altogether 200 people attended the concert, and GH¢ 1400 was collected for the tickets. How many adult were at the concert?

16. A student paid GH¢ 18 for 10 exercise books. Some cost GH¢ 1.50 each and the others cost GH¢ 2 each. How many of each kind did he buy?

17. A man is twice as old as his son. If their combined age is 60, find the age of the father.

18. A father is now three times as old as his son. Eight years ago the father's age was five times that of the son. Find their present ages.

19. A man 40 years old has a son 16 years old. How many years ago was the father five times as old as his son?

20. A 36 years old man has a 6 years old son. In how many years will the father be three times as old as the son?

Motion Problems

Motion problems usually involve distance travelled, speed and time. To solve motion problems, we use the relation

$$s = \frac{d}{t}$$

where s = speed, d = distance and t = time.

Examples

Mensah drove from his house to a village in 5 hours. In coming back heavy traffic slowed his speed by 10 km h^{-1}, and the trip took 6 hr. What was his average speed in each direction?

Let x be Mensah's speed to the village. Then he returns with a speed of $x - 10$. You may find the use of a table helpful in solving motion problems. Here we have

Speed	Time	Distance
x	5	$5x$
$x - 10$	6	$6(x - 10)$

Since the distance is the same each way, we have

$5x = 6(x - 10)$

Solving, we have

$5x = 6x - 60$

$-x = -60$

$x = 60$

So Mensah's speed going to the village was 60 km h^{-1} and he return with a speed of 50 km h^{-1}.

Some students may find sketching the given information in motion problems helpful.

Try this 10

Mensah drove to a village 220 kilometres from Accra. He travelled at a steady speed for 3 hours and then increased his speed by 10 km h^{-1} for the remaining 2 hours of the trip. What was his driving speed for each portion of the trip?

Ofori leaves town A for town B at 10 am, driving at 60 km h^{-1}. At 11 am Addo leaves town B for town A, driving at 65 km h^{-1} along the same route. If the towns are 210 km apart, what time will they meet?

Let's t be the time Ofori left A. Then $(t-1)$ is the time Addo left B. Again we summarized the given information in a table.

Speed	Time	Distance
60	t	$60t$
65	$t-1$	$65(t-1)$

Since the distance travelled by Ofori and the distance travelled by Addo must add to 210 km, we have

$$60t + 65(t-1) = 210$$

Solving, we have

$$60t + 65t - 65 = 210 \qquad \text{Multiply bracket out}$$

$$125t = 375$$

$$t = 3$$

Finally, since Ofori left at 10 am, the two will meet at 1 pm

Try this 11

A bus travelling at the speed of 90 km h^{-1} leaves Accra. Two hours later, a car travelling at the speed of 120 km h^{-1} leaves Accra and travel along the same route. How long will it take the car to catch up with the bus?

Exercise 5.2(d)

1. Efua drove 3 hours to her farm. On the return trip, her speed was 10 km h^{-1} less and the trip took 4 hours. What was her speed each way?

2. A cyclist rode into a village in 5 hours. In returning, his speed was 5 km h^{-1} faster and the trip took 4 hours. What was his speed each way?

3. A car leaves a city and goes north at a rate of 50 km h^{-1} at 3 pm. One hour later a second car leaves, travelling south at a rate of 40 km h^{-1}. At what time will the two cars be 320 km apart?

4. Kweku leaves home at 9 am, cycling at a rate of 36 km h^{-1}. Two hours later, Kojo leaves, driving at the rate of 48 km h^{-1}. At what time will Kojo catch up with Kweku?

5. A bus leaves a station at 1 pm, travelling west at an average rate of 54 km h^{-1}. One hour later a second bus leaves the same station, travelling east at a rate of 68 km h^{-1}. At what time will the two buses be 298 km apart?

6. At 8:00 am Ama leaves on a trip at 63 km h^{-1}. One hour later, Kojo decides to join her and leaves along the same route, travelling at 72 km h^{-1}. When will Kojo catch up with Ama?

7. A train leaves town A for town B, travelling at 45 km h^{-1}. At the same time, a second train leaves town B for town A at 55 km h^{-1}. If the two towns are 300 km apart, how long will it take for the two trains to meet?

8. Two cars are 500 km apart and moving directly towards each other. One car is moving at a speed of 120 km h^{-1} and the other is moving at 80 km h^{-1}. If the cars start moving at the same time, how long does it take for the two cars to meet?

9. Two buses start out 150 km apart and starting moving to the right at the same time. The bus on the left is moving at twice the speed as the bus on the right. Six hours after starting the bus on the left catches up with the bus on the right. How fast was each bus moving?

10. Two cyclists A and B start out at the same time from two villages 105 km apart and travel toward each other. The average speed of B is 5 km h^{-1} more than the average speed of A. If they meet in 3 hours, what are their respective speeds?

11. Two cars travelled along the same road. Car A travelled at 80 km h^{-1} for a certain time. Car B travelled at 60 km h^{-1} for 10 minute less than this time. If the total distance travelled by the two cars is 270 km, how long did each car travel?

12. At 7:00 am, a man starts walking from a village A to another village B at 12 km h^{-1}. At 8:00 am another man left village A for village B walking at 15 km h^{-1}. When did the second man catch up with the first man?

Literal Equations

An equation which has some or all of the numbers represented by letters is known as literal equations. For example, $ax + b = c$ is a literal equation. It is customary to represent variables by the letters of the later part of the alphabet and the known by the early part of the alphabet. Thus, in the equation $ax + b = c$, x is the variable and a, b and c are constants. Literal equations are solved in much the same way that we solve numerical equations.

Examples

Solve $ax - b = c$ for x

$$ax - b = c$$

$ax = c + b$ Add b to both sides

$x = \dfrac{c+b}{a}$ Divide by a

Try this 12

Solve $cx + b = a$

Solve $ax + d = bx + c$ for x

$ax + d = bx + c$

$ax - bx = c - d$ Group like terms

$(a - b)x = c - d$ Factorise the left side

$x = \frac{c-d}{a-b}$ Divide both sides by $a - b$

Try this 13

Solve $bx - c = a - dx$ for x

Exercise 5.2(e)

Solve the following equations for x

1. $cx - d = b$ 2. $bx + a = c$ 3. $\frac{x}{a} + b = c$

4. $\frac{x}{c} + 1 = \frac{x}{b}$ 5. $ex + d = fx$ 6. $a(x + 3) = b(2 - x)$

7. $\frac{1}{a} + \frac{x}{c} = \frac{x}{b}$ 8. $\frac{x-b}{a} = c$ 9. $2a - bx = ax - 3a$

10. $a - c(1 - x) = bx$ 11. $x^2 = b + ax^2$ 12. $ax^3 + 1 = b$

13. $\frac{3-x^2}{g} = \frac{2+x^2}{f}$ 14. $b = \sqrt[3]{\frac{ax}{c}}$ 15. $x = \sqrt{\frac{ax^2+c}{b}}$

Change of the Subject of the Formula

Many real life situations can be modelled by equations with two or more letters known as formulas. For instance, the equation $A = \pi r^2$ gives the area of a circle with radius r. A is called the subject of the formula. A formula is a literal equation, and can be solved as described above. For example, we can solve $A = \pi r^2$ for r as follows:

$$\frac{A}{\pi} = r^2 \qquad \text{Divide both sides by } \pi$$

$$r = \sqrt{\frac{A}{\pi}} \qquad \text{Take positive square root of both sides}$$

Often a formula is given in one form, and you can transform it to an equivalent form. The process of rearranging an equation so that a new variable is left alone on one side is called changing the subject of the formula.

Examples

Make r the subject of the formula $V = \frac{1}{3}\pi r^2 h$

$$V = \frac{1}{3}\pi r^2 h$$

$$3V = \pi r^2 h \qquad \text{Multiply both sides by 3}$$

$$\frac{3V}{\pi h} = r^2 \qquad \text{Divide both sides by } \pi h$$

$$\sqrt{\frac{3V}{\pi h}} = r \qquad \text{Take square root of both sides}$$

Try this 14

Make E the subject of the formula $v = \sqrt{\frac{2E}{m}}$

Make u the subject of the formula $\frac{1}{f} = \frac{1}{u} + \frac{1}{v}$

$$\frac{1}{f} = \frac{1}{u} + \frac{1}{v}$$

$$uv = fv + fu$$ Multiply both sides by fuv

$$uv - fv = fu$$ Group like terms

$$(u - f)v = fu$$ Factorise the left side

$$v = \frac{fu}{u-f}$$ Divide both sides by $u - f$

An alternate solution is shown below:

$$\frac{1}{f} = \frac{1}{u} + \frac{1}{v}$$

$$\frac{1}{f} - \frac{1}{u} = \frac{1}{v}$$

$$\frac{u - f}{fu} = \frac{1}{v}$$

$$v = \frac{fu}{u - fu}$$

Try this 15

Make R_1 the subject of the formula $\frac{1}{R} = \frac{1}{R_1} + \frac{1}{R_2}$

Exercise 5.2(f)

Make the variables indicated the subject of the following formulas

1. $v = u + at$, t 2. $s = \frac{n}{2}(a + l)$, n 3. $E = I(R + r)$, r

4. $s = \frac{a}{1-r}$, r 5. $I = \frac{PRT}{100}$, T 6. $k = \frac{mv^2}{2g}$, v

7. $A = h(R^2 - r^2)$, R 8. $v = \frac{4}{3}\pi r^3$, r 9. $I = \frac{nE}{R+nr}$, n

10.$s = \frac{1}{2}gt^2$, t   11. $T = 2\pi\sqrt{\frac{l}{g}}$, g   12. $\frac{1}{c} = \frac{1}{c_1} + \frac{1}{c_2}$, c_1

13. $E = V + \frac{1}{2}mv^2$, v   14. $k = \sqrt[3]{\frac{mg\cos\alpha}{v}}$, v

15.$R = \sqrt{\frac{m}{P-Q}}$, P

Evaluating a Formula

Consider the formula

$$A = h(R^2 - r^2)$$

Given $A = 160$, $h = 2$ and $R = 12$, we can find the value of r as follows:

We substitute for A, h and R

$160 = 2(12^2 - r^2)$

$\quad 80 = 144 - r^2$ Divide both sides by 2

$\quad r^2 = 144 - 80$

$\quad r^2 = 64$

$\quad r = 8$ Take positive square root

So $r = 8$

You may find it easier to first solve for r before doing the substitution, as illustrated below:

Example

If $I = \frac{nE}{R+nr}$, find n when $I = 8, E = 32, R = 6$ and $r = 2.5$

First, we solve $I = \frac{nE}{R+nr}$ for n

$$I(R + nr) = nE$$

$$IR + Inr = nE$$

$$nE - Inr = IR$$

$$n(E - Ir) = IR$$

$$n = \frac{IR}{E-Ir}$$

Now we substitute values for I, R, E and r

$$n = \frac{8\times6}{32-8\times2.5}$$

$$n = 4$$

Try this 16

Given that $F = \frac{9}{5}C + 32$, find C when $F = 86$

Exercise 5.2(g)

1. If $A = 2\pi rh$, find h when $A = 264$, $r = 4$

2. If $E = \frac{1}{2}mv^2$, find v when $E = 80$ and $m = 10$

3. If $s = ut + \frac{1}{2}ft^2$, find s when $u = 10, t = 5$ and $f = 12$

4. If $A = \frac{1}{2}(a + b)h$, find a when $b = 6, h = 6$ and $A = 30$

5. The time Ts for the simple pendulum is given by the formula

$$T = 2\pi\sqrt{\frac{l}{g}}$$ where g ms^{-2} is the acceleration due to gravity.

(a) Make l the subject of the formula

(b) Find l when $T = 1.2\ s$ and $g = 9.8\ ms^{-2}$

Give your answer to two decimal places

6. When two electrical resistances R_1 and R_2 are wired in parallel
the resulting resistance R is given by the formula

$$\frac{1}{R} = \frac{1}{R_1} + \frac{1}{R_2}$$

Make R_2 the subject of the formula. Hence, find the value of R_2

given that $R = 12$ ohms and $R_1 = 15$ ohms

7. The total mechanical energy (E joules) of a particle of mass m kg

moving at a speed of $v\ ms^{-1}$ is given by $E = V + \frac{1}{2}mv^2$,

where V joules is its potential energy. Find the speed of a

particle of mass 5 kg when its potential energy is 100 joules and

its total mechanical energy is 150 joules

8. The formula gives the amount of heat Q lost through a wall

which is D unit thick and insulated with a material whose

insulation factor is k, if the temperature on the warmer side of

the wall is $T_1°$ and on the cooler side is $T_2°$

$$Q = \frac{k(T_1 - T_2)}{D}$$

Given that $Q = 1800, k = 108, D = 1.5$ and $T_2 = 20°$, find T_1

9. Given that $Q = \dfrac{rk}{P+ms}$

   (a) Make P the subject of the formula

   (b) Find P when $Q = \dfrac{4}{3}, m = 15, s = 0.2, k = 4$ and $r = 10$

10. The formula $\dfrac{1}{f} = \dfrac{1}{v} + \dfrac{1}{u}$ gives the focal length f of a lens

    (a) Make v the subject of the formula

    (b) When $f = 3$ and $u = 12$, find the value of v

5.3 Simple Linear Inequalities

A relation in which one quantity is greater than or is less than another quantity is called an inequality.

The inequality symbols are:

$<$ is less than

$\leq$ is less than or equal to

$>$ is greater than

$\geq$ is greater than or equal to

The pass mark m of a test marked out of 10 is 6. If a student passed then his mark is 6 or more. In symbol, we write $m \geq 6$. A student who failed had less than 6 and we write this as $m < 6$.

Try this 17

Write the following inequalities using symbols

(a) a is not less than 15

(b) b is not greater than 8

(c) c is greater than 12

(d) d is less than 10

Graphs of Inequalities

An inequality has infinitely many solutions. For example, given the inequality $x > 2$, where x is a real number, any number to the right of 2 on a number line is a solution to $x > 2$. Clearly it would be impossible to list all these solutions. The solution can be represented on a number line

Examples

Draw the graph of the inequality $x > -2$, where x is a real number.

The solution set of $x > -2$ are all real numbers greater than -2. This means all numbers to the right of -2. We then start at -2 and draw an arrow extending to the right, as shown in Figure 5.1. The open circle at -2 indicates that the number is not included in the solution.

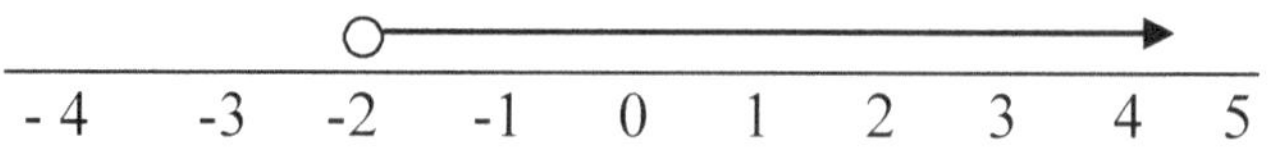

$$-4 \quad -3 \quad -2 \quad -1 \quad 0 \quad 1 \quad 2 \quad 3 \quad 4 \quad 5$$

Figure 5.1

Try this 18

Draw the graph of $x > 2$, where x is a real number

Draw the graph of $x \leq 1$, where x is a real number

The solution set of $x \leq 1$ is as shown in Figure 5.2. The closed circle at 1 indicates that the number is included in the solution.

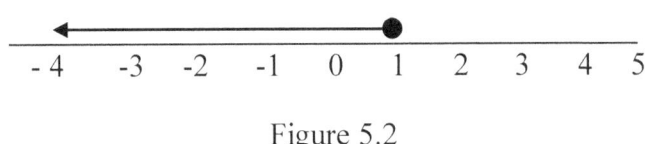

Figure 5.2

Try this 19

Draw the graph of $x \geq -2$, where x is a real number

Exercise 5.3(a)

Draw the graph of the following inequalities

1. $x \geq -1$ 2. $x < 3$ 3. $x \leq -2$

4. $1 < x$ 5. $0 \geq x$ 6. $x > -1\frac{1}{2}$

7. $x < 2\frac{1}{4}$ 8. $x \leq \frac{3}{4}$ 9. $x \leq -\frac{1}{2}$

10. $-1 \leq x < 4$ 11. $0 < x \leq 3$ 12. $-2 \leq x < 5$

Solving Linear Inequalities

Inequalities that have the same solutions are called equivalent inequalities. Solving an inequality is a process of writing equivalent inequalities until you isolate the variable. Each of the operations listed below will produce an equivalent inequality.

1. Adding the same number to both side of the inequality

2. Subtracting the same number from both side of the inequality

3. Multiplying both side by the same positive number

4. Dividing both side by the same positive number

We will consider multiplication and division by the same negative number later in this section

Examples

Solve the inequality $2x - 3 < 5$

$$2x - 3 < 5$$

$2x - 3 + 3 < 5 + 3$ Add 3 to both sides

$2x < 8$ Simplify

$x < 4$ Divide both sides by 2

Try this 20

Solve the inequality $3x + 5 > 8$

Solve the inequality $\frac{2}{3}x + 1 \geq 5$

$$\frac{2}{3}x + 1 \geq 5$$

$2x + 3 \geq 15$ Multiply every term by 3

$2x \geq 12$ Subtract 3 from both sides

$x \geq 6$ Divide both sides by 2

Try this 21

Solve the inequality $\frac{3}{4}x - 2 \leq 1$

Solve the inequality $5x + 7 \leq 1 + 2x$

$$5x + 7 \leq 1 + 2x$$

$$5x - 2x \leq 1 - 7 \qquad \text{Group like terms}$$

$$3x \leq -6 \qquad \text{Combine like terms}$$

$$x \leq -2 \qquad \text{Divide both sides by 3}$$

Notice that if you multiply or divide each side of an inequality by the same positive number, the inequality symbol stays the same.

Try this 22

Solve the inequality $3x - 7 > 3 - 2x$

Exercise 5.3(b)

Solve the following inequalities

1. $2x + 1 > 7$

2. $3x - 2 < 4$

3. $5x \leq 3x + 4$

4. $2x + 1 \leq 16 - 3x$

5. $4x + 9 \geq 1 + 2x$

6. $3(x - 1) > 12$

7. $5 \leq 2(x + 1)$

8. $\frac{1}{3}(x + 2) \geq \frac{1}{2}(3 - x)$

9. $5 < 3x + 8$

10. $1 \geq 2x - 3$

11. $7x - 3 > 3(x + 1)$

12. $\frac{1}{3}(x - 4) + 3 < \frac{1}{4}x$

Multiplication and Division by Negative Numbers

Consider the inequality

$$3 < 5$$

If we multiply both sides by -2, we get

$$-6 < -10$$

which is a false inequality. However, if we reverse the inequality symbol in $-6 < -10$, we get

$-6 > -10,$

which is true.

Similarly, if we multiply

$-3 > -7$

by -1 we get $3 > 7$ which is false. However, if we reverse the inequality symbol we get

$3 < 7,$

 which is true.

In general, if you multiply or divide both sides of an inequality by the same negative number the inequality symbol is reversed. Note that the statement also holds for $\leq$ and $\geq$

Examples

Solve the inequality $1 - 2x < 9$

$1 - 2x < 9$

$-2x < 8$ Subtract 1 from both sides

$x > -4$ Divide both sides by -2

Try this 23

Solve the inequality $2 - 3x > 5$

Solve the inequality $2 - \frac{1}{3}x \geq 4$

$$2 - \frac{1}{3}x \geq 4$$

$$-\frac{1}{3}x \geq 2 \qquad \text{Subtract 2 from both sides}$$

$$x \leq -6 \qquad \text{Multiply both sides by } -3$$

Try this 24

Solve the inequality $3 - \frac{1}{2}x < 5$

Solve the inequality $\frac{1}{4}(x + 5) < \frac{1}{3}(x + 2)$

$$\frac{1}{4}(x + 5) < \frac{1}{3}(x + 2)$$

$$3(x + 5) < 4(x + 2) \qquad \text{Multiply both sides by 12}$$

$$3x + 15 < 4x + 8 \qquad \text{Expand both sides}$$

$$3x - 4x < 8 - 15 \qquad \text{Group like terms}$$

$$-x < -7 \qquad \text{Combine like terms}$$

$$x > 7 \qquad \text{Multiply both sides by } -1$$

Try this 25

Solve the inequality $3x - 4 < 8x + 6$

Exercise 5.3(c)

Solve the following inequalities

1. $3 - 4x < 7$ 2. $5 - 3x > -4$ 3. $-3x + 6 \leq 2$

4. $x > 4 + 3x$ 5. $1 - \frac{1}{2}x \leq 5$ 6. $2(3 - x) < 3$

7. $\frac{1}{4}(x + 1) \geq \frac{1}{3}(x + 2)$ 8. $-4x > 27 + 5x$

9. $5(x + 3) < 3(2x + 4)$ 10. $3(x + 5) - 5(x + 4) \geq 0$

11. $1 - 3(x + 2) < 10$ 12. $4x + 3 \leq 7x + 9$

Solving Double Inequality

Examples

Solve the inequality $-3 < 2x - 5 < 1$

$\qquad -3 < 2x - 5 < 1$

$\qquad -3 + 5 < 2x - 5 + 5 < 1 + 5$ Add 5 to all three parts

$\qquad 2 < 2x < 6$ Combine like terms

$\qquad 1 < x < 3$ Divide each part by 2

The inequality can also be solved as follows:

$\qquad -3 < 2x - 5$ and $2x - 5 < 1$

$\qquad 2 < 2x$ $\qquad 2x < 6$ Add 5 to both sides

$\qquad 1 < x$ $\qquad x < 3$ Divide both sides by 2

The solution set consists of all real numbers that satisfy both inequalities, that is all numbers greater than 1 and less than 3. This is written as $1 < x < 3$.

Solve the inequality $-3 \leq 1 - 2x < 7$

$$-3 \leq 1 - 2x < 7$$

$-3 - 1 \leq 1 - 2x - 1 < 7 - 1$ Subtract 1 from all parts

$-4 \leq -2x < 6$ Combine like terms

$2 \geq x > -3$ Divide each part by -2

We can also write $-3 < x \leq 2$

Try this 26

Solve the inequality $1 < 3x + 4 \leq 7$

Exercise 5.3(d)

Solve the following inequalities

1. $5 < x - 3 < 7$

2. $0 \leq 2x + 4 < 2$

3. $-1 < 3 - 2x \leq 9$

4. $-8 < 1 - 3(x + 2) < 10$

5. $-2 < \frac{1}{2}(x + 1) < 6$

6. $-2 < x + 2 < 3$

7. $5 < 2 + 3x < 11$

8. $1 \leq 5 - x \leq 3$

9. $-2 < \frac{1}{3}(x + 2) - \frac{1}{4}(x + 3) < -1$ 10. $4 \leq 2(1 - x) + 3x < 7$

Problems in Inequalities

The steps for solving problems discussed in Section 5.2 can be used to solve problems involving inequalities. The table below lists some important phrases that are translated to inequalities.

Important words	Examples	Symbols
At least	b is at least 2	$b \geq 2$
At most	s is at least 5	$s \leq 5$
Cannot exceed	t cannot exceed 3	$t < 3$
Is less than	i is less than 200	$i < 200$
Is more than	t is more than 2	$t > 2$
Is between	n is between 30 and 60	$30 < n < 60$
Not more than	d is not more than 3	$d \leq 3$
Not less than	b is not less than 6	$b \geq 6$

Examples

When two times a certain number is added to 3, the result is greater than 15. What are the possible numbers?

Let x represent the number

Therefore $2x + 3 > 15$

$$2x > 12$$

$$x > 6$$

The number must be at least 6

Try this 27

Three times the sum of a number and 2 is at most 18. What are all possible numbers?

A ticket for a concert cost GH¢ 20 for adults and GH¢ 10 for children. Three times as many adults as children attended the concert. If the gate proceeds were not more than GH¢ 3,500, find the maximum number of children at the concert.

Let x represent the number of children at the concert. Then $3x$ adult attended the concert.

Therefore $20(3x) + 10(x) \leq 3500$

$$60x + 10x \leq 3500$$

$$70x \leq 3500$$

$$x \leq 50$$

Therefore there were at most 50 children at the concert

Try this 28

The length of a rectangular garden is 5 metres longer than it is wide. The perimeter of the garden is at most 68 metres. What is the greatest possible width for the garden?

Exercise 5.3(e)

Solve the following inequality problems

1. When three times a certain number is added to 2, the result is less than 8. What are the possible numbers?

2. When 5 is subtracted from a number and the result doubled, the answer is not more than 4. What are the possible numbers?

3. When 7 is added to two times a certain number the result is not less than the number added to 3. What are the possible numbers?

4. The sum of three consecutive even numbers is not less than 36. Find the smallest number.

5. The width of a rectangle is fixed at 8 cm. What lengths will make the perimeter at least 200 cm?

6. A rectangle is twice as long as it is wide. If its perimeter is not greater than 72 cm, find the possible range of values of the width.

7. A rectangle has length x cm, and width 6 cm. The perimeter is less than 48 cm, but greater than 32 cm. What are the possible length of the rectangle?

8. A certain number of equal squares sides 4 cm are placed side by side to form a rectangle two squares in width. If the area of the rectangle is at most 128 cm^2, what is the maximum number of squares used?

9. A man is four times as old as his son, and his daughter is 5 years younger than her brother. If their combined ages is not less than 67 years, find the minimum age of the son?

10. A boy's test grades are 73, 75, 89 and 91. What scores on a fifth test will make his average test grade at least 85?

11. A student took three papers in a physics examination. His marks for two of the papers were 68 and 60 respectively. To obtain a distinction an average of not less than 70 is needed over the three papers. How many marks must he obtain in the third paper?

12. A woman bought a certain quantity of oranges at 25 Gp each and twice as many pine apples at GH¢ 1.00 each. If she did not spend more than GH¢ 45 altogether, what are the maximum quantity of oranges bought?

Review exercise 5

1. Solve:

(a) $5x + 3 = 18$ (b) $3x + 6 = 21$ (c) $7t - 8 = 6$

(d) $6x - 3 = 15$ (e) $-19 = 1 + 4x$ (f) $7 - 3x = 19$

2. Solve:

(a) $5x - 3 = 7 + 3x$ (b) $3x - 2 = 8 - x$

(c) $5x + 3 = 2x + 15$ (d) $4x + 3 = 2x - 5$

(e) $5 - 2x = 25 - 4x$ (f) $10 - 3x = 40 - 6x$

3. Solve:

(a) $7(2x - 1) = 21$ (b) $35 = 5(3x + 1)$

(c) $13 - 3(2x - 1) = 4$ (d) $6x - (3x + 8) = 16$

(e) $5(x + 4) = 7(x - 2)$ (f) $7x - (2x + 8) = 32$

4. Solve each formula for the letter indicated:

(a) $A = \frac{1}{2}h(a + b)$, a (b) $A = P + Prt$, P

(c) $E = mc^2$, c (d) $A = 2\pi r(r + h)$, h

(e) $V = \frac{4}{3}\pi r^3$, r (f) $c = \frac{2ad}{a-d}$, d

5. The formula for the total resistance, R in a parallel circuit is given by the formula

$$\frac{1}{R} = \frac{1}{R_1} + \frac{1}{R_2}$$

Find the total resistance if $R_1 = 6$ ohms and $R_2 = 10$ ohms

6. The perimeter of a rectangle of length l and width w is given by the formula $P = 2l + 2w$. Find the width when $P = 30$ cm and $l = 10$ cm.

7. The formula that relates Celsius and Fahrenheit temperature is $F = \frac{9}{5}C + 32$. If the temperature of the day is $20°C$, what is the Fahrenheit temperature?

8. Translate each statement to an algebraic equation. Let x represent the number in each case.

(a) 5 more than a number is 9

(b) 7 less than a number is 15

(c) 4 less than 3 times a number is twice that same number

(d) 2 times the sum of a number and 3 is 12 more than that same number

9. Solve the following word problems

(a) The sum of twice a number and 7 is 33. What is the number?

(b) 4 times a number, decreased by 20, is 44. What is the number?

(c) The sum of three consecutive integers is 63. What are the three integers?

(d) The sum of three consecutive odd integers is 105. What are the three integers?

(e) The sum of three consecutive even integers is 126. What are the three integers?

(f) In an election, the winning candidate had 160 more votes than the loser. If the total number of votes cast was 3260, how many votes did each candidate receive?

(g) Kofi is 1 year less than twice as old as his sister. If the sum of their ages is 14 years, how old is Kofi?

(h) On her vacation in Europe, Ama expenses for food and lodging were £60 less than twice as much as her airfare. If she spent £2400 in all, what was the cost of her airfare?

10. Draw the graph of each of the following inequalities:

(a) $x \geq -4$ (b) $x \leq 3$ (c) $x < -2$

(d) $x > -4$ (e) $x \leq -3$ (f) $x > -\frac{3}{4}$

11. Solve each of the following inequalities:

(a) $3x \geq 2x - 4$ (b) $5x < 4x + 7$

(c) $6x - 8 < 5x$ (d) $4x - 3 > 3x + 5$

(e) $5x - 3 > 3x + 15$ (f) $5x + 7 \leq 8x - 17$

(g) $3x - 2 > 5x + 3$ (h) $4(x + 7) < 2x + 31$

(i) $2(x - 7) < 5x - 12$

12. Translate the following statements into inequalities. Let x represent the number in each case.

(a) 5 more than three times a number is less than 7

(b) 3 less than a number is greater than 5

(c) 7 more than twice a number is greater or equal to 12

(d) Between 60 and 80 students attended the concert

(e) At least 45 students passed the test

(f) At most 1,200 teachers were interviewed.

(g) A man's weekly wage is not to exceed GH¢ 100

(h) The cost of bread is not less than GH¢ 2.50

13. Solve the following problems

(a) A plumber charges GH¢ 25 plus GH¢ 30 per hour for emergency service. A man was billed over GH¢ 100 for an emergency call. How long was plumber there?

(b) A father cannot spend more than GH¢ 2285 on tuition. If a school charges GH¢ 35 registration fee plus GH¢ 375 per course, what is the greatest number of courses for which he can register?

(c) A student takes mathematics course in which four tests are given. To get a B, he must average at least 80 on the four tests. He scored 82, 76 and 78 on the first three tests What scores on the last test will earn him at least a B?

(d) The perimeter of a rectangular swimming pool is not to exceed 72 metres. The length is to be twice the width. What widths will meet these conditions?

(e) A factory worker earns a daily base pay of GH¢ 15 plus GH¢ 4.50 every hour. How many hours must he work in a day to earn at least GH¢ 42?

(f) A man claims that it cost him at least GH¢ 3.00 to make a call. If a typical call cost 75 Gp plus 45 Gp each minute, how long do his calls typically last?

Chapter Test 5

Take this test as you would take a test in class. After you are done, check you work against the answers in the back of the book.

1. Solve:

(a) $3x + 7 = 2x - 5$

(b) $-3x + 2(x + 5) = 9$

(c) $\frac{2}{3}(x - 1) = \frac{1}{4}(x + 2)$

2. Solve and draw the graph of the solution set for

(a) $7 - 3x < 12 + 2x$

(b) $2(x - 3) \geq 14 - 3x$

(c) $-3 < \frac{2x-3}{2} \leq 1$

3. Solve each formula for the letter indicated:

(a) $V = \frac{1}{3}\pi r^2 h, \quad r$

(b) $S = \frac{360A}{\pi r^2}, \quad A$

(c) $t = \sqrt{\frac{2s}{g}}, \quad s$

4. Given the formula $t = 2\pi\sqrt{\dfrac{l}{g}}$, $t = 1.2$, $\pi = 3.14$ and $g = 9.8$,

find l, correct to one decimal place.

5. The area of a circular ring is given by $A = \pi R^2 - \pi r^2$,where R

and r are the outer and inner radii respectively. If π is taken as $\dfrac{22}{7}$

and $R = 14$ cm, and $r = 13$ cm, find correct to the nearest unit

the value of A.

6. The length of a rectangular flower garden is 6 metres more than

three times the width. The perimeter of the garden is 32 metres.

What is the area of the garden?

7. Tickets sales for a play total GH¢ 2200. There are three times as

many adult tickets sold as children's tickets. The prices of the

ticket for adults and children are GH¢ 6 and GH¢ 4 respectively.

Find the number of children's ticket sold.

8. A bus and a car set out from the same point, headed in the same

direction. The average speed of the car is 40 kilometres per hour

slower than twice the speed of the bus. In three hours, the car is

60 kilometres ahead of the bus, find the speed of the car.

9. A car salesman earns a base pay of GH¢ 2000 plus an 8 %

commission on his sales during the week. What must be the

amount of his sales in one week if he wants his total earnings for the week to be at least GH¢ 4000?

10. A football team can spend at most GH¢ 440 for its annual party at a local restaurant. If the restaurant charges a GH¢ 40 setup fee plus GH¢ 16 per person, at most how many can attend?

6

Simultaneous Linear Equations

Solving Simultaneous Linear Equations

An equation in the form $ax + by = c$, where a, b and c are real numbers, and a, $b \neq 0$ is called a linear equation in two variables. Examples are $-2x + 3y = 5$ and $y = 3x - 2$.

A solution to an equation in two variables is any ordered pair (x, y) that makes the equation true. To find the solution of a linear equation in two variables, you need to solve two equations. The pair of equations that have a common solution are called simultaneous equations.

Consider the two equations below

$$2x + y = 5$$

$$2x - y = -1$$

If we put $x = 1$ and $y = 3$ in the left side of each equation, we have

$$2(1) + 3 = 5$$

$$2(1) - 3 = -1$$

The solution of the pair of equations above is $(1, 3)$. That is $x = 1$ and $y = 3$

In this section, we will look at two algebraic methods of solution. A third method, the graphical method will be discussed in Chapter 9.

6.1 Solving Simultaneous Linear Equations

Method of Elimination

We eliminate one of the variables by either adding or subtracting the two equations.

Examples

Solve the simultaneous equations

$$3x + 2y = 12$$

$$x + 2y = 8$$

The coefficients of the y terms are the same in both equations. So, we can eliminate the y terms. By subtracting each term of the second equation from the first equation, we get

$$2x = 4$$

Dividing both sides of the equation by 2 we have

$$x = 2$$

To find the value of y we substitute $x = 2$ into either of the original equations. By substituting $x = 2$ into the second equation we get

$$2 + 2y = 8$$

$$2y = 6$$

$$y = 3$$

So the two solutions are $x = 2$ and $y = 3$

Try this 1

Solve the simultaneous equations

$$-3x + 2y = 7$$

$$-3x + y = 5$$

Solve the simultaneous equations

$$2x + 3y = 12$$

$$-2x + y = -4$$

Here the coefficients of x in both equations are the same but opposite in sign. So, you can eliminate x by adding the terms of the two equations. The working can be presented briefly as shown below.

First write the two equations and label them

$$2x + 3y = 12 \qquad (1)$$

$$-2x + y = -4 \qquad (2)$$

$(1) + (2) \qquad 4y = 8$

$$y = 2$$

By substituting $y = 2$ into equation (1) we get

$$2x + 3(2) = 12$$

$$2x = 6$$

$$x = 3$$

So, the solutions to the simultaneous equations are $x = 3$ and $y = 2$

Try this 2

Solve the simultaneous equations

$$5x - 3y = 8$$

$$2x + 3y = -1$$

With practice you will noticed that a term with equal coefficient in both equations can be eliminated by:

1. subtracting if they have like signs

2. adding if they have unlike signs

Solve the simultaneous equations

$$3x + 2y = 13$$

$$2x - 5y = -4$$

Adding or subtracting the equations in this form will not eliminate x or y. It will be necessary to multiply one or both of the equations in order to make the coefficients of one variable numerically equal.

$$3x + 2y = 13 \qquad (1)$$

$$2x - 5y = -4 \qquad (2)$$

$(1) \times 5 \qquad 15x + 10y = 65 \qquad (3)$

$(2) \times 2 \qquad 4x - 10y = -8 \qquad (4)$

$(3) + (4) \qquad 19x = 57$

$$x = 3$$

By substituting $x = 3$ into equation (1) we get

$$3(3) + 2y = 13$$

$$2y = 4$$

$$y = 2$$

So, the solutions to the simultaneous equations are $x = 3$ and $y = 2$

Try this 3

Solve the simultaneous equations

$$4x + 3y = 5$$

$$7x - 2y = 16$$

Exercise 6.1(a)

Solve the simultaneous equations

1. $x + 2y = 8$

   $x + y = 5$

2. $3x + 2y = 7$

   $x + 2y = 5$

3. $5a + 2b = 3$

   $4a + 2b = 2$

4. $3y + 5x = 6$

   $3y + 2x = -3$

5. $a + 2b = 9$

   $a + b = 5$

6. $2x + 3y = 12$

   $2x - y = 4$

7. $3n + 4m = 6$

   $-2n + 4m = 16$

8. $-3x + 2y = 5$

   $x + 2y = 9$

9. $2x - 5y = 17$

   $x - 5y = 16$

10. $-2a + 7b = 5$

    $-2a + 5b = 3$

11. $-x + 4y = 10$

    $-x + 3y = 7$

12. $4n - 3m = 10$

    $2n - 3m = 8$

13. $2a - b = 3$

    $a - b = 4$

14. $x - y = 5$

    $3x + y = 7$

15. $3x + 2y = 1$

    $5x - 2y = -9$

16. $2m - n = 10$

    $4m + n = 8$

17. $3a + 2b = 5$

    $3a - 2b = 7$

18. $6x - 4y = 2$

    $3x + 4y = 7$

19. $x + 4y = 11$ 20. $3x + 2y = 4$ 21. $5x + 2y = 6$

  $-x + y = -1$ $2x - y = 5$ $7x - 3y = 20$

22. $3x + 5y = 8$ 23. $2x + 3y = 8$ 24. $2a - 5b = -5$

  $2x + 3y = 5$ $5x + 2y = 9$ $3a - 7b = -6$

25. $3a - 2b = 12$ 26. $-6a + 7b = -9$ 27. $3x - 4y = 5$

  $4a - 3b = 18$ $2a - 5b = 11$ $4x + 2y = 14$

28. $5x - 2y = 7$ 29. $11a - 4b = 10$ 30. $-5x + 4y = 5$

  $2x - 3y = -6$ $6a - 7b = -9$ $-3x + 2y = 2$

Method of Substitution

Choose one of the equations and express one variable in terms of the other. Then substitute the result into the other equation.

Examples

Solve the simultaneous equations

$$3x + 2y = 12$$

$$y = 2x - 1$$

In this case, the second equation has one variable alone on one side. Substitute $y = 2x - 1$ into the first equation.

$$3x + 2(2x - 1) = 12$$

This is an equation in x that you can solve

$$3x + 4x - 2 = 12$$

$$7x = 14$$

$$x = 2$$

To find y we substitute $x = 2$ into either the first or second equation. It will be easier to substitute into the second equation.

$$y = 2(2) - 1$$

$$y = 3$$

So, the solutions to the simultaneous equations are $x = 2$ and $y = 3$

Try this 4

Solve the simultaneous equations

$$3x + 2y = 8$$

$$x = 3y - 1$$

Solve the simultaneous equations

$$5x + 3y = 4$$

$$x + 2y = 5$$

In this case, we solve one equation for one of the variables and then proceed as before. Since the coefficient of x in the second equation is 1, it will be easier to solve that equation for x. Try to avoid fractions if it is possible to do so.

$$x = 5 - 2y$$

Now, substitute this for x in the first equation

$$5(5 - 2y) + 3y = 4$$

Solve this equation for y

$$25 - 10y + 3y = 4$$

$$-7y = -21$$

$$y = 3$$

Finally, substitute $y = 3$ into one of the original equations and solve the resulting equation for x. It will be easier to choose the second equation.

$$x + 2(3) = 5$$

$$x = -1$$

So, the solutions to the simultaneous equations are $x = -1$ and $y = 3$

Try this 5

Solve the simultaneous equations

$$3x - 2y = 8$$

$$2x + y = 3$$

You can use either method to solve simultaneous equations but the method of substitution is usually a good method when one of the variables is express in terms of the other.

Exercise 6.1(b)

Solve the simultaneous equations by substitution

1. $2x + 3y = 4$

   $y = 3 + x$

2. $5a - 2b = 7$

   $b = 13 - 3a$

3. $2s - 5t = 12$ 4. $2m - 3n = 9$

      $s = 7 + 3t$ $n = -3 + 4m$

5. $3a + 2b = 5$ 6. $a - 4b = -5$

      $a + b = 2$ $2a + 5b = 16$

7. $3x + 2y = 6$ 8. $2p - 3q = 9$

      $5x + y = 17$ $4p + q = 11$

9. $5x + 3y = 11$ 10. $7x - 5y = 2$

      $3x = 15 - 6y$ $y = 3 - 2x$

6.2 Application Problems

In this section, you will learn to use two equations in two variables to solve word problems, using the same steps for solving problems discussed in Chapter 5.

Examples

Ama bought 3 pens and 5 pencils and paid a total of GH¢ 3. Kojo bought 5 pens and 8 pencils and paid GH¢ 4.90. Find the cost of a single pen and a single pencil.

Let x = the cost of a pen and y = the cost of a pencil

The total cost of 3 pens is $3x$ and the total cost of 5 pencil is $5y$. Since, the total cost of 3 pens and 5 pencils is GH¢ 3, we have

     $3x + 5y = 3$ (1)

The second equation is obtained in a similar manner.

     $5x + 8y = 4.90$ (2)

Now, solve the simultaneous equation.

You can multiply equation (1) by 5 and equation (2) by 3 to obtain

$$15x + 25y = 15 \qquad (3)$$

$$15x + 24y = 14.70 \qquad (4)$$

Subtracting the terms of equation (4) from equation (3), we have

$$y = 0.30$$

Substituting 0.30 for y in equation (1) gives

$$3x + 5(0.30) = 3$$

$$3x + 1.5 = 3$$

$$x = 0.50$$

So a pen cost 50 Gp and a pencil cost 30 Gp

Try this 6

Kojo bought 4 pen drives and 3 compact disks on sale for GH¢ 26. At the sale, Kweku bought 5 pen drives and 4 compact disks for GH¢ 33. Find the price for one pen drive and the price for one compact disk.

400 tickets for a musical concert were sold out. Adult tickets were GH¢ 25, while children tickets were GH¢ 12. If the total ticket sales were GH¢ 6750, how many of each ticket were sold?

Let x = the number of adult ticket sold and y = the number of children ticket sold. Then

$$x + y = 400 \qquad (1)$$

$$25x + 12y = 6750 \qquad (2)$$

Solve the simultaneous equations.

You can multiply equation (1) by 25

$$25x + 25y = 10{,}000 \qquad\qquad (3)$$

Subtracting the terms of equation (3) from equation (2), we have

$$-13y = -3250$$

$$y = 250$$

Substituting 250 for y in equation (1), we have

$$x + 250 = 400$$

$$x = 150$$

So, 150 adult tickets were sold and 250 children tickets were sold

Try this 7

Kwesi has saved GH¢ 6.25 in 5 Gp and 20 Gp coins. If he has 65 coins, how many of each kind of coin has he?

How much of a 20% acid solution should one mix with a 60% acid solution to produce 200 millilitres of a 44% acid solution?

Let $x =$ the volume of 20% acid solution and $y =$ the volume of 60% acid solution

Note that a 20% acid solution means that 20 % of the solution is pure acid and 80 % is water. So the amount of acid in the solution is percentage of acid $\times$ the volume of the solution. Since, the total volume of the solution is 200 ml, we have

$$x + y = 200 \qquad\qquad (1)$$

Now, the amount of acid in the 20% acid solution is $0.20x$ and that of the 60% acid solution is $0.60y$. The amount of acid in the 200 ml solution is $0.44(200)$ i.e. 88 ml. So

$$0.20x + 0.60y = 88 \qquad (2)$$

Solve the simultaneous equations.

You can multiply the terms of equation (1) by 20 and equation (2) by 100

$$20x + 20y = 4000 \qquad (3)$$

$$20x + 60y = 8800 \qquad (4)$$

Subtracting the terms of equation (4) from equation (3), we have

$$-40y = -4800$$

$$y = 120$$

By substituting $y = 120$ into equation (1), we get

$$x + 120 = 200$$

$$x = 80$$

The amounts to be mixed are 80 ml of the 20% acid solution and 120 ml of the 60% acid solution.

Try this 8

How much of a 30% alcohol solution and a 40% alcohol solution should be mixed to make a 500 ml of a 34% alcohol solution?

Exercise 6.2

1. The sum of two numbers is 25. Their difference is 5. Find the two Numbers

2. The sum of two numbers is 17. Their difference is 1. Find the two numbers.

3. The difference of two numbers is − 1. When twice the second number is subtracted from three times the first number the result is 1. Find the two numbers.

4. Twice a number is five more than thrice another number. The sum of the numbers is 10. Find the numbers.

5. Two pencils and three erasers cost GH¢ 3.50. Five pencils and four erasers cost GH¢ 7.00. How much does each pencil and each eraser cost?

6. Eight compact disks and two pen drives cost GH¢ 2.80. Three Compact disks and four pen drives cost GH¢ 2.35. Find the cost for one of each?

7. A 30 metre rope is cut into two pieces so that one piece is 6 m longer than the other. How long is each piece?

8. The perimeter of a certain rectangle is 26 m. Also the difference between the length of the rectangle and its width is 5 m. Find the length and width of the rectangle.

9. The length of a rectangle is 3 centimetres more than twice its width. If the perimeter of the rectangle is 36 cm, find the dimensions of the rectangle.

10. The sum of the masses of two parcels is 30 kg. Thrice the mass of the heavier parcel is 6 more than four times the mass of the lighter parcel. Find the masses of the parcels.

11. The sum of the ages of a father and his son is 48 years. The father's age is three times that of the son. Find their ages.

12. Three years ago a man was four times as old as his son, but in one year time he will be only three times as old as his son. What are their ages now?

13. Kweku has 22 coins with a total value of GH¢ 2.45. If the coins are all 5 Gp and 20 Gp, how many of each type of coin does he have?

14. Tickets for the school play sell for GH¢ 7 for a student and GH¢ 9 for an adult. One night, 400 people bought tickets. The school took in GH¢ 3100. How many adult tickets and how many students tickets were sold?

15. A concert ticket is GH¢ 10 for adults and GH¢ 5 for children. If the total receipts from 700 tickets were GH¢ 5000, how many adults and how many children attended the concert?

16. A labourer charges a certain fixed amount plus a certain amount per hour for doing a job. If he was paid GH¢ 160 for 3 hours job and GH¢ 200 for 5 hours job, find the fixed charge and the charge per hour.

17. A chemist has a 25% and a 50% acid solution. How much of each solution should be used to form 200 millilitre of a 35% acid solution?

18. You have two alcohol solutions, one a 15% solution and one a 45% solution. How much of each solution should be used to obtain 300 ml of a 25% solution?

19. Kojo has a total of GH¢ 12,000 invested in two accounts. One account pays 8 percent and the other 9 percent. If his interest for 1 year is GH¢ 1010, how much does he have invested at each rate?

20. Esi invests a part of GH¢ 8000 in bonds paying 12 percent interest. The remainder is in a saving account at 8 percent. If she receives GH¢ 840 in interest for 1 year, how much does she have invested at each rate?

Review exercise 6

In Exercises 1 – 10, solve the simultaneous equations using the elimination method

1. $x + y = 5$

   $x - y = 1$

2. $2x + y = 3$

   $2x + 3y = 5$

3. $3x - 2y = 5$

   $3x + y = 11$

4. $x - 3y = 7$

   $2x - 3y = 8$

5. $5x + 3y = 8$

   $-5x + 2y = -3$

6. $4x - 3y = 5$

   $2x - 3y = 1$

7. $2x - 3y = -1$

   $3x + 4y = 24$

8. $2x + 3y = -1$

   $3x + 5y = -2$

9. $2x + 5y = 9$

   $3x - 2y = 4$

10. $3x + 5y = 4$

    $-2x + 3y = 10$

11. $2x + 3y = 8$

    $3x + 4y = 13$

12. $2x - y = 9$

    $3x + 4y = -14$

In Exercises 13 - 18, solve the simultaneous equations using the substitution method

13. $x + y = 5$

    $y = x - 2$

14. $x + y = 9$

    $x = y + 3$

15. $3x + 4y = 9$

   $y = 3x + 1$

16. $x = 7y + 3$

   $2x - 5y = 15$

17. $5x - 4y = 5$

   $y = 4x + 7$

18. $5x - 6y = 21$

   $x = 5 + 2y$

19. The sum of two numbers is 50. The second is 2 more than three times the first. What are the two numbers?

20. The larger of two supplementary angles is 75^0 more than twice the size of the smaller angle. Find size of each angle.

21. An office desk and chair together cost GH¢ 2500. If the desk cost GH¢ 300 less than three times as much as the chair, what did each cost?

22. There are 250 students in the first year class of a school. The number of girls are 20 fewer than twice the number of boys. How many boys and how many girls are in the class?

23. Kojo has 22 coins with a total value of GH¢ 1.70. If the coins are all 10 Gp and 5 Gp, how many of each type of coin does he have?

24. 500 tickets were sold for a concert. The receipts from ticket sales were GH¢ 3100, and the ticket prices were GH¢ 5 and GH¢ 8. How many of each price ticket were sold?

25. The length of a rectangle is 2 cm less than three times its width. If the perimeter of the rectangle is 36 cm, find the dimensions of the rectangle.

26. The perimeter of an isosceles triangle is 19 cm. The lengths of the two equal legs are 3 cm less than twice the length of the base. Find the lengths of the three sides.

27. Ama invests a part of GH¢ 800 in bonds paying 12 percent interest. The remainder is in a savings account at 8 percent. If she receives GH¢ 84 in interest for 1 year, how much does she have invested at each rate?

28. Afua has GH¢ 1200 in savings accounts in two banks. The ratio of money saved in the two banks is 3 to 2. Use simultaneous equations to find how much money is in each bank.

29. A chemist has 15% and a 25% acid solution. How much of each solution should be used to form 500 ml of a 21% acid solution

30. A plane flies 540 km with the wind in 3 hours. Flying back against the wind, the plane takes 9 hours to make the trip. What was the speed of the plane in still air? What was the speed of the plane in the wind?

Chapter Test 6

Take this test as you would take a test in class. After you are done, check your work against the answers in the back of the book.

1 Use the method of elimination to solve

(a) $3x - 4y = -14$
$-3x + y = 8$

(b) $x + y = 7$
$3x - 5y = 15$

(c) $7x + 5y = 2$
$8x - 9y = 17$

2. Use the method of substitution to solve

(a) $x + 3y = 19$

$x = y - 1$

(b) $x + y = 5$

$4x - 3y = 13$

(c) $5x - y = 6$

$4x - 3y = -4$

3. The sum of two numbers is 38. The second number is 2 less than three times the first. Find the two numbers.

4. Two angles are complementary. One angle is $15°$ more than twice the other. Find the size of each angle.

5. 500 tickets were sold for a concert. Two types of tickets were sold. Children tickets cost GH¢ 15 per ticket and adult tickets cost GH¢ 30 per ticket. If the total ticket sales were GH¢ 12,000, how many of each type of ticket were sold?

6. Ofori invests a part of GH¢ 500 in stocks paying 9 percent interest. The remainder is in a savings account at 8 percent. If he receives GH¢ 42 in interest for 1 year, how much does he have invested at each rate?

7. Kojo has a 30% alcohol solution and a 50% alcohol solution. How much of each solution should he combine to make 400 ml of a 45% alcohol solution?

8. A boat can travel 45 kilometres downstream in 3 hours. Coming back upstream the trip takes 5 hours. Find the rate of the boat in still water and the rate of the current.

7

Quadratic Equations

An equation of the form $ax^2 + bx + c = 0$, where a, b and c are numbers and $a \neq 0$, is called a quadratic equation. This is the standard form of quadratic equation in x. Some examples of quadratic equations are:

$$x^2 + 5x - 9 = 0, \ 3x^2 + 8 = 0 \ \text{ and } \ x^2 - 3x = 0.$$

Zero Product Property

If the product of two numbers is zero, then at least one of the numbers is zero. For instance, if a and b are real numbers and $a \cdot b = 0$, then either $a = 0$ or $b = 0$ (or both).

7.1 Solving Quadratic Equations

You can solve quadratic equations algebraically by factorisation, completing the square or by use of the formula.

Solving quadratic equations by factorisation

To solve quadratic equations by factorisation just follow these steps:

1. Write the given quadratic equation in standard form.

2. Factorise the quadratic expression

3. Equate each factor to zero and solve

Examples

Solve $x^2 + 8x + 15 = 0$

$$x^2 + 8x + 15 = 0$$

First, factorise the quadratic expression on the left

$$(x + 3)(x + 5) = 0$$

Next, equate each factor to 0 and solve the resulting equation

$$x + 3 = 0 \quad \text{or} \quad x + 5 = 0$$

$$x = -3 \qquad\qquad x = -5$$

The solutions are -3 and -5

Try this 1

Solve $x^2 - 7x + 10 = 0$

Solve $x^2 + 7x = 30$

$$x^2 + 7x = 30$$

First, write the equation in standard form

$x^2 + 7x - 30 = 0$	Subtract 30 from both sides
$(x - 3)(x + 10) = 0$	Factorise the quadratic expression
$x - 3 = 0 \quad \text{or} \quad x + 10 = 0$	Equate each factor to 0
$x = 3 \qquad\qquad x = -10$	

The solutions are 3 and -10

Try this 2

Solve $x^2 - 6 = x$

Solve $(x + 5)(x - 2) = -6$

$$(x + 5)(x - 2) = -6$$

First multiply out the product on the left

$$x^2 + 3x - 10 = -6$$

Next, write the equation in the standard form

$$x^2 + 3x - 4 = 0 \qquad \text{Add 6 to both sides}$$

$$(x - 1)(x + 4) = 0 \qquad \text{Factorise the quadratic expression}$$

$$x - 1 = 0 \text{ or } x + 4 = 0 \qquad \text{Equate each factor to 0}$$

$$x = 1 \qquad\qquad x = -4$$

The solutions are 1 and – 4

Try this 3

Solve $(x - 7)(x + 1) = -16$

Solve $4x^2 - 9 = 0$

$$4x^2 - 9 = 0$$

In this case, we have a difference of two squares

$$(2x - 3)(2x + 3) = 0 \qquad \text{Factorise the difference of two squares}$$

$$2x - 3 = 0 \quad \text{or} \quad 2x + 3 = 0 \qquad \text{Equate to 0}$$

$$x = \frac{3}{2} = 1\frac{1}{2} \qquad x = -\frac{3}{2} = -1\frac{1}{2}$$

The solutions are $1\frac{1}{2}$ and $-1\frac{1}{2}$

This result can also be obtained as follows:

Rewrite the equation $4x^2 - 9 = 0$ as

$$4x^2 = 9$$

Divide both sides by 4

$$x^2 = \frac{9}{4}$$

Finally, take the square root of each side of the equation

$$\sqrt{x^2} = \pm\sqrt{\frac{9}{4}}$$

$$x = \pm\frac{3}{2}$$

So, we have the two solutions $1\frac{1}{2}$ and $-1\frac{1}{2}$.

Try this 4

Solve $5x^2 - 125 = 0$

Solve $x^2 + 5x = 0$

$$x^2 + 5x = 0$$

First, factorise the expression on the left, by removing the greatest common factor x.

$$x(x + 5) = 0$$

Next, equate each factor to 0 and solve

$$x = 0 \text{ or } x + 5 = 0$$

$$x = -5$$

The solutions are 0 and -5

Try this 5

Solve $3x^2 + 12x = 0$

You may have noticed that quadratic equations usually have two different solutions.

Exercise 7.1(a)

Solve

1. $x^2 + 4x - 12 = 0$ 2. $x^2 - 8x + 15 = 0$

3. $x^2 + 7x - 18 = 0$ 4. $x^2 - x - 30 = 0$

5. $x^2 - x - 20 = 0$ 6. $x^2 + 12x + 32 = 0$

7. $x^2 - 4x - 60 = 0$ 8. $x^2 - 14x + 45 = 0$

9. $x^2 + 3x - 28 = 0$ 10. $x^2 + 15x + 56 = 0$

11. $x^2 - 27x + 50 = 0$ 12. $x^2 - 15x - 54 = 0$

13. $x^2 - 5x - 14 = 0$ 14 $x^2 + 10x - 24 = 0$

15. $x^2 - 17x - 60 = 0$ 16. $42 + x - x^2 = 0$

17. $15 - 2x - x^2 = 0$ 18. $x^2 + 7x = 0$

19. $3x^2 - 4x = 0$ 20. $4x^2 - 25 = 0$

21. $9x^2 - 16 = 0$ 22. $x^2 - 2x = 18 + 5x$

23. $x^2 + 20 = 9x$ 24. $x(x - 7) = -12$

25. $x(x - 3) = 10$ 26. $x(x + 12) = -27$

27. $(x - 2)(x - 5) = 28$ 28. $(x + 2)(x + 3) = 20$

29. $(x - 3)(x - 4) = 42$ 30. $(x - 4)(x + 4) = -6x$

Example

Solve $3x^2 - 5x - 12 = 0$

You may factorise the quadratic expression by the method of grouping. We will not give details of the factorisation.

$3x^2 - 5x - 12 = 0$

$(3x + 4)(x - 3) = 0$ Factorise the quadratic expression

$3x + 4 = 0$ or $x - 3 = 0$ Equate each factor to 0

$x = -\dfrac{4}{3} = -1\dfrac{1}{3}$ $x = 3$

The solutions are $-1\dfrac{1}{3}$ and 3

Try this 6

Solve $2x^2 - x - 6 = 0$

Exercise 7.1(b)

Solve:

1. $3x^2 - 20x - 7 = 0$

2. $8x^2 + 14x + 3 = 0$

3. $6x^2 - 7x - 5 = 0$

4. $3x^2 - 13x + 12 = 0$

5. $6x - 13x - 28 = 0$

6. $7x^2 - 37x + 10 = 0$

7. $2x^2 + 13x - 24 = 0$

8. $4x^2 + 4x - 3 = 0$

9. $5x^2 - 17x + 6 = 0$

10. $3x^2 + 20 = 4x + 35$

11. $x(2x - 17) = -35$

12. $3x^2 - 2x = 5$

13. $(x - 1)(5x + 4) = 2$

14. $(6x + 1)(x + 1) = 21$

15. $30 = 8x(x + 1)$

Solving quadratic equations by completing the square

Consider the two expressions below called the perfect square trinomial:

$$x^2 + 2ax + a^2$$

$$x^2 - 2ax + a^2$$

You may noticed that the last term in each case is the square of half the coefficient of x. In solving quadratic equation by completing the square we use this relationship between the coefficient of x and the constant term to write an equivalent equation that has a perfect square trinomial on one side.

Examples

Solve $x^2 + 8x - 20 = 0$ by completing the square

$$x^2 + 8x - 20 = 0$$

First move the constant term to the right side

$$x^2 + 8x = 20$$

To make the left side a perfect square trinomial, add the square of half the coefficient of x to both sides of the equation. Here dividing 8 by 2 and squaring gives 16.

$$x^2 + 8x + 16 = 36$$

Factorise the expression on the left

$$(x + 4)^2 = 36$$

Taking the square root of both sides gives

$$x + 4 = \pm 6$$

So $x = -4 \pm 6$

Therefore $x = -4 + 6 = 2$ or $x = -4 - 6 = -10$

Try this 7

Solve $x^2 - 8x + 15 = 0$ by completing the square

Solve $2x^2 + x - 6 = 0$ by completing the square

$2x^2 + x - 6 = 0$

Move the constant term to the right side

$2x^2 + x = 6$

Divide both sides by the coefficient of x^2. In this case we divide all the terms by 2.

$$x^2 + \frac{1}{2}x = 3$$

Divide $\frac{1}{2}$ by 2 and then square to get $\frac{1}{16}$.

$$x^2 + \frac{1}{2}x + \frac{1}{16} = \frac{49}{16}$$

Factorise the expression on the left side

$$\left(x + \frac{1}{4}\right)^2 = \frac{49}{16}$$

By taking the square root of each side we get

$$x + \frac{1}{4} = \pm\frac{7}{4}$$

So $x = -\frac{1}{4} \pm \frac{7}{4}$

Therefore $x = -\frac{1}{4} + \frac{7}{4} = 1\frac{1}{2}$ or $x = -\frac{1}{4} - \frac{7}{4} = -2$

Try this 8

Solve $3x^2 + 5x - 2 = 0$ by completing the square

Exercise 7.1(c)

Solve the following quadratic equations by completing the square. Give your answer to two decimal places if possible

1. $x^2 + 2x - 3 = 0$

2. $x^2 - x - 12 = 0$

3. $x^2 - 4x - 6 = 0$

4. $x^2 + 10x + 22 = 0$

5. $x^2 - 3x - 11 = 0$

6. $3x^2 - 6x - 31 = 0$

7. $2x^2 - 3x + 1 = 0$

8. $3x^2 - 9x + 2 = 0$

9. $2x^2 + 10x + 5 = 0$

10. $5x^2 - 7x - 6 = 0$

11. $2x^2 = 7x + 15$

12. $4x(x - 1) = 5$

Solving quadratic equations by use of the formula

The general quadratic equation is $ax^2 + bx + c = 0$, where $a \neq 0$. The solution to this equation is given by the formula below called the quadratic formula.

$$x = \frac{-b \pm \sqrt{b^2 - 4ac}}{2a}$$

Example

Solve $x^2 - 6x - 16 = 0$

Identify the values of a, b and c in the quadratic equation

Here $a = 1$, $b = -6$ and $c = -16$

By substituting these values into the quadratic formula we get

$$x = \frac{6 \pm \sqrt{(-6)^2 - 4(1)(-16)}}{2(1)}$$

$$= \frac{6 \pm \sqrt{100}}{2}$$

$$= \frac{6 \pm 10}{2}$$

So $x = \frac{6+10}{2} = 8$ or $x = \frac{6-10}{2} = -2$

Try this 9

Solve $2x^2 + 7x - 15 = 0$

Solve $3x^2 = -7x + 6$

First write the equation in standard form

$3x^2 + 7x - 6 = 0$

Here $a = 3$, $b = 7$ and $c = -6$

Therefore $x = \frac{-7 \pm \sqrt{7^2 - 4(3)(-6)}}{2(3)}$

$$= \frac{-7 \pm \sqrt{121}}{6}$$

$$= \frac{-7 \pm 11}{6}$$

So $x = \frac{-7+11}{6} = \frac{2}{3}$ or $x = \frac{-7-11}{6} = -3$

Try this 10

Solve $2x^2 = x + 15$

Any quadratic equation can be solved by completing the square or by using the formula. However, if the equation factorises, use the method of factorisation.

Exercise 7.1(d)

Solve the following quadratic equation by use of the formula. Give your answer to two decimal places if possible

1. $x^2 - 4x + 3 = 0$

2. $x^2 - 3x - 10 = 0$

3. $x^2 + 6x + 7 = 0$

4. $x^2 - 5x + 2 = 0$

5. $x^2 + 7x - 30 = 0$

6. $3x^2 - 5x - 2 = 0$

7. $7x^2 - 5x - 2 = 0$

8. $2x^2 + 8x + 3 = 0$

9. $5x^2 - 9x + 3 = 0$

10. $4x^2 = 6 - 5x$

11. $8x^2 + 6x = 3$

12. $3x(x - 2) = 1$

7.2 Application of Quadratic Equations

Some application problems will require us to solve quadratic equations. The two examples below illustrate this

Examples

The length of a rectangular field is 3 kilometres longer than its width. If the area of the field is 108 km², find the dimensions of the field.

Let x be the width of the field, so $(x + 3)$ will be the length of the field.

Hence, the area of the field is $x(x + 3)$. So

$x(x + 3) = 108$ Multiply the bracket out

$x^2 + 3x - 108 = 0$ Subtract 108 from both sides

$(x - 9)(x + 12) = 0$ Factorise the quadratic expression

$x - 9 = 0$ or $x + 12 = 0$ Equate each factor to 0

     $x = 9$ $x = -12$

The solutions of the equation are 9 and − 12. Since the width must be positive, we will reject the negative solution. Therefore, the width of the field is 9 km. The length is 3 km longer than this and so the length of the field is 12 km. Thus, the field is 12 km long and 9 km wide.

Try this 11

The length of a rectangular plot is 2 metres longer than its width. If the area of the plot is 224 square metres, find the length and the width of the plot.

The length of a rectangular picture frame is 1 metre longer than the width. If the diagonal of the frame is 5 m, what are the dimensions of the rectangle?

Let x cm be the width of the frame. So $(x + 1)$ cm is the length of the frame

A sketch of the problem is shown below. Sometimes it may be useful to draw a sketch or diagram of the given information.

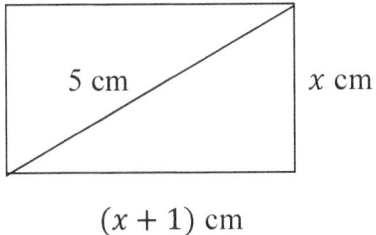

$(x + 1)$ cm

Using the Pythagoras' theorem, we have

$x^2 + (x + 1)^2 = 5^2$

$2x^2 + 2x - 24 = 0$ Simplify the equation

$x^2 + x - 12 = 0$ Divide each term by 2

$(x + 4)(x - 3) = 0$ Factorise the quadratic expression

$x + 4 = 0$ or $x - 3 = 0$ Equate each to 0

$x = -4$ $x = 3$

So, the frame is 3 m wide and 4 m long

Try this 12

A rectangle is 11 metres long and 6 metres wide. If each side is increased by the same amount, the area of the new rectangle is 126 square metres. Find the increase in each side.

Exercise 7.2

1. The product of the page number on two consecutive pages of a

    book is 240. Find the page numbers

2. The product of two consecutive odd integers is 195. Find the

    integers.

3. The product of two consecutive even integers is 168. Find the integers.

4. Twice a number is 8 less than its square. Find all such numbers

5 A rectangular picture frame is twice as long as it is wide. If the area of the frame is 338 cm^2, find its dimensions.

6. One number is 3 more than another. The sum of their squares is 89. What are the numbers?

7. The length of a rectangle is 2 cm longer than its width. If the diagonal of the rectangle is 10 cm, what are the dimensions of the rectangle?

8. The length of the base of a triangle is 3 cm more than the height of the triangle. If the area of the triangle is 35 cm^2, find the height and the length of the base.

9. A triangular traffic island has a base half as long as its height. Find the base and the height if the island has an area of 81 m^2.

10. A 15 metre ladder leaning against a building touches the bottom of a window. The foot of the ladder from the building is 3 m shorter than the height of the window above the ground. Find the height of the window above the ground.

11. A 10 metre rope is fastened from the top of a vertical pole to a peg on the ground. The distance of the peg from the pole is 2

metres longer than the height of the pole. Find the height of the pole.

12. Ama starts at a point and walks north at 2 km h^{-1}. One hour later Kofi starts at the same point and walks east at 3 km h^{-1}. How long after Ama starts' walking does it takes for them to be 5 km apart?

Review exercise 7

Solve:

1.$x^2 - 7x + 6 = 0$ 2.$x^2 + 6x + 5 = 0$

3. $x^2 + 4x - 21 = 0$ 4.$x^2 + 7x - 18 = 0$

5.$x^2 + 9x + 14 = 0$ 6. $x^2 + 8x + 15 = 0$

7. $x^2 + 6x = 0$ 8. $x^2 - 8x = 0$

9. $3x + x^2 = 0$ 10. $4x - x^2 = 0$

11. $x^2 = 2x - 1$ 12. $x^2 + 16 = 8x$

13. $x^2 + 6x = -5$ 14. $(x - 7)(x + 1) = -16$

15. $(x + 2)(x - 7) = -18$

Solve:

16. $3x^2 - 7x = 20$ 17. $3x^2 - 2x = 5$

18. $12x^2 - 5x = 2$ 19. $x(3x + 1) = 2$

20. $(x - 1)(5x + 4) = 2$ 21. $2x^2 + 3x - 20 = 0$

22. $2x^2 + 9x - 35 = 0$ 23. $5x^2 - 26x + 5 = 0$

24. $2x^2 + 5x - 12 = 0$ 25. $3x^2 + 8x - 16 = 0$

26. $4x^2 - x - 14 = 0$ 27. $5x^2 + 9x - 18 = 0$

28. $2x^2 - x - 6 = 0$ 29. $(2x + 1)^2 = 49$

30. $8x^2 - 10x + 3 = 0$

31. The product of two consecutive positive integers is 210. Find the integers

32. The product of two consecutive positive even integers is 224. Find the integers

33. The product of two consecutive negative odd integers is 323. Find the integers

34. The product of twice a number and 3 less than the number is 80. Find the numbers

35. A picture frame is 4 cm taller than it is wide and has an area of 192 cm². What are the dimensions of the picture frame?

36. The height of a triangle is 8 cm less than its base. The area of the triangle is 192 cm². Find the height of the triangle.

37. The length of one leg of a right-angled triangle is 3 cm more than the other. If the length of the hypotenuse is 15 cm. What are the lengths of the two legs?

38. The length of a rectangle is 7 cm longer than its width. If the diagonal of the rectangle is 13 cm, what are the dimensions of the rectangle?

39. Two boys leave from the same point. One boy starts out walking north at 2 km h^{-1}. One hour later the second boy starts walking east at 6 km h^{-1}. How long after the first boy starts walking does it takes for the two boys to be 26 km apart?

40. A 10 metre long ladder leans against a wall. The top of the ladder touches the top of the wall, and the distance of the foot of the ladder from the wall is 2 cm shorter than the height of the wall. Find the height of the wall.

Chapter Test 7

Take this test as you would take a test in class. After you are done, check your work against the answers in the back of the book

1. Solve:

  (a) $x^2 + x - 20 = 0$ (b) $42 + x - x^2 = 0$

2. Solve:

  (a) $4x^2 = 11x + 3$ (b) $8x(x - 1) = 30$

3. Solve the following quadratic equations. Give your answer to two decimal places

(a) $2x^2 - 7x - 10 = 0$ (b) $3x^2 = 4 - 5x$

4. The length of a rectangle is 2 m greater than the width. The area of the rectangle is 48 m². Find the length and the width.

5. The length of a rectangular garden is 7 m greater than the width. If the length of the diagonal is 17 m, find the length and width.

6. The length of the base of a triangular sign is 3 cm less than the height. If the area of the triangular sign is 35 cm², find the height and the length of the base.

7. A rope from a vertical pole is 13 m long. It reaches from the top of the pole to a peg on the ground. If the length of the peg from the base is 7 m shorter than the height of the pole, how tall is the pole?

8. A ball is thrown up, and its height, h metres after time, t seconds is given by $h = 3 + 14t - 5t^2$. How long will it take the ball to hit the ground?

8

Relations and Functions

8.1 Relations

The word relation is used to indicate a relationship between two numbers or objects. Two people belonging to the same family may be related as father and daughter or husband and wife. In a school two people may be related as a student and teacher. In mathematics two numbers may be related by being equal, or by being a multiple or twice of the other.

Definition of Relations

A relation is simply a set of ordered pairs.

Consider the set $\{1, 2, 3, 4\}$.

We can express the relation "a is less than b" by listing all the ordered pairs (a, b) for which the relationship is true. The relation consists of the ordered pairs

$(1, 2), (1,3), (1,4), (2,3), (2,4), (3,4)$

In general, a set of ordered pairs (a, b) can be used to show that some pairs of objects are related. For example, the set $R = \{(4,2), (9,3), (16,4), (25,5), (36,6)\}$ lists a number and its square. This relation can be described as "a is a square of b". In symbol we write aRb, read "a is related to b by R". $4R2$ means 4 is the square of 2.

The first elements in the ordered pairs (the a −values) form the domain of the relation, and the second elements (the b −values) form the range.

Domain

The domain of a relation is the set of all the first components of the ordered pairs

Range

The range of a relation is the set of all the second component of the ordered pairs

Example

Find the domain and range for the relation

$A = \{(2,3), (4,7), (5,8), (6,9), (7,10)\}$

The domain is the set $\{2, 4, 5, 6, 7\}$

The range is the set $\{3, 7, 8, 9, 10\}$

Try this 1

Find the domain and range for the relation

$R = \{(-1,3), (0,5), (1,6), (2,7), (3,12)\}$

Diagrammatic Representation of Relations

A relation can be illustrated by a diagram called an arrow diagram.

Consider the relation $\{(3,4), (-1,3), (2,5), (1,6)\}$

The arrow diagram for this relation is illustrated in Figure 8.1

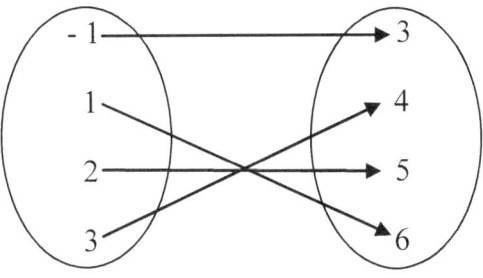

Figure 8.1

The first circle represents the domain of the relation and the second circle represents its range. The arrows drawn to the range represent the relation between any two elements.

Figure 8.2 shows the arrow diagram of a relation

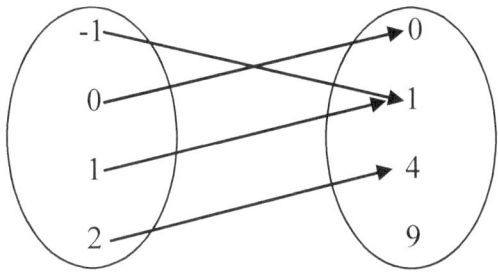

Figure 8.2

This relation consists of the ordered pairs (- 1, 1), (0, 0), (1, 1) and (2, 4). The domain is the set {-1, 0, 1, 2}, and the range is {0, 1, 4}. Note that 9 is not part of the range

Try this 2

Draw an arrow diagram of the relation
$\{(-2,0), (0,1), (1,4), (0,-1)\}$

Inverse of a Relation

The relation obtained by interchanging the first and second components in the ordered pairs of a relation is called the inverse relation. The inverse of a relation R is denoted by R^{-1}.

Example

Find the inverse of the relation $\{(1,3), (2,5), (3,7)\}$

The inverse of the relation is $\{(3,1), (5,2), (7,3)\}$

Try this 3

Find the inverse of the relation $\{(1, -3), (1,2), (2,5), (3, -4)\}$

Exercise 8.1

1. Find the domain and the range of the relation

   (a) $\{(-2, 1), (-1, 2), (0, -1), (3, -2)\}$

   (b) $\{(3, 2), (2, 1), (5, 1), (4, 3)\}$

   (c) $\{(-3, 4), (0, 0), (8, 2), (5, 6), (5, 5)\}$

   (d) $\{(-2, 5), (-2, 2), (-2, 5)\}$

   (e) $\{(2, 1), (4, 2), (6, 3), (8, 4), (10, 5)\}$

2. Draw an arrow diagram to illustrate the following relations

   (a) $\{(-2, 1), (-1, 1), (0, 2), (1, 3)\}$

   (b) $\{(3, 2), (1, 3), (2, 3), (1, 4)\}$

   (c) $\{(2, 2), (2, 4), (3, 4), (3, 6), (4, 5)\}$

   (d) $\{(1, 2), (2, 4), (3, 4), (1, 4), (3, 6)\}$

3. Let $A = \{1, 2, 7\}$ and $B = \{3, 5, 9\}$. The relation R from A to B is defined as 'is less than'. Write the relation as a set of ordered pairs.

4 .Let $A = \{1,2,3,4,5\}$ and $B = \{125,27,64,1,8\}$. The relation R from A to B is defined as is the cube of '. Write the relation as a set of ordered pairs.

5. Let $A = \{8,9,25,49\}$ and $B = \{5,7,3,2\}$. The relation R from A to B is defined as 'is a factor of'. Write the relation as a set of ordered pairs.

6. Find the inverse of each relation

   (a) $\{(1, 1), (1, 2), (2, 5), (3, 7)\}$

   (b) $\{(- 3, -2), (0, -1), (1, 2), (2, 2)\}$

   (c) $\{(2, 1), (- 1, 1), (3, -2), (0,1)\}$

   (d) $\{(0, 0), (1, 1), (-2, 2), (3, -2), (4, 5)\}$

7. R is the relation y is twice as x. Find a number q such that

   (a) $(3,q)$

   (b) $(q,-1)$

   are both in R

8. R is the set of all ordered pairs (x,y) of real numbers, where y is greater than x. Which of the following ordered pairs $(3,5),(4,2),(-5,-1),(-3,-7),(-2.5,0)$ belong to the relation R?

8.2 Mapping

A mapping is a relation in which each member of the domain is related to a member in the range.

Types of Mapping

Many -to -one

Figure 8.3 illustrates a many -to - one mapping

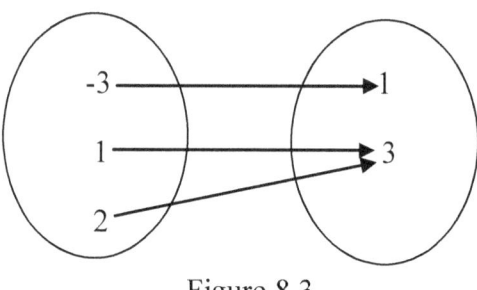

Figure 8.3

The element 3 in the range associates with more than one element in the domain. If two or more elements in the domain are related to one element in the range the mapping is called many-to-one mapping.

One –to- many mapping

. Figure 8.4 illustrates a one -to -many mapping.

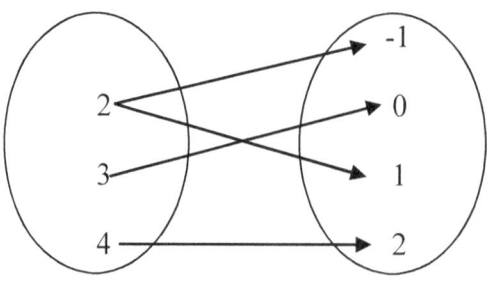

Figure 8.4

The element 2 is linked to two elements in the range. If one element in the domain is related to two or more elements in the range the mapping is called one-to-many.

Many -to -many mapping

Figure 8.5 illustrate many -to - many mapping.

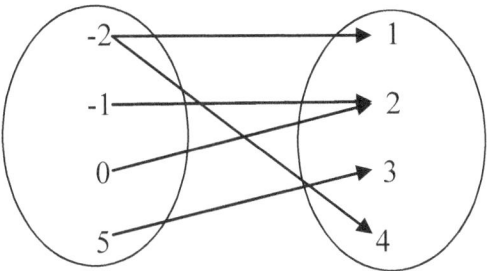

Figure 8.5

Notice that, - 2 in the domain is related to two elements in the range and − 1 and 0 are related to one element. This mapping is called many -to - many.

One –to- one mapping

Figure 8.6 illustrates a one -to -one mapping

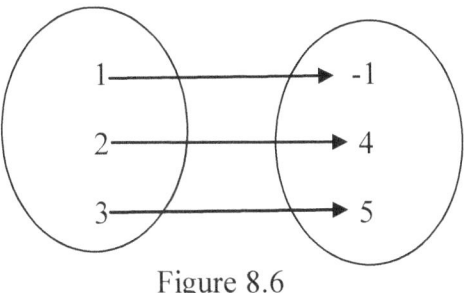

Figure 8.6

The mapping above in which each element of the range is related to exactly one element of the domain is called one -to - one mapping.

Rules for Mappings

You can determine a rule for a mapping from an arrow diagram.

Example

Find the rule for the relation below

$$0 \quad 1 \quad 2 \quad 3 \quad 4 \quad 5$$
$$\downarrow \quad \downarrow \quad \downarrow \quad \downarrow \quad \downarrow \quad \downarrow$$
$$5 \quad 7 \quad 9 \quad 11 \quad 13 \quad 15$$

A careful study of the mapping shows the following results:

You can write 15 as $2 \times 5 + 5$, so we have

$5 \rightarrow 2 \times 5 + 5 = 15$

Similarly, we have

$4 \rightarrow 2 \times 4 + 5 = 13$

$3 \rightarrow 2 \times 3 + 5 = 11$

$2 \rightarrow 2 \times 2 + 5 = 9$

$1 \rightarrow 2 \times 1 + 5 = 7$

It follows that

$0 \rightarrow 2 \times 0 + 5 = 5$

Thus, if x is a number in the domain then $x \rightarrow 2x + 5$. This expression is called the rule of the relation. The rule of a relation shows how the numbers in the domain and the range are related. Also given the rule of a relation and the domain, you can find the range.

Try this 4

Find the rule for the relation

0 1 2 3   4   5
↓ ↓ ↓ ↓   ↓   ↓
2 5 8 11 14 17

Making tables of mappings

Given the domain and the rule of a mapping you can draw the table of the mapping.

Example

Make a table of the mapping defined by the rule $x \rightarrow 2x + 3$ on the domain {0, 1, 2, 3, 4}

You can find the images by replacing x in the rule with each element in the domain

$0 \rightarrow 2(0) + 3 = 3$

$1 \rightarrow 2(1) + 3 = 5$

$2 \rightarrow 2(2) + 3 = 7$

$3 \rightarrow 2(3) + 3 = 9$

$4 \rightarrow 2(4) + 3 = 11$

From the rule we obtain the mapping below

0 1 2 3 4
↓ ↓ ↓ ↓ ↓
3 5 7 9 11

Try this 5

Make a table of the mapping defined by the rule $x \rightarrow 3x - 4$ on the domain $\{0, 1, 2, 3, 4, 5\}$

Exercise 8.2

1. Determine whether the relation is a mapping. Identify the type of

   mapping

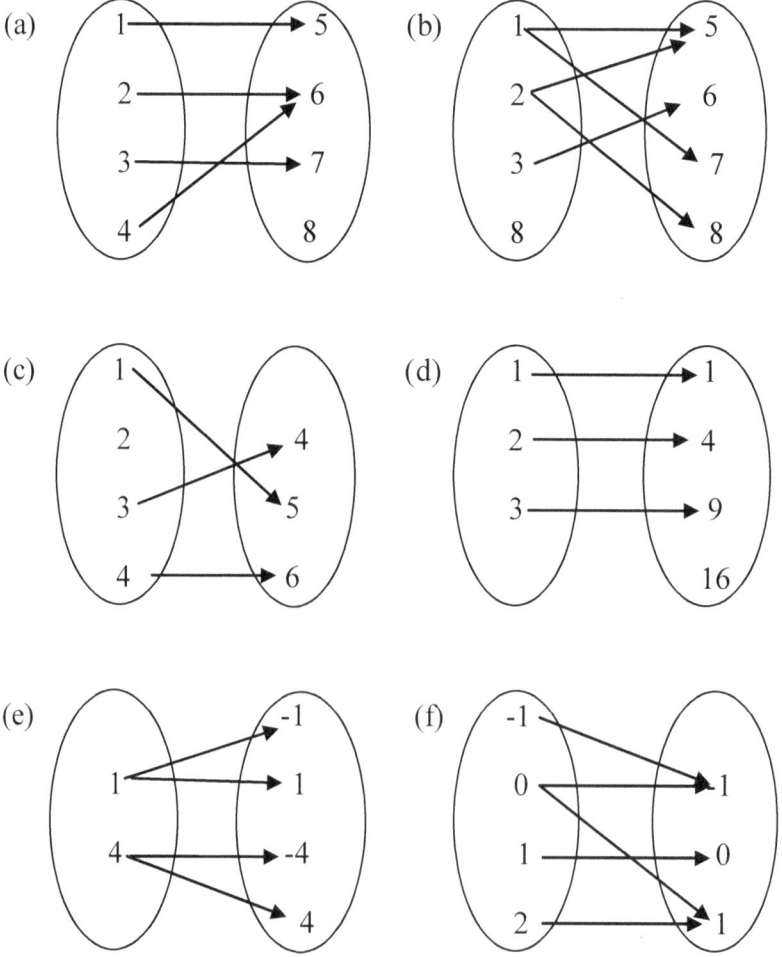

2. Find the rule for each relation

(a)
0	1	2	3	4	5
↓	↓	↓	↓	↓	↓
1	4	7	10	13	16

(b)
0	1	2	3	4	5
↓	↓	↓	↓	↓	↓
1	−1	−3	−5	−7	−9

(c)
0	1	2	3	4	5
↓	↓	↓	↓	↓	↓
−5	−3	−1	1	3	5

(d)
−2	0	2	4	6
↓	↓	↓	↓	↓
2	3	4	5	6

(e)
1	2	3	4	5
↓	↓	↓	↓	↓
1	4	9	16	25

(f)
1	2	3	4	5
↓	↓	↓	↓	↓
2	4	8	16	32

(g)
1	2	3	4	5
↓	↓	↓	↓	↓
3	6	9	12	15

(h)
−1	0	1	2	3
↓	↓	↓	↓	↓
−1	0	1	8	27

3. Use the rule and the stated domain to find the range of each relation.

(a) $x \rightarrow 3x - 2$, $\{-1, 0, 1, 2\}$

(b) $x \rightarrow 3 - 2x$, $\{-2, -1, 0, 1, 3\}$

(c) $x \rightarrow 2x^2$, $\{-3, -2, -1, 1, 2, 3\}$

(d) $x \rightarrow \frac{3}{2}x + 1$, $\{-4, -2, 0, 2, 6\}$

8.3 Functions

A relation is called a function if each member of the domain relates to exactly one member of the range.

A function can be expressed in one of the following forms: a table, a list, an equation, a rule or a graph. Note that while all functions are relations, not all relations are functions. Recall that every member of the domain of a function is assigned exactly one member in the range. Below are some illustrations of this:

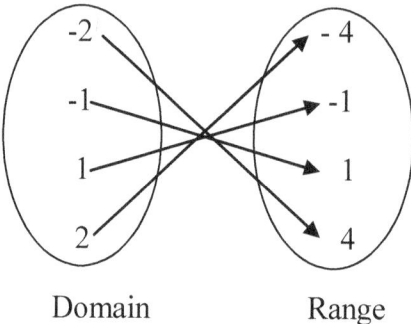

Figure 8.7

The relation shown in Figure 8.7 is a function, since each element in the domain is paired with just one element of the range.

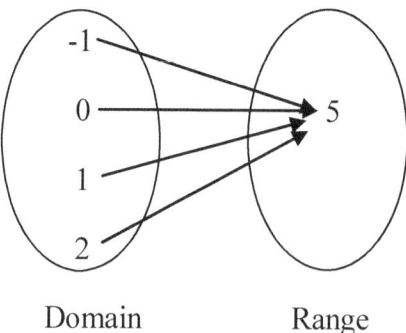

Figure 8.8

The relation shown in Figure 8.8 is a function. Notice that each element in the domain is paired with just one element of the range.

Functions are often denoted by letters such as f, g and h. A function f, that assigns an element of a set A to an element of a set B is written $f: A \rightarrow B$. Set A is called the domain of f and set B is called the co-domain. The range is a subset of the co-domain.

For each element, x, in the domain of a function f, the corresponding element of the range denoted by $f(x)$ is called the image of x or the value of the function at x. The notation $f(x)$ is read " f of x ". The set of all images of the elements of the domain is called the range of the function.

Consider the relation shown in Figure 8.9

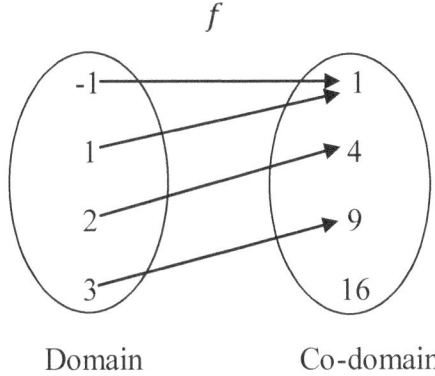

Figure 8.9

Figure 8.9 shows the function f. You can see that $f(2) = 4$. Similarly, $f(-1) = 1$. Also $f(1) = 1$ and $f(3) = 9$. Thus, the range of the function is $\{1, 4, 9\}$.

Try this 6

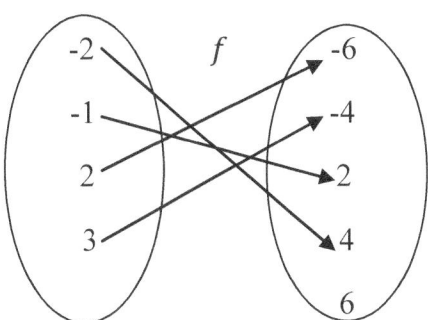

Find

(a) the image of each element in the domain

(b) the range of the function.

Function Notation

You can name a given equation and then write it in function notation. For example, the function $y = 2x + 3$ can be given the name " f " and written in function notation as $f(x) = 2x + 3$ or $f : x \rightarrow 2x + 3$.

Evaluating a Function

The value of a function $f(x)$ for a given value of x, say 3 can be found, if you replace x with 3 and simplify the resulting numerical expression.

Example

Given that $f(x) = 3x + 2$, find (a) $f(2)$ and (b) $f(-3)$

(a) Replacing x with 2, we have

$$f(2) = 3(2) + 2 = 8$$

(b) Replacing x with -3, we have

$$f(-3) = 3(-3) + 2 = -7$$

Try this 7

Given that $g(x) = 3x - 4$, find $g(2)$ and $g(-1)$

Exercise 8.3

1. Determine which sets of ordered pairs represent functions

(a) $\{(7, 4), (6, 3), (5, 2), (4, 1)\}$

(b) $\{(3, 1), (-2, 1), (2, 2), (4, 3)\}$

(c) $\{(2, 3), (1, 4), (2, -1), (4, 3)\}$

(d) $\{(-1, 0), (-2, 0), (1, 2), (1, 3)\}$

(e) $\{(3, 2), (-2, 2), (0, 2), (4, 0)\}$

2. Which of the following arrow diagrams represent functions?

(a) 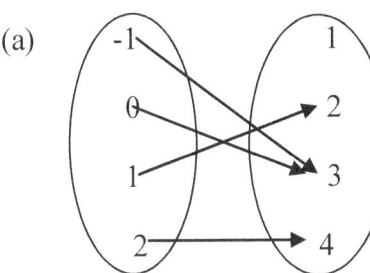   (b)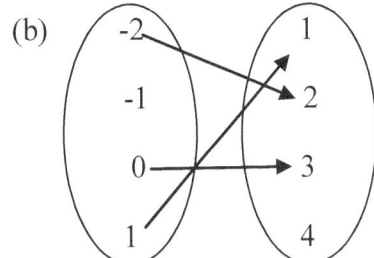

(c) 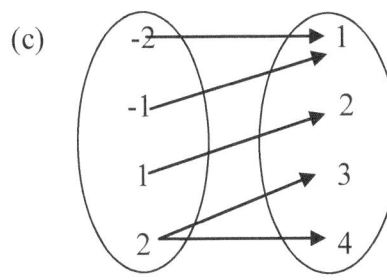   (d)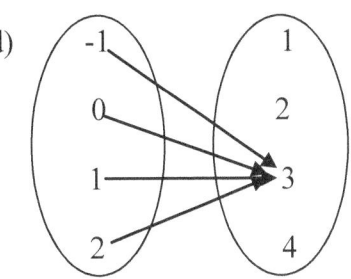

3. Each of the following equations represents a function f. Express each function in a function notation.

(a) $y = 3x + 4$ (b) $y = 2x^2 - 3$ (c) $y = \dfrac{x^2}{x-2}, x \neq 2$

4. Given that $f(x) = 5x - 3$, evaluate

   (a) $f(2)$ (b) $f(-1)$ (c) $f(0) + f(3)$ (d) $f(2) - f(-1)$

5. Given that $g(x) = x^2 - 3$, evaluate

   (a) $g(-2)$ (b) $g(1)$ (c) $-2g(3)$ (d) $2g(-2) + g(0)$

6. Given that $h(x) = 3 - 2x$, evaluate

   (a) $h(-3)$ (b) $h(4)$ (c) $\dfrac{3h(1)}{h(2)}$ (d) $\dfrac{3h(1)-h(-3)}{2h(2)}$

7. A function f is defined on the domain $\{-2, -1, 0, 1, 2, 3\}$ by

   $f: x \rightarrow 2x - 1$, find the range of the function

8. A function g is defined on the set of real numbers by

   $g: x \rightarrow 2 - ax$, where a is a constant, find a if $g(3) = 8$

9. A function h is defined on the set of real numbers by

   $h(x) = a^2 - ax + 3$, where a is a constant, find a if $h(2) = 6$

10. Find the value(s) of x for which the following functions are not

    defined.

    (a) $f: x \rightarrow \dfrac{3}{x+1}$ (b) $g: x \rightarrow \dfrac{2x+1}{x}$ (c) $h: x \rightarrow \dfrac{2x-3}{(x-3)(x+2)}$

    (d) $f: x \rightarrow \dfrac{2x^2+3x+1}{x^2-4}$ (e) $h: x \rightarrow \sqrt{x - 5}$

8.4 Linear Function

The linear function is defined by the general formula $y = mx + c$, where m and c are constants. Equations in the form $y = mx + c$

are said to be linear because the graph of the equation is a line. The value of m gives the gradient and c gives the value of y where the line cuts the y- axis.

Distance between two points

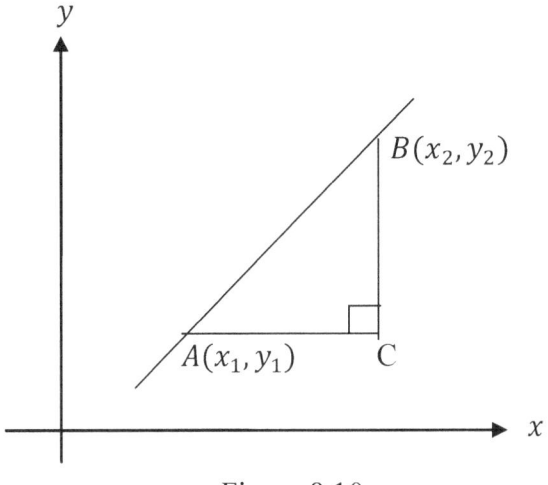

Figure 8.10

Figure 8.10 shows a   line through two points $A(x_1, y_1)$ and $B(x_2, y_2)$. You can see that the coordinate of C is (x_2, y_1). The length of BC is $y_2 - y_1$ and the length of AC is $x_2 - x_1$. By the Pythagoras's theorem, the square of the distance between the point A and B is

$$AB^2 = (x_2 - x_1)^2 + (y_2 - y_1)^2$$

Finding the positive square root, we have

$$AB = \sqrt{(x_2 - x_1)^2 + (y_2 - y_1)^2}$$

Generally, the distance d between two points is

$$d = \sqrt{(x_2 - x_1)^2 + (y_2 - y_1)^2}$$

Example

Find the distance between the points A(3, 5) and B(6, 9).

Let $x_1 = 3$, $y_1 = 5$ and $x_2 = 6$, $y_2 = 9$

Applying the distance formula, we have

$$AB = \sqrt{(6-3)^2 + (9-5)^2}$$

$$= \sqrt{3^2 + 4^2}$$

$$= \sqrt{25}$$

$$= 5$$

Note that it does not matter which point is considered (x_1, y_1) and which is (x_2, y_2), because the result will be the same.

Try this 8

Find the distance between the points P(-1, 2) and Q(2, 4)

Exercise 8.4(a)

Find the distance between each pair of points

1. (2, 3), (14, 8)   2. (4, 3), (1, 7)   3. (4, 2), (10, 10)
4. (3, -2), (8, 3)   5. (2, 1), (2, 5)   6. (4, 3), (-2, -5)

7. (7, -3), (2, 9)   8. (-8, 3), (7, -5)   9. (-9, -4), (6, 4)

10. (5, -7), (-7, 9)   11. (-5, - 1), (1, 2)   12. (-2, 10), (3, -2)

Midpoint of a Line

The midpoint of a line segment that joins two points is the point that divides the line segment into two equal parts. To find the midpoint of the line segment joining the points (x_1, y_1) and (x_2, y_2), we use the formula

$$Midpoint = \left(\frac{x_1 + x_2}{2}, \frac{y_1 + y_2}{2}\right)$$

Example

Find the midpoint of the line segment joining the points A(1, 3) and B(-3, 7)

Letting $x_1 = 1$, $y_1 = 3$ and $x_2 = -3$, $y_2 = 7$, we have

Midpoint of AB $= \left(\frac{1+(-3)}{2}, \frac{3+7}{2}\right) = (-1,5)$

Try this 9

Find the midpoint of the line segment joining the points A(-2,1) and B(4, 5)

Exercise 8.4(b)

Find the midpoint of the line segment joining each pair of points

1. (2, 3), (8, 3)   2. (4, 3), (0, 5)   3. (4, 2), (-10, - 10)

4. (- 1, - 2), (- 3, 4)   5. (3, - 2), (-7, 8)   6. (4, 3), (-2, - 5)

7. (7, -3), (-2, 9)   8. (-8, 3), (7, -5)   9. (- 9, -4), (6, 4)

10. (5, -7), (-7, 9)   11. (1, 4), (7, 2)   12. (- 4, 0), (- 2, - 4)

Gradient of Lines

The measure of the slope of a line is called its gradient. The gradient of a line can be calculated from the fraction

$$Gradient = \frac{change\ in\ y}{change\ in\ x}$$

Figure 8.11 shows the line through the points $A(x_1, y_1)$ and $B(x_2, y_2)$.

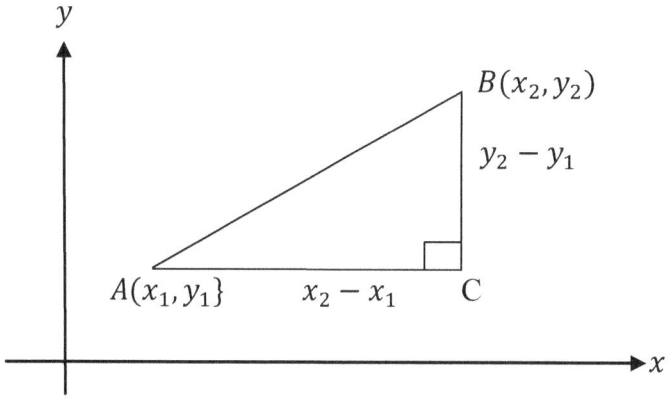

Figure 8.11

The gradient of the line AB is

$$Gradient = \frac{y_2 - y_1}{x_2 - x_1}$$

The order of subtraction is important. The numerator and denominator must be in the same order of subtraction.

The gradient of a line may be positive or negative. A line which slopes up from left to right as shown in Figure 8.12 has a positive gradient.

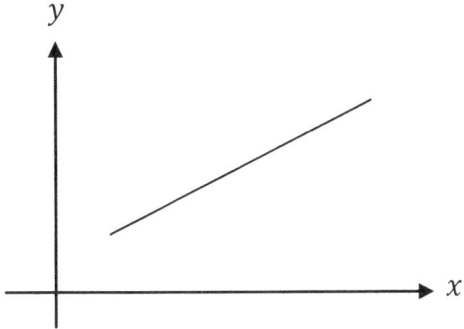

Figure 8.12

A line which slopes down from left to right as shown in Figure 8.13 has a negative gradient.

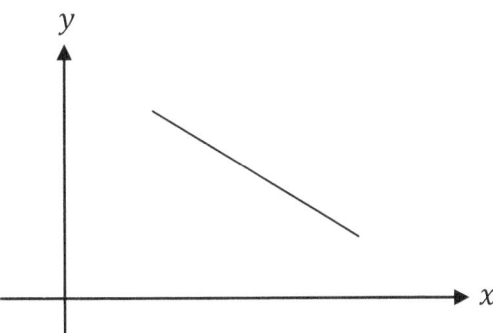

Figure 8.13

Example

Find the gradient of the line passing through the points A(2, 3) and B(5, 9)

Let $x_1 = 2$, $y_1 = 3$ and $x_2 = 5$, $y_2 = 9$

The gradient of AB $= \dfrac{y_2 - y_1}{x_2 - x_1}$

$$= \dfrac{9 - 3}{5 - 2}$$

$$= 2$$

Try this 10

Find the gradient of the line passing through the points (- 2,1) and (1, 4)

Exercise 8.4(c)

Find the gradient of the line passing through each pair of points

1. (3, 4), (1, 8)	2. (1, 7), (2, 10)	3. (- 3, - 2), (1, - 6)
4. (2, - 5), (6, - 3)	5. (1, 1), (- 3, 7)	6. (4, 3), (6, 7)
7. (2, 3), (- 4, - 6)	8. (2, 1), (5, 5)	9. (3, 2), (-2, - 8)
10. (4, 6), (10, 2)	11 (- 1, 1), (5, 3)	12. (- 2, 2),(2, - 6)

Writing Equations of Lines

The equation of a straight line can be written in three forms:

Gradient- intercept form

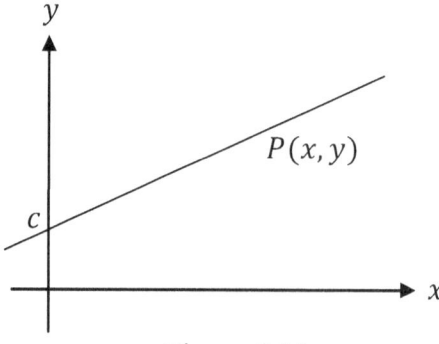

Figure 8.14

Figure 8.14 shows a line that cuts the y-axis at the point $(0,c)$. Let $P(x,y)$ be a point on the line and m the gradient of the line. Then

$$\frac{y-c}{x-0} = m$$

$$y - c = mx$$

$$y = mx + c$$

This form of the equation of a line is called the gradient- intercept form of the equation of a line. The gradient- intercept form of the equation of a line provides the most information about the line. Recall that m is the gradient of the line and the point $(0,c)$ is the intercept on the y-axis.

Example

Find the equation of the line that passes through $(0,-2)$ and has gradient 3

Let (x,y) be a point on the line. Then

$$\frac{y+2}{x} = 3$$

$$y + 2 = 3x$$

$$y = 3x - 2$$

Alternatively, using $y = mx + c$ and replacing m with 3 and c with -2 we have

$$y = 3x - 2.$$

Try this 11

Find the equation of the line that passes through (0, 3) and has gradient − 2.

Exercise 8.4(d)

Find the equation of the lines that cuts the y − axis at the following points and have the specified gradients.

1. (0, - 3), 4   2. (0, 3), 2   3. (0, 5), - 3   4. (0, 6), 4

5. (0, - 4), - 2   6. (0, 1), $\frac{1}{2}$ 7. (0, - 4), $\frac{2}{3}$ 8. (0, 2), $\frac{3}{4}$

Point -gradient form

Figures 8.15 shows a line through the point $A(x_1, y_1)$ with gradient m.

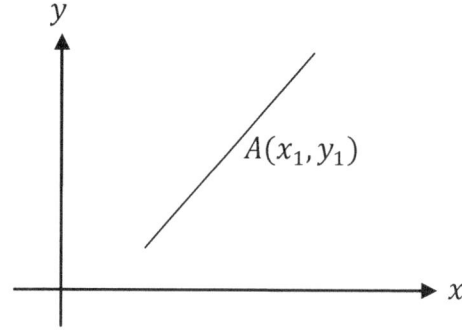

Figure 8.15

Given any other point $P(x, y)$ on the line, we have

$$\frac{y - y_1}{x - x_1} = m$$

$$y - y_1 = m(x - x_1)$$

This form of the equation of a line is called the point- gradient form of the equation of a line.

Example

Find the equation of the line that passes through the point (3, 5) and has gradient 2.

Using the point-gradient form with $x_1 = 3$, $y_1 = 5$ and $m = 2$, we have

$$y - 5 = 2(x - 3)$$

$$y - 5 = 2x - 6$$

$$y = 2x - 1$$

Try this 12

Find the equation of the line that passes through the point (2, - 3) and has gradient – 2.

Exercise 8.4(e)

Find the equation of the lines passing through the following points and has the given gradients.

1. (1, 3), 2   2. (2, - 2), 3   3. (2, - 1), - 3   4. (2, 3), - 1

5. (4, - 5), - 2   6. (1, 1), 4   7. (- 1, 4), 2   8. (- 2, - 3), - 2

9. (2, 3), 5   10. (0, -2), - 4   11. (- 3, 2), -1   12. (- 3, 0), 2

Standard form

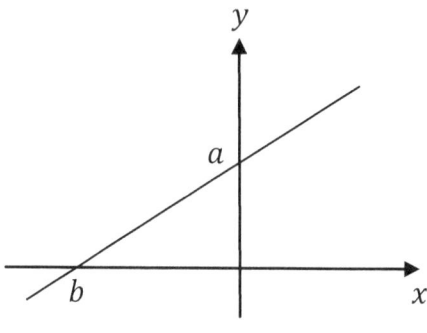

Figure 8.16

Figure 8.16 shows a line that cuts the y-axis at the point $(0, a)$ and the x-axis at the point $(b, 0)$. The gradient of the line is

$$\frac{0 - a}{b - 0} = \frac{-a}{b}$$

Using the point $(0, a)$ and a point (x, y) on the line we have

$$\frac{y - a}{x - 0} = \frac{-a}{b}$$

$$by - ab = -ax$$

$$ax + by - ab = 0$$

Replacing $-ab$ with c, we have

$$ax + by + c = 0$$

This form of the equation of a line is called the standard form.

Example

Find the equation of a line through the point $(1, 2)$ and has gradient $-\frac{3}{4}$

Using the point-gradient form with $x_1 = 1$, $y_1 = 2$ and $m = -\dfrac{3}{4}$, we have

$$y - 2 = -\frac{3}{4}(x - 1)$$

$$4y - 8 = -3x + 3$$

$$3x + 4y - 11 = 0$$

Try this 13

Find the equation of the line through the point $(1, 4)$ and has gradient $\dfrac{2}{3}$

Exercise 8.4(f)

Find the equation in standard form of the lines passing through the following points and has the specified gradient.

1. $(2, 1), \dfrac{2}{3}$ 2. $(3, 2), \dfrac{1}{4}$ 3. $(-1, -2), \dfrac{3}{4}$ 4. $(4, -3), \dfrac{3}{2}$

5. $(-3, 2), -\dfrac{3}{4}$ 6. $(7, 8), -\dfrac{3}{5}$ 7. $((1, -4), \dfrac{4}{3}$ 8. $(-2, -1), \dfrac{5}{3}$

Other Equations of Lines

From the gradient – intercept form of the equation of a line, you can see that:

1. The equation of the line that passes through the origin $(0, 0)$ as shown in Figure 8.17(a) , has the equation of the form $y = mx$.

   Note that $c = 0$

2. A horizontal line has an equation of the form $y = c$, since the gradient of a horizontal line is 0. Notice that each point on a horizontal line through the point $(0, c)$ has a y-coordinate of c (see Figure 8.17(b))

3. The gradient of a vertical line is undefined. Each point on a vertical line through the point $(a, 0)$ has an x-coordinate of a (see Figure 8.17(c)). The vertical line through $(a, 0)$ has an equation of the form $x = a$.

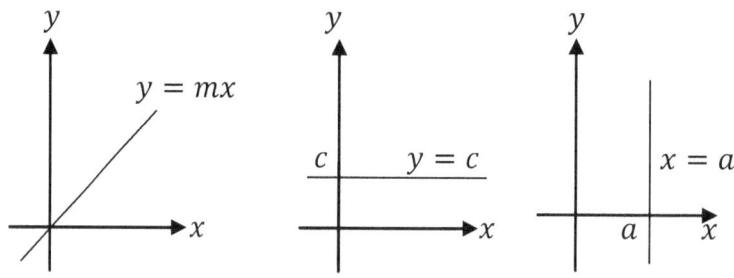

Figure 8.17(a)   Figure 8.17(b)   Figure 8.17(c)

Equation of lines passing through two points

You can use the point- gradient form to find an equation of a line through two points.

Examples

Find the equation of the line that passes through the points (- 2, 1) and (1, - 5).

First determine the gradient of the line. Using the gradient formula with $x_1 = -2$, $y_1 = 1$ $x_2 = 1$ and $y_2 = -5$ we have

$$Gradient = \frac{-5 - 1}{1 + 2} = -2$$

Using the point (- 2, 1) and the point-gradient form, we have

$$y - 1 = -2(x + 2)$$

$$y = -2x - 3$$

You can write an equation of a line in gradient –intercept form or standard form. The standard form of the equation is $2x + y + 3 = 0$

Try this 14

Find the equation of the line that passes through the point (1, - 2) and (3, 2)

Find the equation of the line that passes through the points (5, 8) and (7, 11)

Using $x_1 = 5$, $y_1 = 8$, $x_2 = 7$ and $y_2 = 11$, the gradient of the line is

$$Gradient = \frac{11 - 8}{7 - 5} = \frac{3}{2}$$

Using the point (5, 8) and the point - gradient form, we have

$$y - 8 = \frac{3}{2}(x - 5)$$

$$2(y - 8) = 3(x - 5)$$

$$2y - 16 = 3x - 15$$

$$3x - 2y + 1 = 0$$

Try this 15

Find the equation of the line that passes through the points (-3, 2) and (1, -4).

Exercise 8.4(g)

Find the equation of the line that passes through each pair of points

1. (3, 2), (2, 5)   2. (- 1, 1), (3, 9)   3. (- 5, - 1), (- 2, 5)
4. (5, 6), (6, 8)   5. (- 1, 5), (- 2, 8)   6. (1, 4), (7, 8)

7. (- 2, 4), (- 4, - 6)   8. (3, 2), (- 4, - 6)   9. (4, 8), (10, 2)

10. (5, 4), (2, 2)   11. (6, 2), (2, - 1)   12. (4, 3), (2, 6)

13. (1, 6), (4, 2)   14. (- 2, -1), (1, - 3)   15. (- 3, 1), (7, - 3)
16. (6, - 5), (4, 0)   17. (0, - 3), (4, 2)   18. (3, 4), (7, - 2)

19. (7, - 5), (- 2, - 2)   20. (- 1, 4), (1, 7)

Review exercise 8

1. Find the domain and range of the relation

   (a) {(1, - 2), (2, - 1), (- 1, 0), (- 2, 3)}

   (b) {(2, 3), (1, 2), (1, 0), (3, 4)}

   (c) {(4, - 1), (0, 0), (2, 8), (6, 5), (5, 5)}

   (d) {(5, - 2), (2, - 2), (5, - 3)}

2. Draw an arrow diagram to illustrate each relation

   (a) {(1, 5), (2, 5), (3, 6), (2,4), (4, 5)}

   (b) {(- 1, 3), (- 2, 4), (- 3, 5), (4,4), (5, 6)}

(c) {(4, - 2), (4, 1), (4, - 3), (5, 4)}

(d) {(3, 4), (5, 7), (6, 1), (2, 2), (4, 3)}

3. Find the inverse of the relation

(a) {(2, 5), (3, 5), (3, 6), (2, 4), (5, 5)}

(b) {(- 1, 2), (- 2, 3), (- 2, 4), (3, 4), (5, 6)}

(c) {(3, - 2), (3, 1), (4, - 3), (3, 4)}

(d) {(2, 3), (4, 5), (3, 1), (2, 2), (4, 3)}

4. Determine whether each relation is a mapping. Identify the type

of mapping

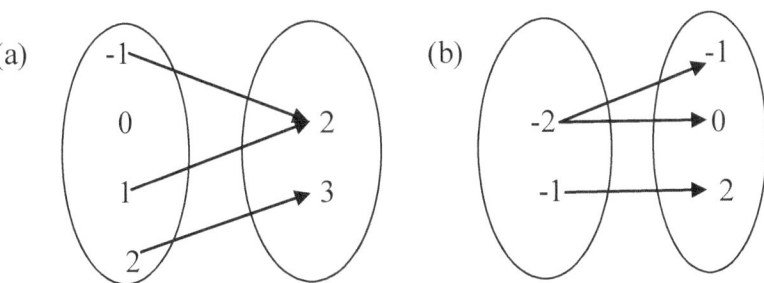

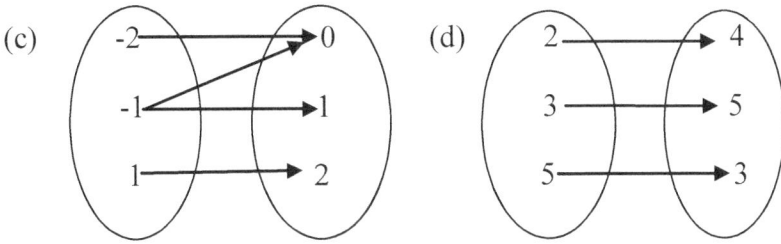

5. Find the rule for each relation

(a)
0	1	2	3	4	5
↓	↓	↓	↓	↓	↓
3	6	9	12	15	18

(b)
0	1	2	3	4	5
↓	↓	↓	↓	↓	↓
0	5	10	15	20	25

6. Use the rule of the relation and the specified domain to find the range

(a) $x \rightarrow 5x - 3$, $\{0, 1, 2, 3, 4, 5\}$

(b) $x \rightarrow 3x^2 + 2$, $(-2, -1, 0, 1, 2, 3\}$

7. Which of the following arrow diagrams represent a function?

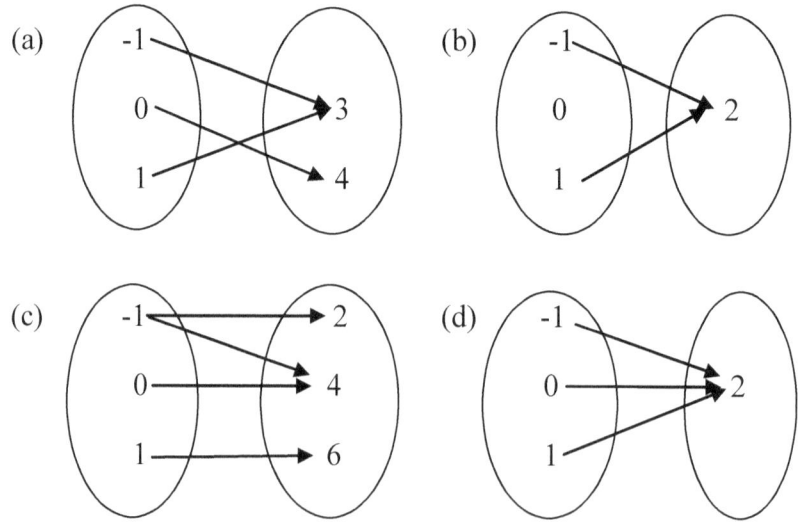

8. Determine which sets of ordered pair represent function

(a) $\{(1, 6), (3, 8), (4, 7)\}$ (b) $\{(1, 3), (1, 2), (1, 1)\}$

(c) $\{(-1, 2), (-2, 3), (0, 3), (1, 0)\}$ (d) $\{(2, -1), (3, 4), (3, -1)\}$

9. Given that $f(x) = 3x + 4$, evaluate

    (a) $f(0)$ (b) $f(1)$ (c) $f(-5)$ (d) $f(-2)$

10. Given that $g(x) = 2x^2 - 3$, evaluate

    (a) $g(-3)$ (b) $g(1)$ (c) $g(0)$ (d) $g(2)$

11. A function f is defined on the set of real numbers by

    $f(x) = a(a + x)$. Find the values of a, if $f(-3) = 10$.

12. Find the distance between each pair of points.

    (a) (7, 3), (5, 4)   (b) (3, 1), (6, 5)

    (c) (10, 3), (5, 15)   (d) (- 5, 0), (- 8, 2)

13. Find the midpoint of the line segment joining each pair of

    points

    (a) (0, - 2), (8, 4)   (b) (- 2, - 3), (2, 7)

    (c) (6, 1), (3, 6)   (d) (7, 2), (- 1, 9)

14. Find the gradient of the line passing through each pair of points.

    (a) (4, 3), (7, - 3)   (b) (5, - 8), (2, 4)

    (c) (- 2, - 3), (4, 6)   (d) (5, 6), (- 3, - 4)

15. Find the equation of the lines that passes through the following

    points and has the given gradient

    (a) (0, -2), 3   (b) (2, - 3), 4   (c) (3, 0), - 2 (d) (- 2, 3), $\frac{3}{4}$

16. Find the equation of the line that passes through each pair of

    points

(a) (1, 4), (5, 6)   (b) (2, 6), (4, 1)

(c) (2, 3), (5, 6)   (d) (1, - 3), (4, 2)

(e) (- 3, 4), (2, 4)   (f) (4, - 5), (2, 1)

(g) (- 2, - 3), (-5, 1)   (h) (5, 3), (8, 5)

Chapter Test 8

Take this test as you would take a test in class. After you are done, check your work against the answers in the back of the book.

1. Find the domain and range of the relation

   (a) {(0, 1), (1, 1), (- 1, 2), (2, 1)}

   (b) {(3, 2), (4, 3), (3, 1), (4, 4)}

2. Which of the following relations are not functions?

   (a) {(-2, 3), (- 1, 1), (0, 5), (1, 3) (3, 3), (4, 2)}

   (b) {(2, 3), (3, 4), (5, 6), (2, 7)}

   (c) {(- 1, - 1), (- 2 , - 2), (0, 0), (1, 1), (2, 2)}

   (d) {(- 2, 4), (3, 4), (5, 6), (- 2, 6)}

3. Determine whether the relation is a mapping. Identify the type of

   mapping

(a) 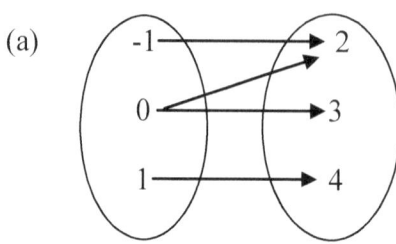   (b)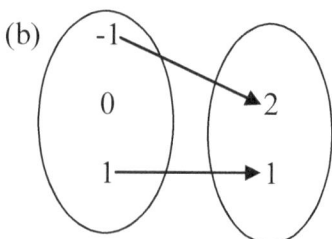

4. Find the rule for each relation

(a)
$$0 \quad 1 \quad 2 \quad 3 \quad 4 \quad 5$$
$$\downarrow \quad \downarrow \quad \downarrow \quad \downarrow \quad \downarrow \quad \downarrow$$
$$7 \quad 9 \quad 11 \quad 13 \quad 15 \quad 17$$

(b)
$$-1 \quad 0 \quad 1 \quad 2 \quad 3$$
$$\downarrow \quad \downarrow \quad \downarrow \quad \downarrow \quad \downarrow$$
$$3 \quad 1 \quad -1 \quad -3 \quad -5$$

5. Which of the following arrow diagrams represent a function?

(a) 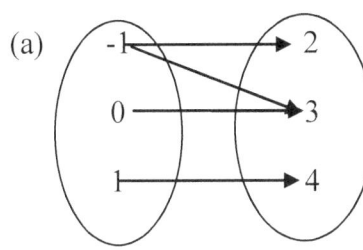   (b)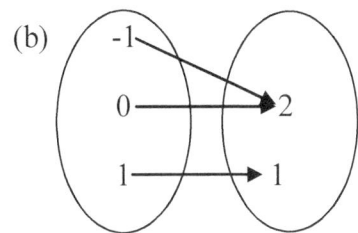

6. Two functions f and g are defined by $f: x \rightarrow x^2 - 3$ and $g: x \rightarrow 2x + 1$, where x is a real number. Evaluate

(a) $f(-2)$ (b) $g(3)$ (c) $\frac{f(2) \cdot g(-3)}{f(3) - g(5)}$

7. A function f is defined on the set of real numbers by $f(x) = ax + b$, where a and b are constants. Given that $f(-1) = 2$ and $f(2) = 11$, find:

(a) a and b

(b) $f\left(\frac{1}{3}\right)$

8. A function f is defined on the set of real numbers, R, by $f: x \rightarrow 2x^2 + 3$. If $f(x - 1) - 2 = x$, find the values of x.

9. Find the value(s) of x for which the following functions are not defined

(a) $f(x) = \dfrac{2x}{3x-2}$ 　　　　 (b) $g(x) = \dfrac{3x+2}{x^2-2x-3}$

10. Find the distance between each pair of points

(a) $(-7, 3), (-5, 4)$ 　　　　 (b) $(3, -1), (6, -5)$

11. Find the midpoint of the line segment joining each pair of

points

(a) $(2, 3), (-2, -7)$ 　　　　 (b) $(-7, 2), (1, -6)$

12. Find the gradient of the line passing through each pair of points

(a) $(4, -3), (7, 3)$ 　　　　 (b) $(-5, -8), (-2, 4)$

13. A line with gradient $-\dfrac{3}{4}$ passes through the point $A(5, -2)$ and

the point $B(-3, n)$. Find the value of n.

14. Find the equations of the lines that passes through the following

points and have the given gradients.

(a) $(3, -1), \dfrac{2}{3}$ 　　　　 (b) $(-2, 3), -3$

15. Find the equation of the line that passes through each pair of

points

(a) $(2, -3), (5, 6)$ 　　　　 (b) $(-1, -2), (-3, 1)$

9

Graphs of Functions

We begin by reviewing some of the basic ideas in drawing graphs.

9.1 Cartesian Plane

You can represent ordered pairs of real numbers by points in a plane. This plane is called the Cartesian plane (or xy-plane), named after the French mathematician Rene' Descartes. The Cartesian plane consists of two perpendicular number lines. Each number line is called an axis (plural axes). The horizontal axis is usually called the x – axis, and the vertical axis is usually called the y -axis. The point of intersection of the two axes is called the origin, and the axes separate the plane into four regions called quadrants.

Each point in the Cartesian Plane is named by an ordered pair of the form (x, y), called the coordinates of the point. For example, the coordinates of the origin are (0, 0). The first number is called the x-coordinate, and the second number is called the y-coordinate of the point.

Note that the x- coordinate measures the distance of the point in the x-direction and the y-coordinate measures the distance of the point in the y-direction. Figure 9.1 shows a Cartesian plane, with few points plotted.

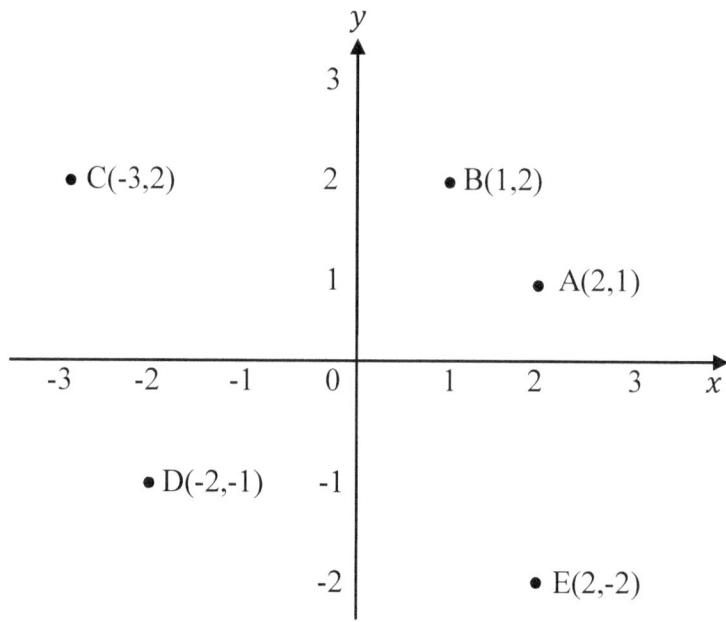

Figure 9.1

Every point on the plane can be represented by only one ordered pair. The point A has coordinates (2, 1), and is two units to the right of the origin and one unit up, while the point B(1, 2) is the point that is one unit to the right of the origin and two units up. Note that (2, 1) and (1, 2) are different points. The order of the coordinates is important.

The point C(- 3, 2) is three units to the left and two units above the origin, the point D(- 2, - 1) is two units to the left and one unit below the origin, and the point E(2, - 2) is two units to the right and two units below the origin. You may have noticed that numbers on the axes to the right and above the origin are positive. The numbers on the axes to the left and below the origin are negative.

Scales

Scales for axes must be chosen carefully. A badly chosen scale may make the plotting and the reading of points on the graph difficult. Choose a scale such that the graph covers as much space as possible on the graph paper. You may use different scales on the two axes.

Choosing a scale

To choose a scale for the axes, find the range of values to be represented on each axis, and the length of axis that is available. In an examination, you would be given 2 cm graph paper. The 2 cm graph paper has 10 big squares (each of length 2 centimetres) along the x -axis and 12 big squares along the y- axis. Suppose you had to draw a graph to represent the following results.

x	- 3	-2	-1	0	1	2	3	4
y	-3	-1	1	3	5	7	9	11

Table 9.1

First, look for the largest and smallest values of both x and y. These values are indicated in the diagram shown in Figure 9.2

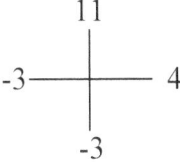

Figure 9.2

You need to represent 7 units on the x – axis. Since you have 10 squares available you can use the scale of 2 cm to represent 1 unit on the x -axis.

You need to represent 14 units on the y- axis. Since there are only 12 squares available, you cannot take the length of each square as 1 unit. Dividing 14 units by 12 gives $\frac{14}{12} = 1.1\dot{6}$. It will not be suitable to represent the length of each square as $1.1\dot{6}$. The most convenient scale for the y- axis will be 2 cm to 2 units. Avoid awkward scales such as 2 cm to 3 units.

The following are illustrations of scales which normally make graphing easy.

2 cm to 0.1 unit; 1 unit; 10 units, 2 cm to 0.2 unit; 2 units; 20 units and 2 cm to 0.5 unit; 5 units; 50 units

Try this 1

Suggest suitable scales to represent the ranges of values for the given number of square.

1(a). Range – 4 to 3; 10 squares   (b). Range – 8 to 6; 10 squares

(c). Range – 15 to 30; 12 squares   (d). Range 0 to 1; 10 squares

(e). Range 0 to 200; 12 squares   (f). Range – 80 to 20; 12 squares

(g). Range – 1 to 2; 10 squares   (h) Range 5 to 40; 12 squares

Drawing Axes

A rough sketch of the axes as illustrated in Figure 9.3 may help you draw the axes on the graph paper.

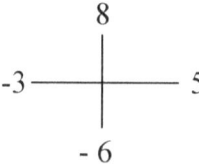

Figure 9.3

You can see that the x – axis extends from – 3 to 5. Using a scale of 2 cm to 1 unit, you require 3 squares to the left of the vertical axis and 5 squares to the right. Counting three squares from the left draw a vertical line to represents the y-axis. You can add one or more squares provided you have enough squares on the right. If you use a scale of 2 cm to 2 units on the y- axis, you will require 3 squares below the x–axis and 4 squares above it. Counting 3 squares from the bottom draw a horizontal line to represent the x -axis. You may add one or more squares provided you have enough squares above. The horizontal axis and the vertical axis are labelled as x and y respectively. The arrows at the ends of each axis denote the positive direction.

Finally, mark off equal length on each axis to show clearly the scale for that axis. Positive x values are marked to the right of the origin and negative to the left. Positive y values are marked above the origin and negative values below. The axes may look as shown in Figure 9.4

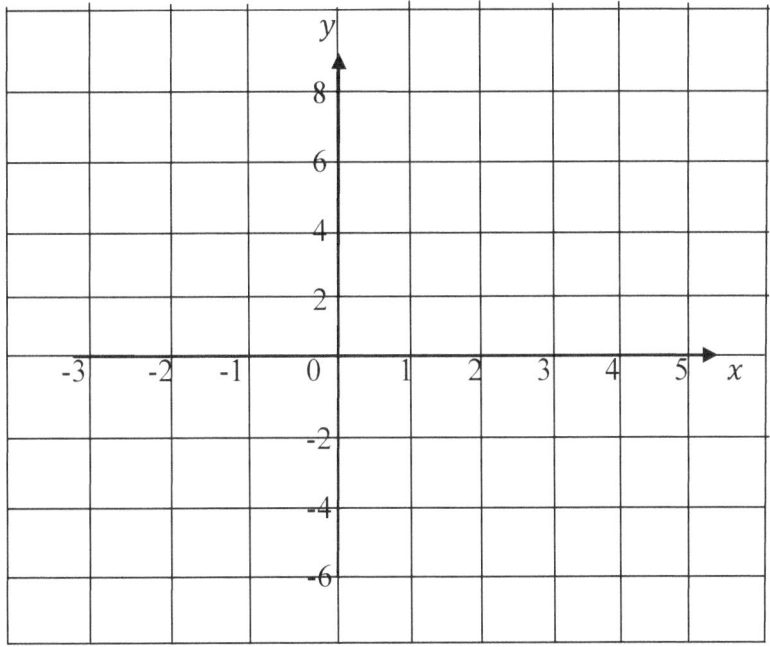

Figure 9.4

Constructing tables of values

To construct a table of values for an equation, we choose values for x and substitute them in the equation to calculate the corresponding values for y.

Example

Construct a table of values for the function $y = x^2 - 2x$.

First we choose a convenient range of values for x. We will choose values of x from -2 to 4. The table of values can be constructed as illustrated in Table 9.2.

x	-2	-1	0	1	2	3	4
x^2	4	1	0	1	4	9	16
$-2x$	4	2	0	-2	-4	-6	-8
y	8	3	0	-1	0	3	8

Table 9.2

The first row gives the values of x picked from -2 to 4 inclusive. The second row gives the square of the x values and the third row is obtained by multiplying the x values by -2. The fourth row is obtained by adding the values in the second and third rows.

Try this 2

Construct a table of values for the function $y = x^2 - 2x - 4$ in the range $-2 \leq x \leq 4$.

Exercise 9.1

Construct a table of values for each of the following functions in the given range.

1. $y = 2x + 5$, $-1 \leq x \leq 4$

2. $y = 3 - 2x$, $-2 \leq x \leq 5$

3. $y = 3x - 7$, $0 \le x \le 6$

4. $y = \frac{1}{2}x - 3$, $-2 \le x \le 4$

5. $y = x^2 - 5x + 6$ $-1 \le x \le 6$

6. $y = 5 + 4x - x^2$ $-2 \le x \le 6$

7. $y = 3x^2 + 5x - 2$ $-3 \le x \le 2$

8. $y = (x - 1)^2 - 9$ $-3 \le x \le 5$

9. $y = (3 - x)(2 + x)$ $-3 \le x \le 4$

10. $y = x^3 - 5x$ $-3 \le x \le 3$

11. $y = x^3 - 3x^2 + 4$ $-2 \le x \le 4$

12. $y = \frac{2x-5}{x+3}$, $x \ne -3$ $0 \le x \le 4$

Plotting Points

The position of a point is located by its co-ordinates. The first co-ordinate called the x co-ordinate measures its distance from the origin measured parallel to the x-axis, and the second co-ordinate called the y co-ordinate measures its distance from the origin parallel to the y-axis.

To plot a point, say A (3, 2), move 3 units to the right of the origin, and then 2 units up. The intersection of the vertical line through 3 on the x -axis and the horizontal line through 2 on the y –axis is the point that corresponds to the point A (3, 2) as shown in Figure 9.5. A cross (or a dot) is generally used to designate a point.

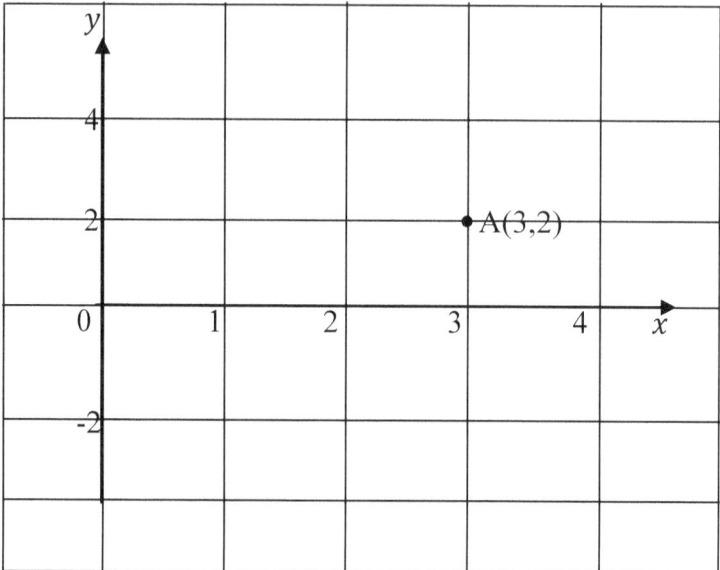

Figure 9.5

Try this 3

Draw two perpendicular axes with x-values from -5 to 5 and y-values from -6 to 6.

Use a scale of 2 cm to represent 1 unit on the x- axis and 2 cm to represent 2 units on the y-axis. Plot the following points.

1. A(3, 5)   2. B(- 3, 6)   3. C(- 4, -3)   4. D(2.5, - 3.2)

5. E(- 4.8, 2.6)   6. F(1.6, 6.4)   7. G(- 2.3, 6)   8. H(4.7, - 5.8)

9.2 Graphs of functions

The graph of an equation, such as $y = x^2 - 3x$ is the set of all points in the Cartesian plane that satisfy the equation. We will consider graphs of linear and quadratic functions but the technique of graphing these functions is exactly the same way that other functions are graph.

Drawing graphs of Linear Functions by using tables

To draw a graph of a function, you can find a set of ordered pairs(x, y), plot them and then connect the points.

Example

Draw the graph of the function $y = 3x + 2$ in the range $-2 \leq x \leq 1$

First construct a table of values as shown in Table 9.3

x	-2	-1	0	1
$3x$	-6	-3	0	3
2	2	2	2	2
y	-4	-1	2	5

Table 9.3

Next plot the points and use a ruler or other straight-edge to draw a straight line through them. The graph is shown in Figure 9.6.

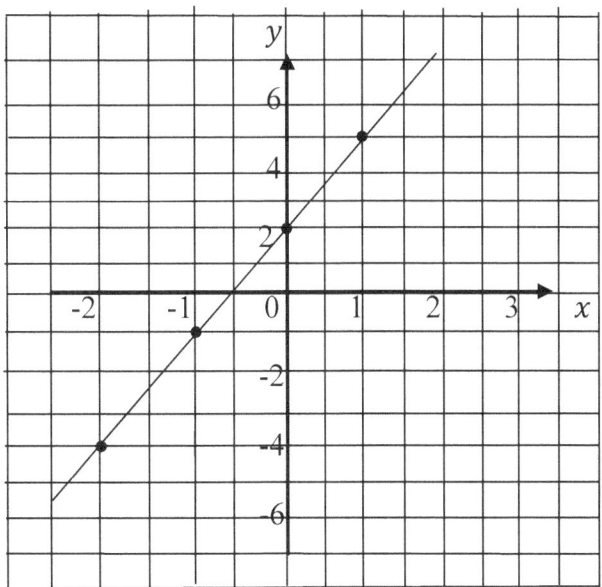

Figure 9.6

Notice that all the points lie on a straight line. The line is said to be the graph of $y = 3x + 2$.

You need only two points to fix the position of a straight line. However, it is advisable to work out three pairs of values. The third

point serves as a check. If you make a mistake the three points will not line up.

The two points often used in drawing graphs of linear equations are the x intercept and the y intercept because they are very easy to find.

Drawing graphs of Linear Equations using the Intercepts

Example

Draw the graph of $3x + 2y = 6$ in the range $-2 \leq x \leq 4$

To find the x intercept set $y = 0$, and then solve for x

$$3x + 2(0) = 6$$

$$3x = 6$$

$$x = 2$$

So, the x- intercept is (2, 0).

To find the y intercept by set $x = 0$, and then solve for y

$$3(0) + 2y = 6$$

$$2y = 6$$

$$y = 3$$

So, the y -intercept is (0, 3).

Plot the points (2, 0) and (0, 3). Then draw a straight line through the points, as shown in Figure 9.7.

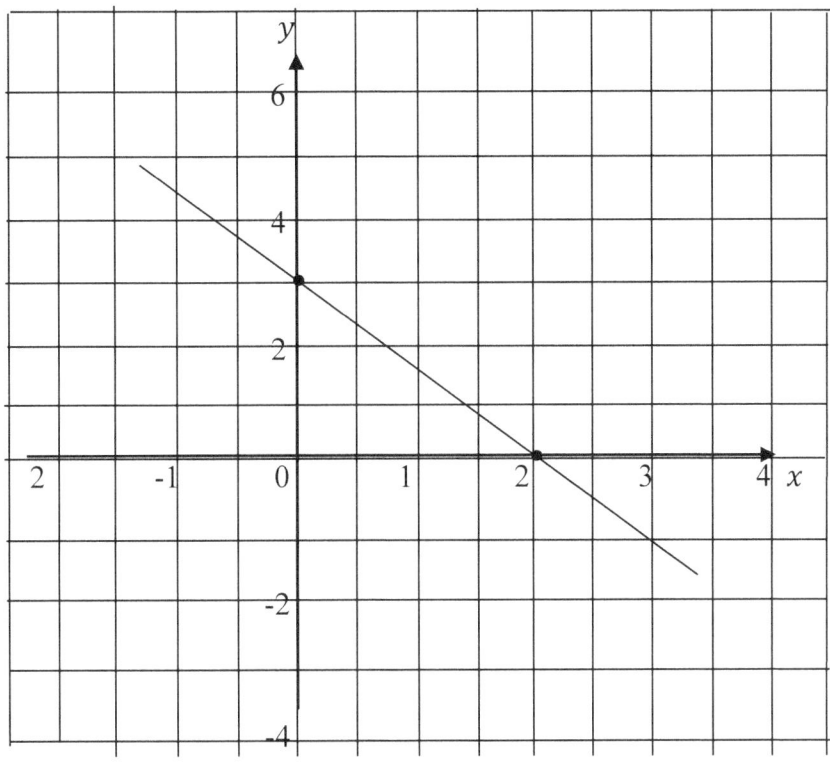

Figure 9.7

Try this 4

Draw the graph of $y = 2x - 5$ in the range $-2 \leq x \leq 4$

Exercise 9.2(a)

Draw the graph of each of the following functions in the given range

1. $y = 2x + 5$, $-4 \leq x \leq 3$

2. $y = 4 - 3x$, $-3 \leq x \leq 5$

3. $y = 5x - 3,$ $-2 \leq x \leq 4$

4. $y = 2 - x,$ $-5 \leq x \leq 3$

5. $y = 4x + 7,$ $0 \leq x \leq 5$

6. $3x + y = 7,$ $-1 \leq x \leq 5$

7. $2x - y = 3,$ $-2 \leq x \leq 5$

8. $x + 2y = 6,$ $-4 \leq x \leq 8$

Using Graphs to Solve Simultaneous Linear Equations

Recall that the solutions to two linear simultaneous equations are the values of x and y which satisfy both equations. These values of x and y will be the co-ordinates of a point which lies on both graphs. The only point which lies on both graphs is their point of intersection.

Simultaneous linear equations in two unknowns can be solved graphically by drawing the graphs of the two equations on the same set of axes and noting their point of intersection. The x and the y coordinates of the point of intersection of the two graphs gives the solution to the equation.

Example

Solve the simultaneous equations by plotting graphs

$2x + y = 8$

$x + y = 5$

Use one of the methods described above to draw the graph of $2x + y = 8$ and $x + y = 5$ on the same axes. The two graphs are shown in Figure 9.8.

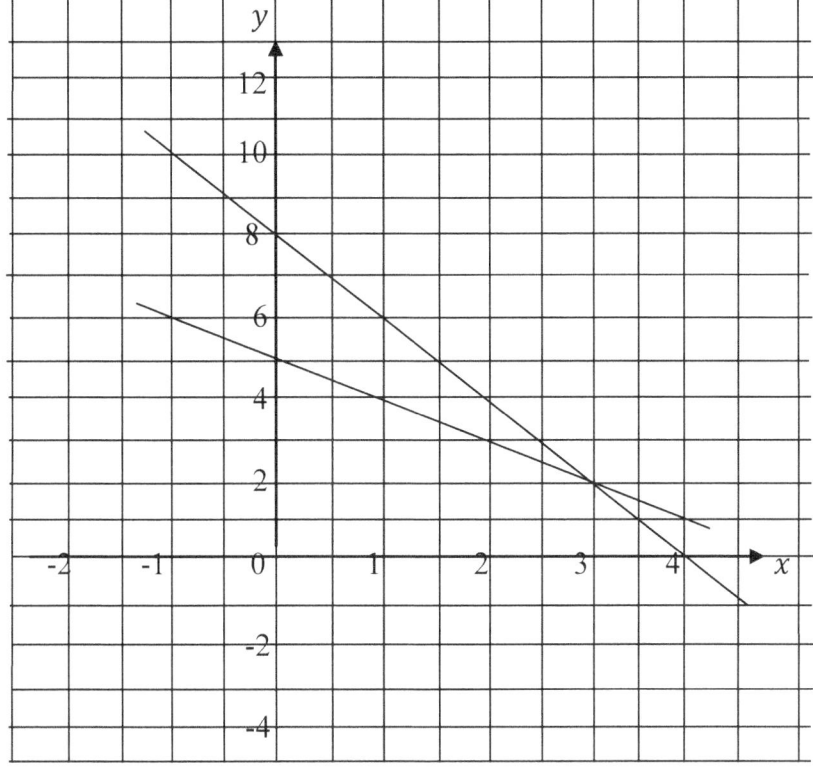

Figure 9.8

Read off the co-ordinates of the point of intersection of the two graphs. The point of intersection is (3, 2), so the solutions of the simultaneous equations are $x = 3$ and $y = 2$.

Try this 5

Solve the simultaneous equations by plotting graphs

$3x + 2y = 8$

$2x - 3y = 1$

Exercise 9.2(b)

Solve each of the following simultaneous equations by plotting graphs

1. $2x - y = 3$ 2. $2x + y = 2$ 3. $x + 4y = 4$

   $x + y = 6$ $x + y = 5$ $-x + y = 1$

4. $2x + 3y = -3$ 5. $3x + y = 1$ 6. $2x + y = 3$

   $2x + y = 3$ $x - 2y = 5$ $x + 3y = 4$

Graphs of Quadratic Functions

The general quadratic function is defined by the equation $y = ax^2 + bx + c$, where a, b and c are constants and a is not equal to zero.

The graph of a quadratic function may have the shape illustrated in either Figure 9.9(a) or Figure 9.9(b). The graph opens upward when the coefficient of the squared term, a, is positive $(a > 0)$ as shown in Figure 9.9(a). The graph opens downward when the coefficient, a, of the squared term is negative $(a < 0)$ as shown in Figure 9.9(b).

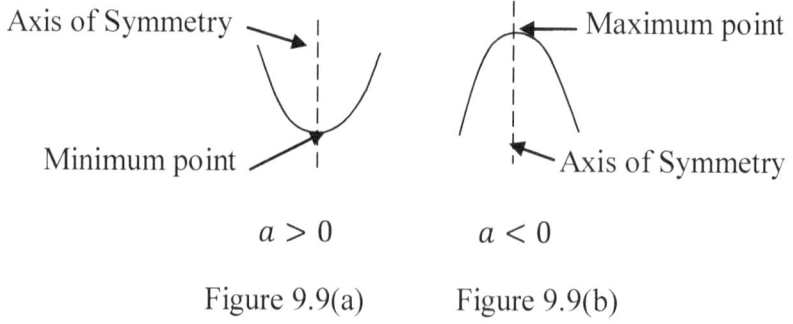

Axis of Symmetry   Maximum point

Minimum point   Axis of Symmetry

$a > 0$ $a < 0$

Figure 9.9(a)   Figure 9.9(b)

The graph of a quadratic function has a lowest point, called the minimum point if the graph opens upward. The y-coordinate of the minimum point is called the minimum value of the function. The graph reaches a highest point, called the maximum point if the graph opens downward. The y-coordinate of the maximum point is called the maximum value of the function. The highest or the lowest point on the curve is called the vertex (or the turning point).

Every quadratic graph is symmetric with respect to a vertical line through its maximum or minimum point. The line is called the axis of symmetry, and its equation is $x = -\dfrac{b}{2a}$..

To find the vertex, we use $-\dfrac{b}{2a}$ to find the x- coordinate and then substitute the value of x to find the y-coordinate. For instance, if $y = x^2 - 6x + 5$ then the x-coordinate of the vertex is $\dfrac{6}{2} = 3$ and $y = 3^2 - 6(3) + 5 = -4$. Therefore the coordinates of the vertex is (3, - 4).

Drawing Graphs of Quadratic Functions

Example

Draw the graph of the function $y = x^2 - 3x - 8$ in the range $-3 \leq x \leq 5$.

A table of values is drawn up as shown in Table 9.4

x	-3	-2	-1	0	1	2	3	4	5
x^2	9	4	1	0	1	4	9	16	25
$-2x$	6	4	2	0	-2	-4	-6	-8	-10
-6	-6	-6	-6	-6	-6	-6	-6	-6	-6
y	9	2	-3	-6	-7	-6	-3	2	9

Table 9.4

Plot the ordered pairs and join the points by a smooth curve drawn freehand as shown in Figure 9.10. The smooth curve shown below is called a parabola.

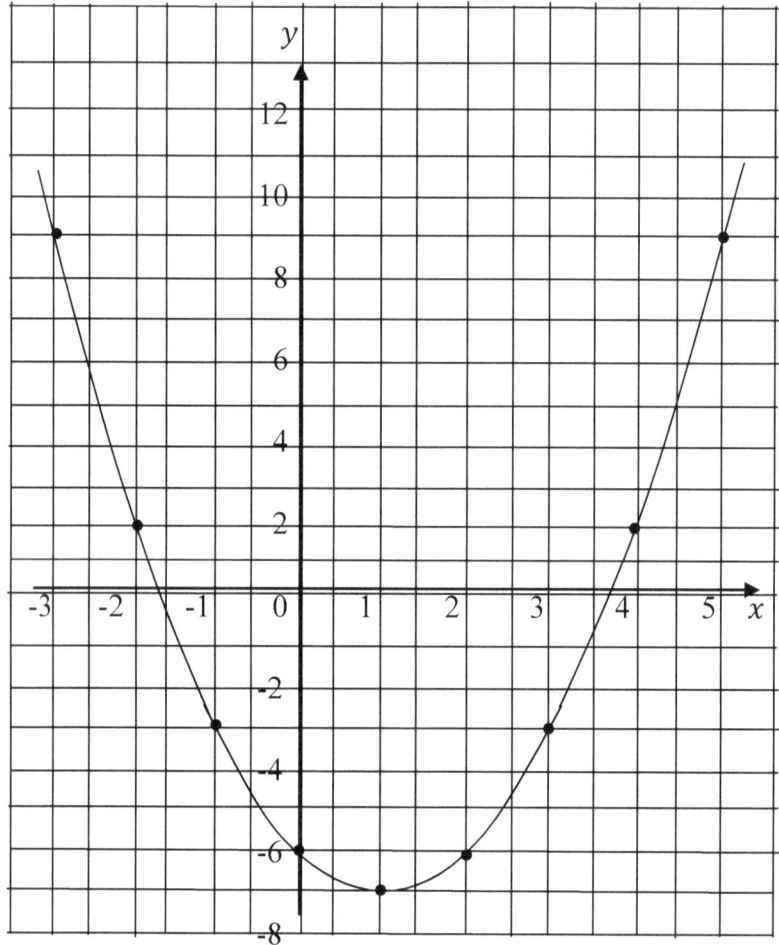

Figure 9.10

Try this 6

Draw the graph of $y = 3 - 2x - x^2$ in the range $-5 \leq x \leq 3$

Exercise 9.2(c)

Draw the graph of each function in the given range

1. $y = x^2 - 2x, -2 \leq x \leq 4$ 2. $y = x^2 - 4x + 3, -1 \leq x \leq 5$

3. $y = 6 + 3x - 2x^2$, $-2 \leq x \leq 4$

4. $y = (3 - x)(2 + x)$, $-4 \leq x \leq 5$

Solving Quadratic Equations with Graphs

You can obtain the solution of the quadratic equation $ax^2 + bx + c = 0$ from the graph of $y = ax^2 + bx + c$. The solutions of the equation are the x co-ordinates of the points where the graph crosses the x-axis.

To draw the graph of $y = ax^2 + bx + c$ you will have to construct a table of values and then plot the points. Choose the x coordinate of the vertex and some x-values on both sides of the vertex when constructing the table of values.

The number of solutions of a quadratic equation depends on how many times the curve cuts the x-axis. If the graph cuts the x-axis at two points then we have two solutions. If the graph touches the x-axis at one point then we have one solution. If the graph does not intersect with the x-axis then the equation has no real solution.

Examples

Draw the graph of $y = x^2 - 5x + 3$ in the range $-1 \leq x \leq 6$. Use your graph to solve

(a) $x^2 - 5x + 3 = 0$ (b) $x^2 - 5x + 4 = 0$

A table of values is drawn up as shown in Table 9.5

x	-1	0	1	2	3	4	5	6
x^2	1	0	1	4	9	16	25	36
$-5x$	5	0	-5	-10	-15	-20	-25	-30
3	3	3	3	3	3	3	3	3
y	9	3	-1	-3	-3	-1	3	9

Table 9.5

A graph of $y = x^2 - 5x + 3$ is shown in Figure 9.11

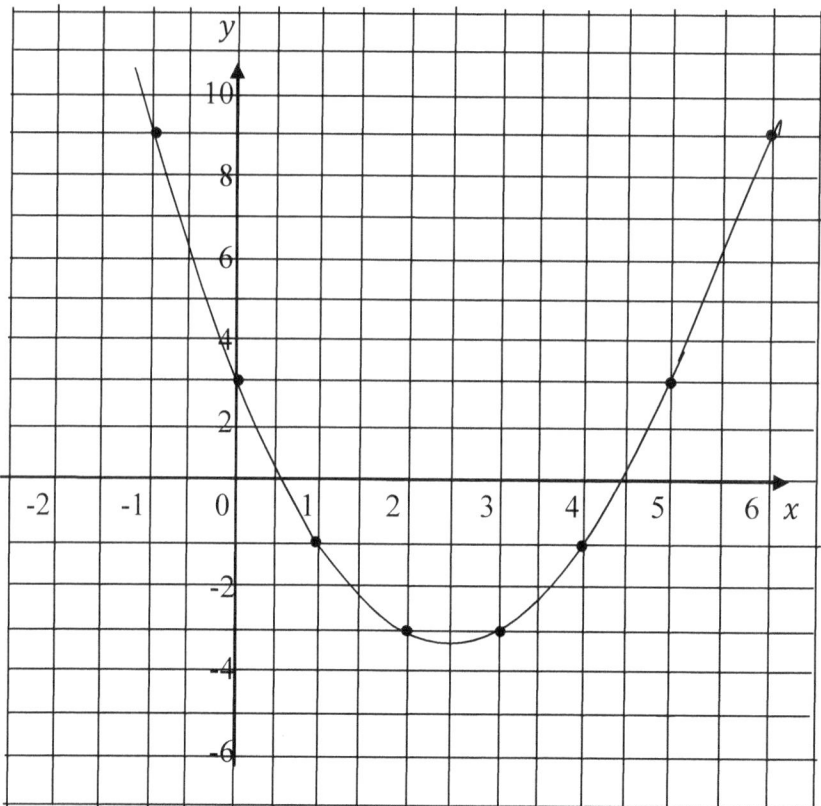

Figure 9.11

(a) You are to find the values of x when $y = 0$. These values occur at the point the graph crosses the x-axis. The graph cuts the x-axis at $(0.7, 0)$ and $(4.3, 0)$, so the solutions to the equation $x^2 - 5x + 3 = 0$ are $x = 0.7$ and $x = 4.3$.

(b) Rewrite $x^2 - 5x + 4 = 0$ so that the left hand side will be the same as the right hand side of $y = x^2 - 5x + 3$. Adding $- 1$ to both sides gives $x^2 - 5x + 3 = -1$. Draw the line $y = -1$ on the same graph.

The solutions are given by the x-co-ordinates of the points where the line $y = -1$ crosses the curve $y = x^2 - 5x + 3$. You can see from Figure 9.11, that the graphs cross at $(1, -1)$ and $(4, -1)$. Therefore the solutions to $x^2 - 5x + 4 = 0$ are $x = 1$ and $x = 4$.

Try this 7

Draw the graph of $y = x^2 - 3x - 5$ in the range $-2 \leq x \leq 5$. Use your graph to solve

(a) $x^2 - 3x - 5 = 0$

(b) $x^2 - 3x - 7 = 0$

Draw the graph of $y = 4 + 2x - x^2$ in the range $-3 \leq x \leq 5$. Use your graph to solve $3 + 3x - x^2 = 0$. Find the equation of the axis of symmetry of the curve.

A table of values is drawn up as shown in Table 9.6

x	-3	-2	-1	0	1	2	3	4	5
4	4	4	4	4	4	4	4	4	4
$2x$	-6	-4	-2	0	2	4	6	8	10
$-x^2$	-9	-4	-1	0	-1	-4	-9	-16	-25
y	-11	-4	1	4	5	4	1	-4	-11

Table 9.6

A graph of $y = 4 + 2x - x^2$ is shown in Figure 9.12

Rewrite the equation $3 + 3x - x^2 = 0$ as $4 + 2x - x^2 = 1 - x$. Draw the graph of $y = 1 - x$ on the same axes, using three points. Here, we plot the points $(-3, 4)$, $(0, 1)$ and $(5, -4)$. Use a straight edge to draw a line through the points (see Figure 9.12).

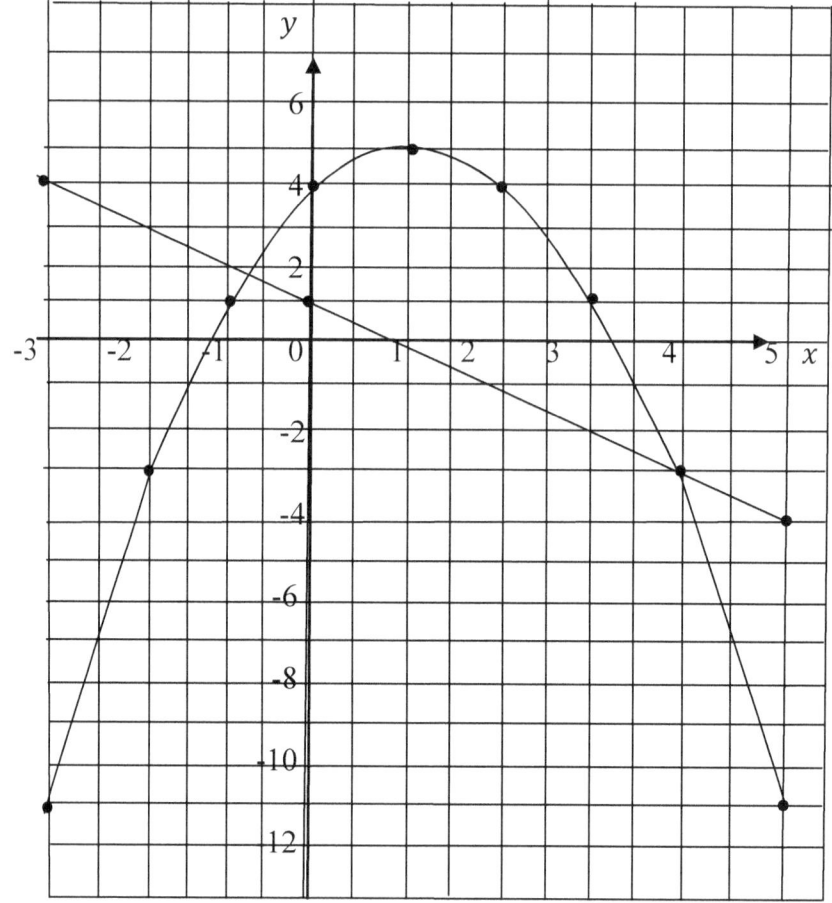

Figure 9.12

The solutions of the given equation correspond to the points where the two graphs intersect. The solutions of the equation are $x = -0.8$ and $x = 3.7$.

The vertex of the curve is $(1, 5)$. The equation of the axis of symmetry is $x =$ 'the $x -$ coordinate of the vertex', so the equation is $x = 1$.

Try this 8

Draw the graph of $y = 2x^2 - 3x - 4$ in the range $-2 \leq x \leq 4$. Use your graph to solve $2x^2 - 5x - 7 = 0$

An alternative graphical method of solving the quadratic equation $ax^2 + bx + c = 0$ is to rearrange the equation as $ax^2 = -bx - c$ and then plot the graphs of $y = ax^2$ and $y = -bx - c$ on the

same axes. The points of intersection of the two graphs give the solutions to the equation $ax^2 + bx + c = 0$.

Example

Solve the quadratic equation $2x^2 + 3x - 4 = 0$ given that the solutions lie between $x = -3$ and $x = 3$.

Begin by rearranging the equation $2x^2 + 3x - 4 = 0$ as $2x^2 = -3x + 4$.

Then let $y = 2x^2$ and $y = -3x + 4$.

Draw up a table of values for $y = 2x^2$ as shown in Table 9.7

x	-3	-2	-1	0	1	2	3
y	18	8	2	0	2	8	18

Table 9.7

Plot the points and join them by a smooth curve.

Next draw up tables of values for $y = -3x + 4$ as shown in Table 9.8.

x	-3	0	2
y	13	4	-2

Table 9.8

Plot these points and join them with a straight edge.

The two graphs are shown in Figure 9.13.

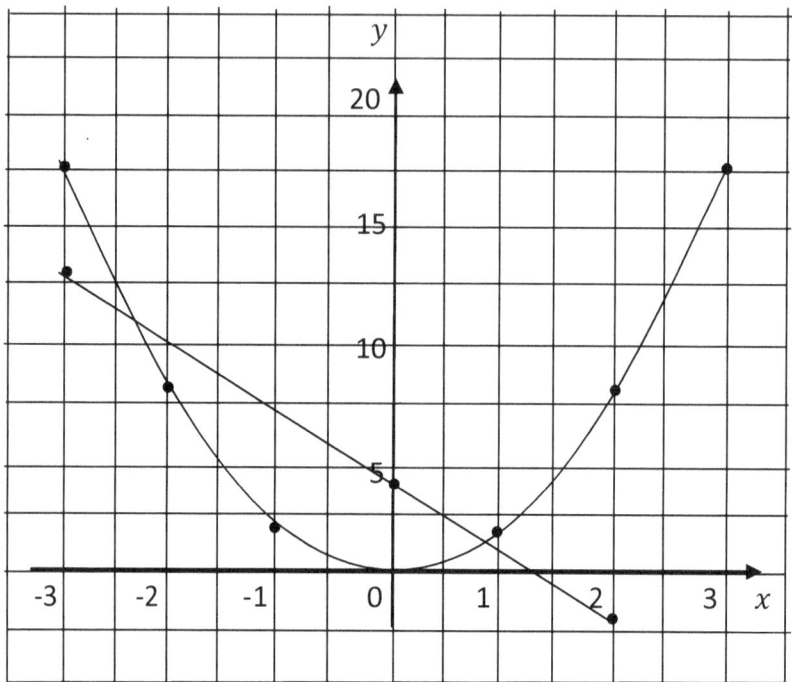

Figure 9.13

The solution of the equation $2x^2 + 3x - 4 = 0$ is obtained from the points of intersection of the two graphs. The line and the curve intersect at (-2.3, 11) and (0.8, 1.5). Hence the solutions of $2x^2 + 3x - 4 = 0$ are $x = -2.3$ and $x = 0.8$.

Try this 9

Draw the graph of $y = 2x^2$ in the interval $-2 \leq x \leq 2$ and hence solve

(a) $2x^2 - x - 3 = 0$

(b) $2x^2 - 5 = 0$

Estimating the gradient of a curve at a given point

The gradient of a curve at a given point is defined as the gradient of the tangent to the curve at that point. The tangent to the curve

at a point is a straight line which touches the curve at that point as illustrated in Figure 9.14

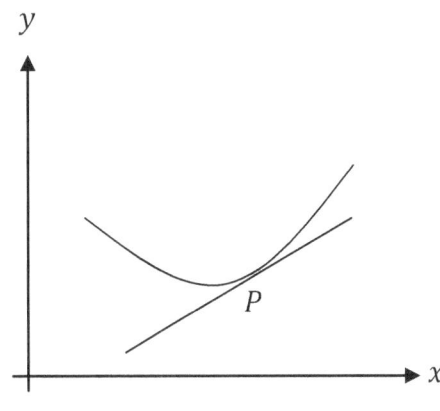

Figure 9.14

You can find the gradient of a curve at a given point by drawing a tangent to the curve at that point and then use the gradient formula to calculate the gradient of the tangent.

Example

Draw the graph of $y = 2x^2 - 6x + 5$ in the interval $-1 \leq x \leq 4$, and find the gradient of the curve at the point (2, 1)

The graph of $y = 2x^2 - 6x + 5$ is shown in Figure 9.15.

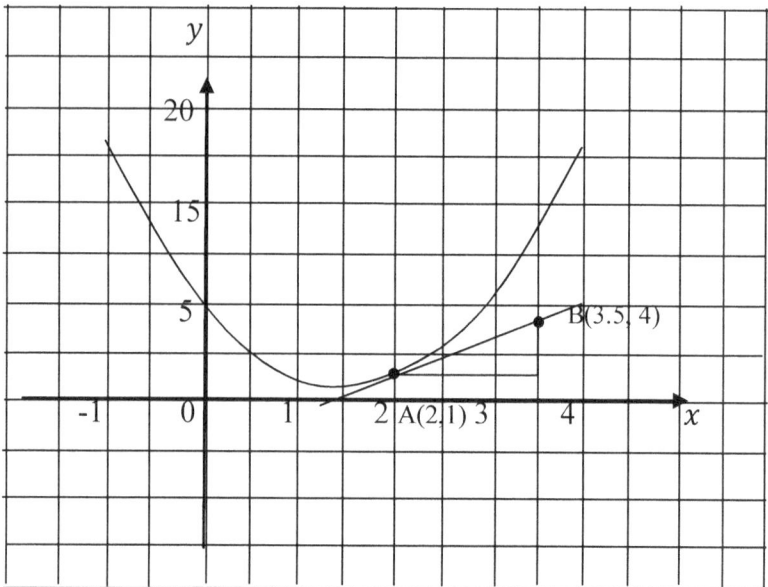

Figure 9.15

First draw a straight line segment to touch the curve at A (2, 1) as shown in Figure 9.15.

Next select another point say B on the tangent. The coordinates of B are (3.5, 4).

Finally, calculate the gradient of the tangent. The gradient of the curve at A (2, 1) is equal to the gradient of the tangent to the curve at A.

Here $x_1 = 2$, $y_1 = 1$, $x_2 = 3.5$ and $y_2 = 4$.

$$Gradient = \frac{y_2 - y_1}{x_2 - x_1}$$

$$= \frac{4-1}{3.5-2}$$

$$= 2$$

Therefore the gradient of the curve at (2, 1) is 2.

Try this 10

Draw the graph of $y = 2x^2 - 3x + 4$ in the interval $-2 \leq x \leq 4$, and find the gradient of the curve at the point (2, 6)

Exercise 9.2(d)

1. Draw the graph of $y = x^2 + 2x - 3$ for values of x from -4 to

   2. Use your graph to find

   (a) the solution of the equation $x^2 + 2x - 5 = 0$

   (b) the set of values of x for which $y = x^2 + 2x - 3$ is negative

   (c) the least point on the graph

2. Draw the graph of $y = x^2 - 3x + 5$ for values of x from -2 to

   5. Use your graph to find

   (a) the solution of the equation $x^2 - 4x - 3 = 0$

   (b) the equation of the axis of symmetry of the graph

3. Draw the graph of $y = 2x^2 - 5x - 3$ for values of x from -2 to

   5. Use your graph to find

   (a) the solution of the equation $2x^2 - 5x - 10 = 0$

   (b) the least value of $y = 2x^2 - 5x - 3$ and the corresponding

      value of x

   (c) in the given interval the set of values of x for which y

      decreases as x increases

4. Draw the graph of $y = 3 + 2x - x^2$ for values of x from -3 to

   5. Use your graph to find

   (a) the solution of the equation $x^2 - 2x - 5 = 0$

   (b) the greatest value of $y = 3 + 2x - x^2$

5(a) Copy and complete the following table of values for

   $y = (x - 3)^2 - 1$

x	-1	0	1	2	3	4	5	6	7
y	15				-1		3		

   (b) Using a scale of 2 cm to represent 1 unit on the x-axis and

   2 cm to represent 2 units on the y-axis, draw the graph of

   $y = (x - 3)^2 - 1$ for values of x from -1 to 7.

   (c) Use our graph to solve

   (i) $(x - 3)^2 - 1 = 0$

   (ii) $(x - 3)^2 - 5 = 0$

6(a) Copy and complete the following table of values for

   $y = 6 + 3x - 2x^2$

x	-3	-2	-1	0	1	2	3	4	5
y			1		7		-3		

   (b) Using a scale of 2 cm to represent 1 unit on the x-axis and

   2 cm to represent 5 units on the y-axis, draw the graph of

$$y = 6 + 3x - 2x^2$$

(c) Use your graph to solve

(i) $6 + 3x - 2x^2 = 0$

(ii) $5 + x - 2x^2 = 0$

7(a) Copy and complete the following table of values for

$$y = x^2 - 2x - 3$$

x	-2	-1.5	-1	0	0.5	1	2	3	3.5	4
y			0		-3.75		-3			

(b) Using a scale of 1 cm to represent 0.5 unit on each axis, draw

the graph of $y = x^2 - 2x - 3$ for values of x from – 2 to 4

(c) Use your graph to solve

(i) $x^2 - 2x - 5 = 0$

(ii) $x^2 - x - 5 = 0$

8(a) Copy and complete the following table of values for

$$y = 2 + 2x - x^2$$

x	-1	-0.5	0	0.5	1	1.5	2	2.5	3
y			2			2.75			

(b) Using a scale of 4 cm to represent 1 unit on each axis, draw the

graph of $y = 2 + 2x - x^2$ for values of x from – 1 to 3

(c) Use your graph to solve

(i) $1 + 2x - x^2 = 0$

(ii) $3x - x^2 = 1$

9. Draw the graph of $y = 3x^2$ for $-3 \leq x \leq 3$. Use your graph to

solve

(a) $3x^2 - 8 = 0$

(b) $x^2 - x - 3 = 0$

10. Draw the graph of $y = x^2 - 3x - 4$ for $-2 \leq x \leq 5$, and find

the gradient of the curve at $(3, -4)$

11. Draw the graph of $y = 3x^2 - x - 7$ for $-3 \leq x \leq 3$, and find

the gradient of the curve at $(-2, 7)$

12(a) Copy and complete the following table of values for $y = \dfrac{8}{x}$ in

the range $-3 \leq x \leq 3$

x	-3	-2	-1	-0.5	0.5	1	2	3
y	-2.7		-8		16		4	

(b) Draw the graph of $y = \dfrac{8}{x}$ in the range $-3 \leq x \leq 3$

(c) Use your graph to solve the equation $2x^2 - x = 8$

Review exercise 9

In Exercises 1 – 4, construct a table of values for each of the
following functions in the given range

1. $y = 3x - 2$, $-3 \leq x \leq 2$

2. $y = 5 + 3x$, $-4 \leq x \leq 1$

3. $y = x^2 + 3x - 2$, $-5 \leq x \leq 2$

4. $y = 5 + 3x - 2x^2$, $-3 \leq x \leq 4$

In Exercises 5 – 8, draw the graph of each of the following functions in the given range

5. $y = 2x + 5$, $-3 \leq x \leq 3$

6. $2x + 3y = 6$, $-1 \leq x \leq 4$

7. $y = 2x^2 - 3x - 4$, $-2 \leq x \leq 4$

8. $y = 6 + 3x - 2x^2$, $-2 \leq x \leq 4$

In Exercises 9 – 12, solve each of the following simultaneous equations by plotting graphs

9. $2x - y = 4$ 10. $3x + 2y = 8$

   $x + y = 5$ $2x - y = 3$

11. $2x - 3y = 0$ 12. $y = 2x + 7$

   $x + 3y = 9$ $2y + 3x = 0$

13. Find the coordinates of the vertex of each of the following

   quadratic functions

   (a) $y = x^2 - 4x - 3$ (b) $y = x^2 + 2x + 5$

   (c) $y = 2x^2 + 4x - 3$ (d) $y = 3x^2 - 8x + 5$

14. Draw the graph of $y = x^2 - 3x + 5$ for values of x from -2 to

   5. Use your graph to solve

   (a) $x^2 - 3x - 5 = 0$ (b) $2x^2 - 3x - 4 = 0$

15. Draw the graph of $y = 3 + 6x - 2x^2$ for values of x from -2 to 5.

   (a) Use your graph to solve $5 + 3x - x^2 = 0$

   (b) Find the greatest value of $y = 3 + 6x - 2x^2$

   (c) Find in the given range the values of x for which y decreases as x increases.

16(a) Copy and complete the following table of values for $y = 2x^2 - x - 7$

x	-4	-3	-2	-1	0	1	2	3	4
y		14			-7				21

   (b) Using a scale of 2 cm to represent 1 unit on the x-axis and 2 cm to represent 5 units on the y-axis, draw the graph of $y = 2x^2 - x - 7$

   (c) Use your graph to solve

   (i) $2x^2 - x - 7 = 0$ (ii) $2x^2 - 4x - 9 = 0$

17. Draw the graph of $y = 2x^2$ for $-3 \leq x \leq 3$. Use your graph to solve

   (a) $2x^2 + 3x = 0$ (b) $2x^2 - 7 = 0$

18. Draw the graph of $y = x^2 + 3x - 2$ for $-2 \leq x \leq 5$, and find the gradient of the curve at (-3, -2)

19. Draw the graph of $y = 2x^2 + 4x - 7$ for $-5 \leq x \leq 3$, and find

   the gradient of the curve at (1, - 1)

20(a) Draw the graph of $y = \frac{2}{x}, x \neq 0$ for $-4 \leq x \leq 4$.

   (b) Use your graph to solve the equation $x^2 = 2 - x$

Chapter Test 9

Take this test as you would take a test in a class. After you are
done, check your work against the answers in the back of the book.

1(a) Copy and complete the following table of values for

   $y = x^2 - 4x + 8$

x	-2	-1	0	1	2	3	4	5	6
y			8		4			13	

   (b) Using a scale of 2 cm to represent 1 unit on the x −axis and 4

   cm to represent 5 units on the y −axis, draw the graph of

   $y = x^2 - 4x + 8$

   (c) Use your graph to solve

   (i) $x^2 - 7x + 6 = 0$ (ii) $x^2 - 4x - 2 = 0$

2(a) Copy and complete the following table of values for

   $y = 3 + 2x - x^2$

x	-2	-1	0	1	1.5	2	2.5	3	3.5	4
y			3		3.75			0		-5

(b) Using a scale of 1 cm to represent 0.5 unit on both the x −axis and the y −axis, draw the graph of $y = 3 + 2x - x^2$ for values of x from − 2 to 4

(c) Use your graph to solve $4 + 2x - x^2 = 0$

(d) Find the greatest value of $y = 3 + 2x - x^2$

3. A missile is projected vertically into the air and its height, h metres after t seconds is given by the formula $h = 112t - 16t^2$

(a) Draw the graph of $h = 112t - 16t^2$ for values of t from 0 to 7.

(b) Find, from your graph

   (i) the greatest height it reaches

   (ii) the time it takes to reach the greatest height

   (iii) the time it takes to reach 160 metres after projection

   (iv) the time it takes to hit the ground

4. Draw the graph of $y = 2x^2$ in the interval $-3 \le x \le 3$ and hence

   (a) solve $2x^2 - 2x - 7 = 0$ (b) find the value of $\sqrt{5}$

5(a) Draw the graph of $y = \frac{6}{x}, x \ne 0$ for $-5 \le x \le 5$.

   (b) Use your graph to solve the equation $x = 1 + \frac{6}{x}$

10

Modular Arithmetic

One use of modular arithmetic is in the 12 – hour clock, in which the day is divided into two 12 – hour periods. If it is 6 am, what time will it be 9 hours from now? In ordinary arithmetic $6 + 9$ equals 15, but there is no 15 on the clock. We only use 12 numbers: 1, 2, 3, 4, 5, 6, 7, 8, 9, 10, 11 and 12 to tell standard time. If we start at 6 and add 9, we end at 12 and then start over at 1. So, what would have been 13 would be 1, what would be 14 would be 2 and what would be 15 would be 3. The time will be 3 pm. We might think of this operation as adding 6 and 9, dividing the sum by 12, and using the remainder, 3 as our sum i.e. $5 + 9 = 3$.

Generally, when we are only interested in the remainder left after division by a certain number, we are doing modular arithmetic. The number we divide by is called the modulus. For instance, the clock arithmetic is arithmetic modulo 12, abbreviated mod 12. 12 is the modulus.

We will replace 12 with 0, since the numbers we use are those numbers that we obtain as remainders when we divide by 12. The remainders start at 0 and increases by 1 each time, until the number reaches one less than 12.

The set of elements $\{0, 1, 2, 3, 4, 5, 6, 7, 8, 9, 10, 11\}$ together with a binary operation is called modulo 12 or mod 12 system. Generally, the modulo n system consists of n elements $0, 1, 2, 3, \cdots . n - 1$, and a binary operation. For example, by mod 5, we mean a system in which there are five members: 0, 1, 2, 3, 4 When two numbers a and b leave the same remainder after division by n, we say that they are equal in that modulus, and we write $a = b(\bmod n)$. For example, because

$7 \div 4 = 1$ rem. 3

and $11 \div 4 = 2$ rem. 3,

7 and 11 are 'equal modulo 4'. We write

$7 = 11 = 3$ (mod 4)

Notice that 3(mod 4) is the remainder when 7 or 11 is divided by 4. 3(mod 4) is read '3 modulo 4'.

We can reduce any number to a whole number less than the modulus, called its simplest form. So 3 is the simplest form of 27 (mod 4).

10.1 Addition in Modular Arithmetic

A method of determining the sum, say $9 + 12$ in arithmetic modulo 8 is to divide the sum, 21 by 8 and observe the remainder.

$21 \div 8 = 2$, remainder 5

The remainder, 5 is the sum of $9 + 12$ in modulo 8. So, we write

$9 + 12 = 5$ (mod 8)

Example

Find $13 + 32$ in modulus 6

$13 + 32 = 45$ Add the numbers together

$\qquad = 3(\text{mod } 6)$ Remainder after dividing the sum by 6

You can reduced 13 and 32 to their simplest form and then add.

$13 + 32 = 1 + 2$ (mod 6)

$\qquad = 3$ (mod 6)

Try this 1

Find $24 + 17$ in modulo 5

Exercise 10.1(a)

1. Write down the set of numbers for each of the following modular

   system

   (a) mod 5   (b) mod 6   (c) mod 8   (d) mod 9

2. Reduce each of the following numbers to their simplest form in

   the modulo given

   (a) 17 (mod 5)   (b) 23 (mod 4)   (c) 30 (mod 6)

   (d) 125 (mod 12)   (e) 97 (mod 8)   (f) 158 (mod 9)

   (g) 238 (mod 7)   (h) 102 (mod 13)   (i) 245 (mod 12)

   (j) 190 (mod 3)   (k) 345 (mod 5)   (l) 769 (mod 13)

3. Work out each of the following additions in the given modulo

   (a) $18 + 24$ (mod 6)   (b) $18 + 19$ (mod 4)

   (c) $125 + 36$ (mod 8)   (d) $96 + 84$ (mod 5)

   (e) $63 + 75$ (mod 9)   (f) $210 + 136$ (mod 7)

   (g) $67 + 95$ (mod 8)   (h) $146 + 253$ (mod 12)

   (i) $100 + 57$ (mod 6)   (j) $48 + 99$ (mod 5)

Constructing Addition Table in a given Modulo

Example

Construct addition table in modulo 5. Use your table to find n if

(a) $2 + n = 0$

(b) $n + 3 = 2$

(c) $n + n = 1$

We set up the addition table as shown below. The sum of, say $3 + 4$ is 2 in modulo 5. 2 is put in a cell right of 3 in the column to the extreme left and below 4 in the row at the top.

+	0	1	2	3	4
0	0	1	2	3	4
1	1	2	3	4	0
2	2	3	4	0	1
3	3	4	0	1	2
4	4	0	1	2	3

(a) Moving across the fourth row, we find 0 in the fifth column.

Thus, $n = 3$

(b) Moving down the fifth column we find 2 in the sixth row. Thus,

$n = 4$

(c) $3 + 3 = 1$, therefore $n = 3$

Try this 2

Construct the addition table in modulo 8

Exercise 10.1(b)

1. Construct addition table in

   (a) modulo 4   (b) modulo 6   (c) modulo 9

2. Construct the addition table in modulo 12 on the set

   $P = \{4, 7, 9, 11\}$

3. Construct the addition table for arithmetic modulo 7. Use your

   table to evaluate

   (a) $(4 + 5) + 6$ (b) $(2 + 4) + (3 + 4)$

4. Construct the addition table on the set $S = \{1, 3, 5, 7\}$ in

   arithmetic modulo 8. Use your table to find n if

   (a) $n + n = 6$ (b) $5 + n = 2$ (c) $n + 3 = 0$

10.2 Multiplication in Modular Arithmetic

Example

Find 13×26 in modulo 4

Multiply the two numbers and then divide the result by 4

$13 \times 26 = 338$

$$= 2 \ (\text{mod } 4)$$

You can also obtain this result by reducing 13 and 26 to their simplest form before you multiply:

$13 \times 26 = 1 \times 2 \ (\text{mod } 4)$

$$= 2 (\text{mod } 4)$$

Try this 3

Find 75×30 in modulo 5

Exercise 10.2(a)

Work out each of the following multiplications in the modulo given

1. 6×5 (mod 4)   2. 8×9 (mod 5)

3. 15×17 (mod 6)   4. 28×12 (mod 8)

5. 32×46 (mod 9)   6. 50×62 (mod 12)

7. 19×13 (mod 7)   8. 21×18 (mod 5)

9. 76×14 (mod 6)   10. 82×11 (mod 7)

11. 83×37 (mod 7)   12. 75×15 (mod 6)

Constructing Multiplication Table in a given Modulo

Example

Construct the multiplication table in modulo 4

We set up the multiplication table as shown below.

×	0	1	2	3
0	0	0	0	0
1	0	1	2	3
2	0	2	0	2
3	0	3	2	1

Try this 4

Construct the multiplication table in modulo 7

Exercise 10.2(b)

1. Construct the multiplication table in the following modulo

    (a) modulo 5   (b) modulo 6   (c) modulo 8

2. Construct the multiplication table in modulo 9 on the set

    $P = \{2, 3, 5, 6\}$. Use your table to evaluate

    (a) $(3 \times 5) \times 2$ (b) $(6 \times 5) \times (2 \times 6)$

3. Construct the multiplication table in modulo 12 on the set

    $S = \{3, 5, 7, 9\}$. Use your table to find n if

    (a) $9 \times n = 3$ (b) $n \times n = 1$

4. Construct the multiplication table in modulo 9 on the set

    $\{1, 3, 5, 7\}$. Use your table to find n if:

    (a) $3 \times n = 0$ (b) $n \times (n \times 5) = 8$

10.3 Using Modular Arithmetic

Some problems are easily solved using modular arithmetic.

Example

If today is Tuesday, what day of the week will it be in 25 days?

25 days is 3 weeks plus 4 days, so the day of the week will be four days later than Tuesday which is Saturday.

You can arrive at the answer by dividing 25 by 7 and observing the remainder. $25 \div 7$, will give a remainder of 4. The remainder of 4 indicates the answer will be four days later than Tuesday.

An alternate solution is given below.

You may represent the days of the week by the members 0, 1, 2, 3, 4, 5, 6, with 0 representing Sunday, 1 representing Monday, and so on. Tuesday is represented by 2, and then the day of the week 25 days later is represented by

$2 + 25 = 6(\text{mod } 7).$

The remainder of 6 indicates the answer will be Saturday.

Try this 5

In a particular year 7th January was a Monday, what day of the week will 7th January be the following year, assuming it is not a leap year?

Exercise 10.3

1. If today is Tuesday, find the day of the week it will be in

    (a) 50 days   (b) 161 days

2. If the time now is 7 am, find the time it will be in

    (a) 15 hours   (b) 23 hours

3. In a particular year, 7 th December was a Friday. What day of the week will 7 th December be in the following year, assuming it is not a leap year?

4. In 1954, 27 th September was a Monday. Find the day of the week it will be in 1960.

5. If this month is February, find the month it will be in

   (a) 38 months   (b) 3 years 5 months

6. If this month is September, find the month it will be in

   (a) 83 month   (b) 8 years

7. A man took 17 days to walk from a town A to a town B. If he left town A on a Thursday, what day of the week did he arrive in Town B?

8. A painter works from 8 am to 12 noon and from 1 pm to 5 pm every day. He started a job at 8 a.m. and finished it after working for 31 hours. At what time did he finish it?

Review exercise 10

1. Reduce the following numbers to their simplest form in the given modulo

   (a) 37 (mod 5)   (b) 16 (mod 4)   (c) 72 (mod 6)

   (d) 237 (mod 12)   (e) 87 (mod 8)   (f) 168 (mod 9)

2. Work out each of the following additions in the given modulo

   (a) $28 + 34$ (mod 6)   (b) $28 + 49$ (mod 4)

   (c) $225 + 136$ (mod 8)   (d) $163 + 175$ (mod 9)

(e) $217 + 154$ (mod 7)   (f) $246 + 153$ (mod 12)

3. Work out each of the following multiplications in the modulo given

(a) 16×25 (mod 4)   (b) 28×38 (mod 5)

(c) 45×67 (mod 6)   (d) 42×36 (mod 9)

(e) 60×72 (mod 12)   (f) 29×23 (mod 7)

4. Construct the (i) addition and (ii) multiplication tables in modulo

5. Use your tables to the truth set of

(a) $(n \times 4) + (n \times 4) = 1$ (b) $n \times n = 4$

5. Construct the (i) addition and (ii) multiplication tables for the set $\{5, 7, 9, 11\}$ in arithmetic modulo 13. Using your table

(a) evaluate $7 + (7 \times 9)$

(b) find the truth set of $n \times n = 3$

6. Construct a table for multiplication modulo 8 on the set $S = \{2, 3, 5, 6\}$. Use your table to find on S the truth set of:

(a) $n \times n = 1$

(b) $n \times (n \times 3) = 4$

7. It is 8 a.m., what will the big hand of a clock be in 29 hours?

8. A flight which was schedule to arrive at 9 p.m. was delayed for 17 hours. What time will it land?

Chapter Test 10

Take this test as you would take a test in a class. After you are done, check your work against the answers in the back of the book.

1. Reduce the following numbers to their simplest forms in the

   given modulo

   (a) 73 (mod 5)   (b) 130 (mod 6)   (c) 243 (mod 8)

2. Work out each of the following additions in the given modulo

   (a) $32 + 71$(mod 5)   (b) $157 + 36$ (mod 6)

   (c) $83 + 17$ (mod 8)

3. Work out each of the following multiplications in the given

   modulo

   (a) 54×80 (mod 6)   (b) 132×47 (mod 9)

   (c) 76×32 (mod 12)

4. Construct (i) the addition and (ii) the multiplication tables for the

   set $\{1, 5, 7, 11\}$ in arithmetic modulo 12. Using your tables

   (a) evaluate $(7 \times 11) + 7$

   (b) find the truth set of $n \times n = 1$

5. Construct (i) the addition and (ii) the multiplication tables for the

   set $\{1, 3, 5, 7\}$ in arithmetic modulo 9. From your tables

   (a) evaluate $(3 \times 7) + (5 \times 5)$

   (b) find the truth set of $n \times 3 = 3$

6. A shop has 50 boxes of can drinks for sale. There are 127 drinks in each box. The shop plans to pack the drinks into cases of 6 cans. After making as many complete cases as possible, how many drinks will be left over?

11

Variation

There are four types of variation problems: direct, inverse, joint and combined variation.

11. 1 Direct Variation

The table below shows the cost of items and the number of items bought. Study the table.

Number of items	1	2	3	4	5
Cost(in GH¢)	60	120	180	240	300

Notice that if you divide the cost by the number of items, you will obtain GH¢ 60 in each case. For example,

$$\frac{120}{2} = 60$$

Also

$$\frac{180}{3} = 60$$

and so on. If the cost of x items is y cedis then

$$\frac{y}{x} = 60$$

So, $y = 60x$

In this example, the cost is a constant multiple of the number of items bought. So we say that there is direct variation and that the cost varies directly as number of items bought or the cost is proportional to number of items bought. This can be written in symbol as

$$y \propto x$$

The equation of variation is $y = 60x$.

If we replace 60 with k we obtain the general equation

$$y = kx$$

k is called the constant of proportionality or constant a of variation.

In most variation problems, the constant of variation k, may not be known. In such cases, you can found k if one pair of values of x and y is known.

There are many examples of equations of variation that we are familiar with. For example, the volume of a sphere of radius r, is $V = \frac{4}{3}\pi r^3$. Here, we say that V varies directly as the cube of r. Notice that in this example the constant of variation is $\frac{4}{3}\pi$. The area A of a circle is given by the formula $A = \pi r^2$. Since, π is a constant we say that A varies directly as the square of r.

Try this 1

(a) Write the following in symbols

(i) p varies directly as q

(ii) y varies directly as the square of x

(iii) p varies directly as the cube of r

(iv) p varies directly as the cube root of q

(v) y varies directly as the square root of x

(b) Write down the equations for the following statements

(i) y varies directly as the cube of x

(ii) p varies directly as the fourth root of q

(iii) s varies directly as the square root of t

(iv) The force, f, acting on a particle varies directly as the acceleration, a.

(v) The energy, E, of a moving particle varies directly as the square of the velocity, v.

(vi) The length, l, that a spring will stretch varies directly as the weight, w.

(vii) The distance, d, a car travelled varies as directly as time, t.

Finding the constant of variation

Example

y varies directly as the square of x. If y is 80 when x is 4, find the constant of variation

$$y \propto x^2$$

The general equation is

$$y = kx^2$$

Substitute 80 for y and 4 for x, and solve for k

$$80 = k \times 4^2$$

$$80 = 16k$$

$$5 = k$$ Divide both sides by 16

The constant of variation is 5.

Try this 2

y varies directly as the square root of x. If $y = 45$ when $x = 25$, find the constant of variation

Solving Direct Variation

You may solve direct variation problems using the method illustrated in the examples below.

Examples

y varies directly as x. If $y = 2$ when $x = 4$, find y when $x = 10$

$$y \propto x$$

Write this as

$$y = kx$$

To find the value of k, substitute 2 for y and 4 for x.

$$2 = 4k$$

$$\frac{1}{2} = k$$ Divide both sides by 4

Write the equation of variation.

$$y = \frac{1}{2}x$$ Substitute $\frac{1}{2}$ for k

When $x = 10$, we have

$$y = \frac{1}{2}(10)$$

$$y = 5$$

So y is 5 when $x = 10$

Try this 3

y varies directly as of x. If $y = 30$ when $x = 6$, find y when $x = 9$

p varies directly as the square root of r. If $p = 60$ when $r = 25$, find r when $p = 36$.

$$p \propto \sqrt{r}$$

Write this as

$$p = k\sqrt{r}$$

$\quad 60 = k\sqrt{25}$ Substitute 60 for p and 25 for r

$\quad 60 = 5k$

$\quad 12 = k$ Divide both sides by 5

So, $p = 12\sqrt{r}$

Substituting 36 for p, we have

$\quad 36 = 12\sqrt{r}$

$\quad\quad 3 = \sqrt{r}$ Divide both sides by 12

$\quad\quad 9 = r$ Square both sides

So r is 9 when $p = 36$

Try this 4

y varies directly as the cube root of x. If $y = 12$ when $x = 8$, find y when $x = 125$

Exercise 11.1

1. y varies directly as x. If $y = 40$ when $x = 8$, find the equation of variation

2. p varies directly as the square of r. If $p = 100$ when $r = 5$, find the equation of variation.

3. y varies directly as the square root of x. If $y = 150$ when $x = 36$, find the equation of variation.

4. y varies directly as the cube of x. If $y = 48$ when $x = 4$, find the equation of variation.

5. p varies directly as the cube root of r. If $p = 135$ when $r = 27$, find the equation of variation.

6. y varies directly as x. If $y = 12$ when $x = 4$, find y when $x = 2$

7. p varies directly as q. If $p = 18$ when $q = 3$, find p when $q = 5$

8. y varies directly as x. If $y = 20$ when $x = 5$, find y when $x = 6$

9. y varies directly as x. If $y = 30$ when $x = 6$, find x when $y = 25$

10. y varies directly as the square of x. If $y = 32$ when $x = 4$, find

   y when $x = 3$

11. p varies directly as the square of q. If $p = 81$ when $q = 3$, find

   q when $p = 36$

12. y varies directly as the square root of x. If $y = 30$ when

   $x = 25$, find y when $x = 16$

13. y varies directly as the square root of x. If $y = 40$ when

   $x = 25$, find x when $y = 24$

14. q varies directly as the cube root of p, If $q = 6$ when $p = 27$,

   find p when $q = 10$

15. y varies directly as the cube of x. If $y = 40$ when $x = 2$, find x

   when $y = 135$

16. The weight of a steel bar, W grams, varies directly as its length,

   L centimetres. A 15 centimetre steel bar weighs 180 grams.

   Find the weight of a steel bar 26 centimetres long

17. The cost, GH¢ C of producing a booklet varies directly as the

   square root of the number N of pages it contains. A booklet

   with 25 pages cost GH¢ 2.40 to produce. Find the cost of

   producing a booklet with 36 pages.

18. The volume of juice in a can V cm^3, varies directly as its

   height, H cm. The volume is 150 cm^3 when the height is 6 cm.

Calculate the volume when the height is 9 cm.

11.2 Inverse Variation

If y varies directly as $\frac{1}{x}$, we say that y varies inversely as x.

Try this 5

1. Write the following in symbols

(a) p varies inversely as q

(b) y varies inversely as the square of x

(c) s varies inversely as the cube of t

(d) z varies inversely as the square root of y

(e) p varies inversely as the cube root of r

2. Write an equation for each of the following statement

(a) p varies inversely as the square of q

(b) s varies inversely as the cube root of r

(c) p varies inversely as the cube of r

(d) F varies inversely as the square root of r

Examples

y varies inversely as the square of x, and $y = 9$ when $x = 2$. Find y when $x = 3$.

The method of solution is exactly the same as in the examples on direct variation.

$$y \propto \frac{1}{x^2}$$

The general equation is

$y = \frac{k}{x^2}$, where k is the variation constant

Substituting 2 for x and 9 for y, we have

$$9 = \frac{k}{4}$$

$$36 = k \qquad\qquad \text{Multiply both sides by 4}$$

So, $y = \frac{36}{x^2}$

Substituting 3 for x, we have

$$y = \frac{36}{9}$$

$$y = 4$$

Try this 6

y varies inversely as x, and $y = 15$ when $x = 3$. Find y when $x = 9$

p varies inversely as the cube of r, and $p = 1$ when $r = 2$. Find r when $p = 27$,

$$p \propto \frac{1}{r^3}$$

Write this as

$$p = \frac{k}{r^3}$$

$$1 = \frac{k}{8} \qquad\qquad \text{Substitute 1 for } p \text{ and 2 for } r$$

$$8 = k \qquad\qquad \text{Multiple both sides by 8}$$

So, $p = \dfrac{8}{r^3}$

$$27 = \dfrac{8}{r^3} \qquad\qquad \text{Substitute 27 for } p$$

$$r^3 = \dfrac{8}{27}$$

$$r = \dfrac{2}{3} \qquad\qquad \text{Take cube root of both sides}$$

Try this 7

y varies inversely as the square root of x, and $y = 8$ when $x = 36$. Find x when $y = 12$.

Exercise 11.2

1. y varies inversely as x, and $y = 15$ when $x = 2$. Find y when

   $x = 6$.

2. p varies inversely as q, and $p = 8$ when $q = 3$. Find q when

   $p = 6$.

3. y varies inversely as x, and $y = 6$ when $x = 2$. Find x when

   $y = 3$.

4. p varies inversely as the square of q, and $p = 5$ when $q = 3$.

   Find q when $p = 125$.

5. y varies inversely as the square of x, and $y = 5$ when $x = 2$.

   Find x when $y = 45$.

6. y varies inversely as the square root of x, and $y = 20$ when

   $x = 25$. Find y when $x = 16$.

7. y varies as inversely as the square root of x, and $y = 6$ when

   $x = 25$. Find x when $y = 15$.

8. p varies inversely as the square root of q, and $p = 9$ when

   $q = 25$. Find q when $p = 10$.

9. p varies inversely as the cube of r, and $p = 4$ when $r = 3$. Find

   r when $p = 32$.

10. p varies inversely as the cube root of r, and $p = 9$ when $r = 8$.

    Find p when $r = 27$.

11.3 Joint Variation

If y varies directly as the product of x and z, we say that y varies jointly as x and z.

Try this 8

Write an equation for each of the following statement

(a) y varies jointly as z and the square of x

(b) p varies jointly as q and the cube root of r

(c) y varies jointly as x and the square root of z

(d). p varies jointly as q and the cube of r

Examples

z varies jointly as x and y, and $z = 36$ when $x = 3$ and $y = 6$. Find z when $x = 4$ and $y = 5$.

$z \propto xy$

This can be written as

$z = kxy$

$36 = k \times 3 \times 6$ Substitute 3 for x and 6 for y

$36 = 18k$

$2 = k$ Divide both sides by 18

$z = 2xy$

$z = 2 \times 4 \times 5$ Substitute 4 for x and 5 for y

$z = 40$

p varies jointly as q and the square root of r, and $p = 60$ when $q = 3$ and $r = 25$. Find r when $p = 80$ and $q = 5$.

$p \propto q\sqrt{r}$

We write this as

$p = kq\sqrt{r}$

$60 = k \times 3 \times \sqrt{25}$ Substitute 3 for q and 25 for r

$60 = 15k$

$4 = k$ Divide both sides by 15

$$p = 4q\sqrt{r}$$

$$80 = 4 \times 5 \times \sqrt{r}$$ Substitute 80 for p and 5 for q

$$80 = 20\sqrt{r}$$

$$4 = \sqrt{r}$$ Divide both sides by 20

$$16 = r$$ Square both sides

Try this 9

y varies jointly as x and z, and $y = 60$ when $x = 5$ and $z = 6$. Find y when $x = 4$ and $z = 7$.

Exercise 11.3

1. y varies jointly as x and z, and $y = 36$ when $x = 3$ and $z = 4$.

  Find y when $x = 5$ and $z = 2$.

2. z varies jointly as x and y, and $z = 60$ when $x = 3$ and $y = 4$.

  Find z when $x = 6$ and $y = 3$.

3. y varies jointly as x and z, and $y = 45$ when $x = 3$ and $z = 5$.

  Find z when $x = 2$ and $y = 60$.

4. y varies jointly as x and the square of z, and $y = 100$ when

  $x = 2$ and $z = 5$. Find z when $y = 36$ and $x = 2$.

5. p varies jointly as q and the square root of r, and $p = 96$ when

  $q = 4$ and $r = 36$. Find r when $p = 40$ and $q = 2$.

6. r varies jointly as s and the cube of t, and $r = 48$ when $s = 8$

   and $t = 3$. Find t when $s = 5$ and $r = 240$.

7. p varies jointly as q and the square of r, and $p = 60$ when $r = 2$

   and $q = 5$. Find q when $p = 300$ and $r = 5$.

8. z varies jointly as x and the square root of y, and $z = 70$ when

   $x = 7$ and $y = 25$. Find y when $x = 4$ and $z = 32$.

11.4 Combined Variation

Combined variation involves a combination of direct or joint variation and inverse variation.

Examples

y varies directly as x and inversely as z, and $y = 2$ when $x = 3$ and $z = 9$. Find y when $x = 8$ and $z = 4$.

$$y \propto \frac{x}{z}$$

Write this as

$$y = \frac{kx}{z}$$

$$2 = \frac{k \times 3}{9}$$ Substitute 2 for y, 3 for x and 9 for z

$$2 = \frac{1}{3}k$$

$$6 = k$$ Multiple both sides by 3

$$y = \frac{6x}{z}$$ Substitute 6 for k

$$y = \frac{6 \times 8}{4}$$ 　　　　　Substitute 8 for x and 4 for z

$$y = 12$$

Try this 10

y varies directly as x and inversely as z, and $y = 10$ when $x = 5$ and $z = 3$. Find x when $y = 14$ and $z = 15$.

p varies directly as q and inversely as the square root of r, and $p = 12$ when $q = 4$ and $r = 25$. Find r when $p = 30$ and $q = 6$.

$$p \propto \frac{q}{\sqrt{r}}$$

Write this as

$$p = \frac{kq}{\sqrt{r}}$$

$$12 = \frac{k \times 4}{\sqrt{25}}$$ 　　　　　Substitute 12 for p, 4 for q and 25 for r

$$12 = \frac{4}{5}k$$

$$15 = k$$

$$p = \frac{15q}{\sqrt{r}}$$ 　　　　　Substitute 15 for k

$$30 = \frac{15 \times 6}{\sqrt{r}}$$ 　　　　　Substitute 30 for p and 6 for q

$$\sqrt{r} = 3$$

$$r = 9$$ 　　　　　Square both sides

Try this 11

y varies directly as the square of x and inversely as z, and $y = 18$ when $x = 3$ and $z = 2$. Find x when $y = 27$ and $z = 12$.

Exercise 11.4

1. y varies directly as x and inversely as z, and $y = 12$ when

   $x = 6$ and $z = 2$. Find y when $x = 8$ and $z = 4$.

2. z varies directly as x and inversely as the square of y, and $z = 9$

   when $x = 3$ and $y = 2$. Find z when $x = 6$ and $y = 3$.

3. p varies directly as q and inversely as the square root of r, and

   $p = 6$ when $q = 3$ and $r = 25$. Find p when $q = 6$ and $r = 16$.

4. y varies directly as x and inversely as the square of z, and

   $y = 16$ when $x = 8$ and $z = 2$. Find z when $x = 4$ and $y = 2$.

5. p varies directly as q and inversely as the square root of r, and

   $p = 4$ when $q = 2$ and $r = 9$. Find r when $p = 12$ and $q = 8$.

6. p varies directly as q and inversely as the square root of r, and

   $p = 6$ when $q = 8$ and $r = 16$. Find r when $p = 15$ and $q = 35$.

7. z varies directly as x and inversely as y, and $z = 36$ when $x = 3$

   and $y = 6$. Find x when $z = 9$ and $y = 32$.

8. r varies directly as the square of s and inversely as t, and $r = 18$

   when $s = 3$ and $t = 2$. Find s when $r = 9$ and $t = 16$.

9. y varies directly as z and inversely as the square of x,and

   $y = 10$ when $z = 5$ and $x = 2$. Find x when $y = 9$ and $z = 18$.

10. w varies jointly as x and y, and inversely as z, and $w = 12$

    when $x = 2$, $y = 3$ and $z = 4$. Find w when $x = 9$, $y = 5$ and

$z = 6.$

11.5 Partial Variation

A man pays GH¢20 plus GH¢12 for each unit of electricity used in a month. If he uses x units of electricity in a particular month, the total amount y, paid by him will be

$y = 20 + 12x$

This equation is of the form

$y = a + bx$

Here, we say that y is partly a constant and partly varies directly as x. Notice that there are two constants of variation in this case.

Try this 12

Write an equation for each of the following statements

(a) y is partly constant and partly varies directly as the square of x

(b) p is partly constant and partly varies inversely as the cube of q

(c) The distance, d, moved by a particle is partly constant and partly varies as the square of the time t

(d) y is partly constant and partly jointly as x and the square of z

(e) p is partly constant and partly varies directly as q and inversely as the square root of r

Examples

y is partly constant and partly varies directly as x, and $y = 8$ when $x = 2$ and $y = 12$ when $x = 4$. Find y when $x = 3$.

Begin by writing

$y = a + bx$

$8 = a + 2b$ (1)   Substitute 8 for y and 2 for x

$12 = a + 4b$ (2)   Substitute 12 for y and 4 for x

Subtracting equation (2) from equation (1), we have

$-4 = -2b$

$2 = b$ Divide both sides by -2

Substitute $b = 2$ into (1)

$8 = a + 2(2)$

$4 = a$

$y = 4 + 2x$

So, $y = 4 + 2(3)$ Substitute 3 for x

$y = 10$

Try this 13

y is partly constant and partly varies as the square of x, and $y = 39$ when $x = 4$ and $y = 25$ when $x = 3$. Find y when $x = 5$.

y is partly constant and partly varies inversely as the square root of x, and $y = 8$ when $x = 4$ and $y = 6$ when $x = 9$. Find y when $x = 16$.

Begin by writing

$$y = a + \frac{b}{\sqrt{x}}$$

$$8 = a + \frac{b}{2} \qquad (1)$$

$$6 = a + \frac{b}{3} \qquad (2)$$

$(1) - (2) \quad 2 = \frac{b}{6}$

$\qquad 12 = b$ \qquad Multiply both sides by 6

Substitute $b = 12$ in (1)

$$8 = a + \frac{12}{2}$$

$$a = 2$$

$$y = 2 + \frac{12}{\sqrt{x}}$$

So, $y = 2 + \frac{12}{\sqrt{16}}$ \qquad Substitute 16 for x

$$y = 5$$

Try this 14

y is partly constant and partly varies inversely as x, and $y = 6$ when $x = 3$ and $y = 8$ when $x = 2$. Find x when $y = 5$.

Exercise 11.5

1. p is partly constant and partly varies directly as q, and $p = 15$ when $q = 3$ and $p = 12$ when $q = 2$. Find p when $q = 4$.

2. y is partly constant and partly varies as the square of x, and $y = 19$ when $x = 2$ and $y = 39$ when $x = 3$. Find x when $y = 67$.

3. p is partly constant and partly varies as the square root of r. If $p = 33$ when $r = 36$ and $p = 38$ when $r = 49$, find

(a) p when $r = 9$

(b) r when $p = 23$

4. y is partly constant and partly varies inversely as the square of x, and $y = 24$ when $x = 2$ and $y = 34$ when $x = 3$. Find y when $x = 6$.

5. p is partly constant and partly varies as the square of q, and $p = 30$ when $q = 3$ and $p = 10$ when $q = 2$. Find p when $q = 4$.

6. y is partly constant and partly varies directly as x and inversely as z, and $y = 12$ when $x = 6$, $z = 3$ and $y = 15$ when $x = 12$, $z = 4$. Find y when $x = 2$ and $z = 3$.

7. y is partly constant and partly varies directly as x and inversely as the square root of z. $y = 9$ when $x = 10$ and $z = 25$, and $y = 13$ when $x = 12$ and $z = 9$. Find y when $x = 9$ and $z = 36$.

8. y is partly constant and partly varies directly as the square of x and inversely as z. $y = 16$ when $x = 6$ and $z = 9$, and $y = 40$ when $x = 8$ and $z = 4$. Find x when $y = 11$ and $z = 6$.

9. y is partly constant and partly varies jointly as x and z. $y = 26$ when $x = 3$ and $z = 2$, and $y = 44$ when $x = 4$ and $z = 3$. Find y when $x = 2$ and $z = 4$.

10. y is partly constant and partly varies jointly as x and the square root of z. $y = 30$ when $x = 4$ and $z = 9$, and $y = 46$ when $x = 4$ and $z = 25$. Find z, when $y = 36$ and $x = 10$.

Review exercise 11

1. y varies directly as x. If $y = 30$ when $x = 15$, find the constant of variation

2. w varies directly as x. If $w = 18$ when $x = 2$, find the equation of variation.

3. z varies inversely as y. If $z = 5$ when $y = 6$, find the constant of variation.

4. p varies inversely as the square root of r. If $p = 8$ when $r = 36$, find the equation of variation.

5. y varies jointly as x and z. If $y = 40$ when $x = 4$ and $z = 5$, find the constant of variation

6. w varies jointly as x^2 and y. If $w = 72$ when $x = 3$ and $y = 2$, find the equation of variation.

7. y varies directly as x and inversely as z. If $y = 8$ when $x = 2$ and $z = 3$, find the constant of variation.

8. p varies directly as q and inversely as the square root of r. If $p = 6$ when $q = 3$ and $r = 25$, find the equation of variation.

9. w varies jointly as x and y, and inversely as z. If $w = 50$ when $x = 2, y = 5$ and $z = 3$, find the equation of variation.

10. r varies directly as s. If $r = 12$ when $s = 4$, find r when $s = 11$

11. z varies directly as the square of y. If $z = 32$ when $y = 4$, find z when $y = 5$

12. x varies inversely as y. If $y = 10$ when $x = \frac{1}{2}$, find x when $y = 15$

13. p varies inversely as the square root of r. If $r = 25$ when $p = 12$, find r when $p = 15$

14. z varies jointly as w and y. If $z = 12$ when $w = 9$ and $y = 4$, find z when $w = 50$ and $y = 6$

15. z varies jointly as x and the square of y. If $z = 135$ when $x = 5$ and $y = 3$, find y when $z = 192$ and $x = 4$

16. z varies directly as x and inversely as y. If $z = 8$ when $x = 4$

and $y = 5$, find z when $x = 6$ and $y = 15$

17. w varies jointly as x and y and inversely as the square of z. If

$w = 24$ when $x = 4$, $y = 6$ and $z = 3$, find z when

$w = 15, x = 27$ and $y = 5$

18. y is partly constant and partly varies as the square of x. If

$y = 14$ when $x = 2$ and $y = 29$ when $x = 3$, find x when

$y = 77$

19. y is partly constant and partly varies inversely as the square of

x. If $y = 24$ when $x = 2$ and $y = 19$ when $x = 3$, find y when

$x = 6$.

20. z is partly constant and partly varies jointly as x and y. If

$z = 27$ when $x = 3$ and $y = 4$, and $z = 43$ when $x = 4$ and

$y = 5$, find x when $z = 39$ and $y = 3$

Chapter Test 11

Take this test as you would take a test in a class. After you are
done, check your work against the answers in the back of the book.

1. Find the constant of variation for each variation

(a) y varies directly as the square of x. $y = 36$ when $x = 3$

(b) z varies jointly as x and y. $z = 42$ when $x = 2$ and $y = 7$

(c) w varies directly as x and inversely as y. $w = 15$ when $x = 6$

and $y = 8$

2. Write an equation for each variation

   (a) y varies directly as x. $y = 28$ when $x = 7$

   (b) z varies inversely as x. $z = 8$ when $x = 3$

   (c) w varies jointly as x and y and inversely as z. $w = 15$ when

      $x = 2$, $y = 9$ and $z = 6$

3. y varies directly as the square of x. If $y = 72$ when $x = 6$, find x

   when $y = 18$

4. z varies inversely as the square root of y. If $z = 12$ when

   $y = 25$, find y when $z = 90$

5. z varies jointly as y and the square of x. If $z = 60$ when $x = 2$

   and $y = 3$, find x when $z = 90$ and $y = 8$

6. z varies directly as the square of x and inversely as y. If $z = 56$

   when $x = 8$ and $y = 16$, find z when $x = 5$ and $y = 10$

7. y is partly constant and partly varies as the square of x. If

   $y = 6.3$ when $x = 3$ and $y = 9.5$ when $x = 5$, find y when

   $x = 10$

8. C is partly constant and partly varies inversely as n. If $C = 15$

   when $n = 3$ and $C = 18$ when $n = 2$, find C when $n = 6$

12

Indices and Logarithm

12. 1 Indices

$9 = 3 \times 3$. In brief we write 3×3 as 3^2. The number 2 is called the index or the exponent, and 3 is called the base. Notice that the index indicates how many times 3 is multiplied by itself. 3^2 is read "3 squared" or "3 to the second power".

In general, if n is any positive integer then for any whole number a where $a \neq 0$

$$a^n = \underbrace{a \cdot a \cdot a \cdots a}_{n\ factors}$$

For example, $2^5 = 2 \cdot 2 \cdot 2 \cdot 2 \cdot 2$

In this example the number 2 is raised to the power 5.

The plural of index is indices

Laws of Indices

We develop some laws for working with indices

Multiplying expressions with like base

Let us consider the product

$a^5 \cdot a^3$

Recall that the expression a^5 means $a \cdot a \cdot a \cdot a \cdot a$. Also a^3 means $a \cdot a \cdot a$

So $a^5 \cdot a^3 = (a \cdot a \cdot a \cdot a \cdot a) \cdot (a \cdot a \cdot a)$

$$= a \cdot a \cdot a \cdot a \cdot a \cdot a \cdot a \cdot a$$

$$= a^8$$

Notice that the index in a^8 is the sum of the indices in $a^5 \cdot a^3$. That is, $5 + 3 = 8$.

In general, if m and n are positive integers then

$$a^m \cdot a^n = a^{m+n}$$

Notice that to multiply expressions with the same base, we keep the base and add the indices.

Example

Simplify:

(a) $2^4 \cdot 2^5$ (b) $3^2 \cdot 3^6$

(a) $2^4 \cdot 2^5 = 2^9$

(b) $3^2 \cdot 3^6 = 3^8$

Try this 1

Simplify:

(a) $5^7 \cdot 5^5$ (b) $7^3 \cdot 7^4$ (c) $3^2 \cdot 3^4 \cdot 3^5$

Dividing expressions with like base

Consider the division $a^6 \div a^4$

Now $a^6 \div a^4 = \dfrac{a^6}{a^4}$

$$= \frac{a \cdot a \cdot a \cdot a \cdot a \cdot a}{a \cdot a \cdot a \cdot a}$$

$$= a \cdot a$$

$$= a^2$$

Notice that the index in a^2 is the difference of the indices in $a^6 \div a^4$.

Generally, when m and n are positive integers then

$$a^m \div a^n = a^{m-n}$$

Example

Simplify:

(a) $5^8 \div 5^5$ (b) $2^6 \div 2^4$

(a) $5^8 \div 5^5 = 5^3$

(b) $2^6 \div 2^4 = 2^2$

Try this 2

Simplify:

(a) $2^7 \div 2^4$ (b) $3^{10} \div 3^6$ (c) $7^3 \div 7$

Raising a Power to a Power

Consider an expression like $(a^2)^3$

$$(a^2)^3 = a^2 \cdot a^2 \cdot a^2$$

$$= (a \cdot a) \cdot (a \cdot a) \cdot (a \cdot a)$$

$$= a \cdot a \cdot a \cdot a \cdot a \cdot a$$

$$= a^6$$

Notice that the index in a^6 is the product of the indices in $(a^2)^3$

In general, if m and n are positive integers then

$(a^m)^n = a^{mn}$

Example

Simplify:

(a) $(3^2)^4$ (b) $(5^3)^4$

(a) $(3^2)^4 = 3^8$

(b) $(5^3)^4 = 5^{12}$

Try this 3

Simplify:

(a) $(7^5)^2$ (b) $(3^3)^2$ (c) $(5^2)^6$

It is also possible for an index to be 0, a negative integer or a fraction. Expressions containing fractional and negative indices are defined in a manner that ensures that the laws of indices remain true for all numbers.

Fractional Index

Now $a^{\frac{1}{2}} \cdot a^{\frac{1}{2}} = a^{\frac{1}{2}+\frac{1}{2}}$

$= a^1$

$= a$

So, the square of $a^{\frac{1}{2}}$ is a which suggest that the positive square root of a is $a^{\frac{1}{2}}$

Thus $a^{\frac{1}{2}} = \sqrt{a}$

In general, $a^{\frac{1}{n}} = \sqrt[n]{a}$

Example

Find the value of:

(a) $16^{\frac{1}{4}}$ (b) $27^{\frac{1}{3}}$

(a) $16^{\frac{1}{4}} = \sqrt[4]{16}$

$\quad\quad = 2$

This result can also be obtained in the following way:

$16^{\frac{1}{4}} = (2^4)^{\frac{1}{4}}$ Write 16 as 2 raised to the power 4

$\quad\quad = 2$ Multiply the indices

(b) $27^{\frac{1}{3}} = \sqrt[3]{27}$

$\quad\quad\quad = 3$

Try this 4

Find the value of:

(a) $100^{\frac{1}{2}}$ (b) $64^{\frac{1}{3}}$ (c) $81^{\frac{1}{4}}$

Now let us consider the product $a^{\frac{2}{3}} \cdot a^{\frac{2}{3}} \cdot a^{\frac{2}{3}}$

$a^{\frac{2}{3}} \cdot a^{\frac{2}{3}} \cdot a^{\frac{2}{3}} = a^{\frac{2}{3}+\frac{2}{3}+\frac{2}{3}}$

$\quad\quad\quad\quad = a^2$

So $a^{\frac{2}{3}} = \sqrt[3]{a^2}$

Notice that $a^{\frac{2}{3}} = \left(a^{\frac{1}{3}}\right)^2$

It follows that $a^{\frac{2}{3}} = \left(\sqrt[3]{a}\right)^2$

In general, $a^{\frac{m}{n}} = \sqrt[n]{a^m} = \left(\sqrt[n]{a}\right)^m$

Example

Find the value of

(a) $16^{\frac{3}{4}}$ (b) $125^{\frac{2}{3}}$

(a) $16^{\frac{3}{4}} = \left(\sqrt[4]{16}\right)^3$

$\qquad = 2^3$

$\qquad = 8$

This result can also be obtained in the following way:

$16^{\frac{3}{4}} = (2^4)^{\frac{3}{4}} = 2^3 = 8$

(b) $125^{\frac{2}{3}} = \left(\sqrt[3]{125}\right)^2$

$\qquad = 5^2$

$\qquad = 25$

Try this 5

Find the value of

(a) $27^{\frac{2}{3}}$ (b) $8^{\frac{4}{3}}$ (c) $25^{\frac{3}{2}}$

The zero Index

For any number a, where $a \neq 0$

$$a^0 \cdot a^2 = a^2$$

Dividing both sides by a^2, we have

$$a^0 = 1$$

Thus, for any nonzero number a, $a^0 = 1$.

For example, $1000^0 = 1$ and $\left(\frac{1}{2}\right)^0 = 1$

Negative Indices

$$a^{-2} \cdot a^2 = a^{-2+2}$$
$$= a^0$$

Dividing both sides by a^2, we have

$$a^{-2} = \frac{1}{a^2}, \text{ since } a^0 = 1$$

In general, $a^{-n} = \frac{1}{a^n}$ for any nonzero number a and any integer n

Example

Find the value of:

(a) 2^{-5} (b) $8^{-\frac{2}{3}}$

(a) $2^{-5} = \frac{1}{2^5} = \frac{1}{32}$

(b) $8^{-\frac{2}{3}} = \frac{1}{8^{\frac{2}{3}}} = \frac{1}{\left(\sqrt[3]{8}\right)^2} = \frac{1}{2^2} = \frac{1}{4}$

This result can also be obtained in the following way:

$$8^{-\frac{2}{3}} = (2^3)^{-\frac{2}{3}}$$ Write 8 as 2 raised to the power 3

$$= 2^{-2}$$ Multiply the indices

$$= \frac{1}{4}$$

Try this 6

Find the value of:

(a) 3^{-4} (b) $32^{-\frac{2}{5}}$ (c) $\dfrac{1}{5^{-3}}$ (d) $\dfrac{1}{16^{-\frac{3}{4}}}$

We summarized the laws of indices:

For any nonzero number a and any integers m and n we have

1. $a^m \cdot a^n = a^{m+n}$

2. $a^m \div a^n = a^{m-n}$

3. $(a^m)^n = a^{mn}$

4. $a^{\frac{1}{n}} = \sqrt[n]{a}$

5. $a^{\frac{m}{n}} = \sqrt[n]{a^m} = \left(\sqrt[n]{a}\right)^m$

6. $a^{-n} = \dfrac{1}{a^n}$

7. $a^0 = 1$

Exercise 12.1(a)

Simplify the following and write your answers in index form

1. (a) $2^3 \cdot 2^4$ (b) $3^3 \cdot 3^2$ (c) $5^7 \cdot 5$ (d) $6^5 \cdot 6^4$

2. (a) $2^{-2} \cdot 2^{-3}$ (b) $3^{-5} \cdot 3^2$ (c) $5^{-1} \cdot 5^4$ (d) $7^6 \cdot 7^{-8}$

3. (a) $3^5 \div 3^2$ (b) $2^8 \div 2^6$ (c) $7^4 \div 7$ (d) $8^7 \div 8^2$

4. (a) $2^2 \div 2^4$ (b) $3^3 \div 3^{-2}$ (c) $5^{-3} \div 5^{-5}$ (d) $6^{-3} \div 6^5$

5. (a) $(4^3)^2$ (b) $(2^4)^3$ (c) $(3^2)^5$ (d) $(7^5)^4$

6. (a) $(2^{-3})^2$ (b) $(3^2)^{-4}$ (c) $(4^{-3})^{-2}$ (d) $(5^{-1})^3$

7. (a) $(3^{-4})^{\frac{3}{2}}$ (b) $(2^6)^{-\frac{2}{3}}$ (c) $(4^{-8})^{-\frac{1}{4}}$ (d) $(5^3)^{-\frac{2}{3}}$

Work out the value of the following

8. (a) $8^{\frac{1}{3}}$ (b) $25^{\frac{1}{2}}$ (c) $256^{\frac{1}{4}}$ (d) $32^{\frac{1}{5}}$

9. (a) $81^{\frac{3}{4}}$ (b) $8^{\frac{2}{3}}$ (c) $625^{\frac{3}{4}}$ (d) $16^{\frac{5}{4}}$

10. (a) $8^{-\frac{1}{3}}$ (b) $9^{-\frac{1}{2}}$ (c) 5^{-2} (d) 2^{-3}

11. (a) $\left(-\frac{1}{3}\right)^0$ (b) $\left(\frac{1}{4}\right)^{-2}$ (c) $\left(\frac{1}{2}\right)^{-1}$ (d) 6^0

12. Simplify the following, writing your answer in index form

(a) $\dfrac{3^{-4} \times 3^6}{3^{-2}}$ (b) $\dfrac{(2^2)^3 \times 2^4}{2^7}$ (c) $\dfrac{(5^3)^4}{5^{-7} \times (5^8)^2}$

(d) $\dfrac{(4^3)^4 \times 4^3 \times 3^{10}}{(4^5)^2 \times (3^2)^4}$ (e) $\dfrac{25^{-4} \times 5^6 \times 25^2}{25^{-3} \times 5^6}$ (f) $\dfrac{(9^2)^3 \times 81^{-\frac{3}{4}}}{27^{\frac{1}{3}} \times (3^2)^3}$

13. Without using a calculator, evaluate the following:

(a) $\dfrac{3^4 \times 3^{-5}}{3^{-2}}$ (b) $\dfrac{2^{-5}}{2^3 \times (2^{-3})^4}$ (c) $6^{-6} \times 12^3$

(d) $\dfrac{9^{\frac{1}{2}} \times 8^{-2}}{12^{-4}}$ (e) $\dfrac{8^{\frac{1}{3}} \times 6^2}{36^{\frac{1}{2}}}$ (f) $\dfrac{2^{-3} \times 32^{\frac{2}{5}}}{4^{\frac{3}{2}} \times 8^{-\frac{1}{3}}}$

Exponential Equations

An equation such as $2^x = 8$ is known as exponential equation. The example below illustrates the method used in solving exponential equations

Examples

If $2^x = 16$ find x

We express 16 in index form as 2^4. Hence,

$2^x = 2^4$

Since the two numbers have equal base their exponents must be equal

So, $x = 4$

Try this 7

Solve

(a) $5^x = 25$ (b) $9^x = 27$

Solve $3^y = \dfrac{1}{81}$

$3^y = \dfrac{1}{81}$

$\quad = \dfrac{1}{3^4}$ Write 81 as 3 raised to the power 4

$\quad = 3^{-4}$

So, $y = -4$

Try this 8

Solve

(a) $2^x = \dfrac{1}{32}$ (b) $\dfrac{1}{3^{-x}} = 27$

Exercise 12.1(b)

Find the value of x in each equation

1. $2^x = 32$ 2. $3^x = 27$ 3. $2^x = 8$

4. $32^x = 2$ 5. $3^{x+1} = 81$ 6. $125^{x+3} = 5$

7. $5^{2x-1} = 125$ 8. $2^{x-3} = \dfrac{1}{32}$ 9. $8^x = \dfrac{1}{2}$

10. $27^x = 81$ 11. $25^x = 1$ 12. $\left(\dfrac{1}{4}\right)^x = 16$

12.2 Standard Forms

When working with very large or very small numbers standard form provides a useful way of writing the numbers in terms of powers of 10.

Let us consider the number 4,500,000,000. Now

$4{,}500{,}000{,}000 = 4.5 \times 1{,}000{,}000{,}000 = 4.5 \times 10^9.$

We have expressed 4,500,000,000 as a product of two numbers, the first number is a number that is more than 1, but is less than 10 and the second number is an integer power of 10, expressed in index form. The index on 10 indicates the number of places we must move the decimal point so that the first number will be a number between 1 and 10.

Any number written in the form $a \times 10^n$ where the index n is an integer and a lies in the range $1 \leq a < 10$, is said to be in standard form. Standard form is also known as scientific notation.

Example

Write (a) 734000 and (b) 0.00057 in standard form

(a) Move the decimal point to the left until you obtain a number greater than or equal to 1 and less than 10. Raised 10 to the number of places moved. In this case move the decimal point 5 places to the left

$$734000. = 7.34 \times 10^5$$

(b) Move the decimal place to the right until you obtain a number greater than or equal to 1 and less than 10. Raised 10 to the number of places moved. In this case move the decimal point 4 places to the right

$$0.00057 = 5.7 \times 10^{-4}$$

Try this 9

Write (a) 6500 and (b) 0.000000735 in standard form

Exercise 12.2(a)

Write the following numbers in standard form

1. 4700000   2. 86000   3. 230

4. 76   5. 3640000000   6. 124000

7. 27.5   8. 572   9. 627

10. 9.5   11. 10500   12. 4156

13. 0.0056   14. 0.24   15. 0.0000431

16. 0.000703   17. 0.0643   18. 0.0000085

19. 0.009   20. 0.126   21. 0.03164

22. 0.00002   23. 0.0605   24. 0.00000057

Using Standard Form

In finding the product of two or more numbers we can sometimes simplify our calculations by rewriting the given numbers in standard form before multiplying.

Example

Calculate the following and write your answers in standard form

(a) 180000×0.000002

(b) $\dfrac{24000 \times 0.003}{0.12}$

Begin by writing each number as a whole number multiplied by a power of 10.

(a) $180000 \times 0.000002 = 18 \times 10^4 \times 2 \times 10^{-6}$

$$= 36 \times 10^{-2}$$

Our answer is not yet in standard form because 36 is not a number between 1 and 10.

Now, $36 = 3.6 \times 10^1$. Substituting 3.6×10^1 for 36, we have

$$180000 \times 0.000002 = 3.6 \times 10^1 \times 10^{-2}$$

$$= 3.6 \times 10^{-1}$$

(b) $\dfrac{24000 \times 0.003}{0.12} = \dfrac{24 \times 10^3 \times 3 \times 10^{-3}}{12 \times 10^{-2}}$

$$= \dfrac{6 \times 10^0}{10^{-2}}$$

$$= 6 \times 10^2$$

Try this 10

Evaluate the following and write your answer in standard form

(a) 4000×0.000032

(b) $\dfrac{8100 \times 0.00008}{0.000054}$

Exercise 12.2(b)

Work out the following and write your answers in standard form

1. 250000×16000

2. 0.00081×0.005

3. 75000×0.003

4. $12100000 \times 0.0000045$

5. $6800 \times 12000 \times 0.00003$

6. $9700000 \times 0.05 \times 0.00012$

7. $\dfrac{0.008 \times 90000}{0.12}$

8. $\dfrac{36000 \times 3.2}{0.06 \times 0.08}$

9. $\dfrac{0.125 \times 0.026}{0.025 \times 0.0013}$

10. $\dfrac{35000 \times 5.60}{0.005 \times 140000}$

11. $\dfrac{169 \times 7.5 \times 0.036}{3900 \times 0.18}$

12. $\dfrac{45000 \times 1600 \times 81}{18000 \times 5400}$

12.3 Logarithms

The logarithm of a number to a given base is the power to which the base must be raised in order to give the number. For example $8 = 2^3$, so 3 is the logarithm of 8 to base 2. In brief, we write

$\log_2 8 = 3$. Similarly, $\frac{1}{9} = 3^{-2}$, thus $\log_3 \frac{1}{9} = -2$.

Notice that the logarithm (to a given base) of any number is the power to which the base must be raised to give the number.

Examples

Write (a) $125 = 5^3$ and (b) $2^5 = 32$ in logarithm form

(a) $125 = 5^3$ We write $\log_5 125 = 3$

(b) $2^5 = 32$ We write $5 = \log_2 32$

Try this 11

Write the following in logarithm form

(a) $10000 = 10^5$ (b) $2^{-5} = \frac{1}{32}$

Write $\log_2 128 = 7$ in index form

We write $\log_2 128 = 7$ in index form as

$128 = 2^7$

Try this 12

Write the following in index form

(a) $\log_2 64 = 6$ (b) $\log_2 \frac{1}{8} = -3$

We can find the logarithm of some numbers by changing the logarithm to the index form. Then solve the resulting exponential equation.

Evaluate (a) $\log_3 243$ and (b) $\log_2 \frac{1}{128}$

(a) Let $\log_3 243 = x$

Begin by writing $\log_3 243$ in index form

$$243 = 3^x$$

$$3^5 = 3^x$$

$$5 = x$$

So $\log_3 243 = 5$

(b) Let $\log_2 \frac{1}{128} = x$

Write the logarithm in index form

$$\frac{1}{128} = 2^x$$

$$2^{-7} = 2^x$$

$$-7 = x$$

So $\log_2 \frac{1}{128} = -7$

Try this 13

Evaluate:

(a) $\log_3 27$ (b) $\log_2 \frac{1}{16}$

Exercise 12.3(a)

Write each of the following in logarithm form

1. $16 = 2^4$ 2. $27 = 3^3$ 3. $81 = 3^4$ 4. $25 = 5^2$

5. $64 = 4^3$ 6. $\frac{1}{8} = 2^{-3}$ 7. $\sqrt[5]{32} = 2$ 8. $\frac{1}{36} = 6^{-2}$

Write each of the following in index form

9. $\log_2 64 = 6$ 10. $\log_3 729 = 6$ 11. $\log_5 625 = 4$

12. $\log_{27} 3 = \frac{1}{3}$ 13. $\log_3 1 = 0$ 14. $\log_5 5 = 1$

Without calculator evaluate the following logarithms:

15. $\log_4 64$ 16. $\log_2 32$ 17. $\log_9 81$ 18. $\log_5 125$

19. $\log_{10} 10000$ 20. $\log_7 1$ 21. $\log_2 \frac{1}{16}$ 22. $\log_{27} 9$

23. $\log_3 \frac{1}{243}$ 24. $\log_8 32$

Laws of Logarithms

Logarithms obey laws which are related to the laws of indices. We will derive one of the laws of logarithm.

Let $X = a^p$ and $Y = a^q$, where $a \neq 0$

Now $X \cdot Y = a^{p+q}$

In logarithm form we have

$\log_a X \cdot Y = p + q$

But $\log_a X = p$ and $\log_a Y = q$

So $\log_a XY = \log_a X + \log_a y$

We state the rest of the laws

1. $\log_a xy = \log_a x + \log_a y$

2. $\log_a \dfrac{x}{y} = \log_a x - \log_a y$

3. $\log_a x^n = n \log_a x$

4. $\log_a a = 1$

5. $\log_a 1 = 0$

Using the Laws of Logarithms

Examples

Simplify:

(a) $\log_3 2 + \log_3 5$ (b) $\log_2 3 + \log_2 6 - 1$

(a) $\log_3 2 + \log_3 5 = \log_3(2 \times 5)$

$$= \log_3 10$$

(b) $\log_2 3 + \log_2 6 - 1 = \log_2(3 \times 6) - \log_2 2$

$$= \log_2 \frac{18}{2}$$

$$= \log_2 9$$

Try this 14

Simplify:

(a) $\log_5 3 + \log_5 8$ (b) $\log_2 27 - \log_2 9$

Simplify:

(a) $\log_2 8$ (b) $\dfrac{\log_4 81}{\log_4 9}$

(a) $\log_2 8 = \log_2 2^3$

$$= 3 \log_2 2$$

$$= 3$$

(b) $\dfrac{\log_4 81}{\log_4 9} = \dfrac{\log_4 3^4}{\log_4 3^2}$

$$= \dfrac{4 \log_4 3}{2 \log_4 3}$$

$$= 2$$

Try this 15

Simplify:

(a) $\log_4 64$ (b) $\dfrac{\log_3 32}{\log_3 8}$

Given that $\log_{10} 2 = 0.3010$ and $\log_{10} 3 = 0.4771$, find

(a) $\log_{10} 12$ (b) $\log_{10} 15$

(a) $\log_{10} 12 = \log_{10}(4 \times 3)$

$$= \log_{10} 2^2 + \log_{10} 3$$

$$= 2 \log_{10} 2 + \log_{10} 3$$

$$= 2(0.3010) + 0.4771$$

$$= 1.0791$$

(b) $\log_{10} 15 = \log_{10} \frac{30}{2}$

$$= \log_{10} 3 + \log_{10} 10 - \log_{10} 2$$

$$= 0.4771 + 1 - 0.3010$$

$$= 1.1761$$

Try this 16

Given $\log_{10} 2 = 0.3010$ and $\log_{10} 3 = 0.4771$, find

(a) $\log_{10} 24$ (b) $\log_{10} 5$

Exercise 12.3(b)

Write each of the following as a single logarithm

1. $\log_{10} 5 + \log_{10} 3$ 2. $\log_5 8 + \log_5 2$

3. $\log_3 32 - \log_3 4$ 4. $\log_8 1000 - \log_8 40$

5. $\log_3 8 - \log_3 128$ 6. $\frac{1}{2}\log_4 729 - \log_4 9$

7. $\log_2 48 - \log_2 16$ 8. $2\log_{10} 4 + 3\log_{10} 3 - \log_{10} 36$

9. $1 + \log_{10} 90 - \log_{10} 60$ 10. $\frac{1}{2}\log_{10} 81 + \frac{1}{3}\log_{10} 8$

Without a calculator evaluate:

11. $\log_3 27$ 12. $\log_2 32$ 13. $\log_5 625$

14. $\frac{1}{3}\log_6 36$ 15. $\frac{3}{4}\log_2 64$ 16. $\log_8 2$

17. $\log_{\frac{1}{2}} 16$ 18. $\dfrac{\log_3 64}{\log_3 16}$ 19. $\dfrac{\log_2 36}{\log_2 216}$

20. $\dfrac{\log_5 125}{\log_5 25}$ 　　　21. $\dfrac{\log_{10} 9}{\log_{10} 81}$ 　　　22. $\dfrac{\log_{10} 256}{\log_{10} 64}$

Given $\log_{10} 2 = 0.3010$, $\log_{10} 3 = 0.4771$ and $\log_{10} 5 = 0.6990$, find

23. $\log_{10} 6$ 　　　24. $\log_{10} 18$ 　　　25. $\log_{10} 20$

26. $\log_{10} 60$ 　　　27. $\log_{10} 45$ 　　　28. $\log_{10} 25$

29. $\log_{10} 81$ 　　　30. $\log_{10} 75$ 　　　31. $\log_{10} 1.5$

32. $\log_{10} \dfrac{1}{8}$ 　　　33. $\log_{10} \sqrt[3]{27}$ 　　　34. $\log_{10} \dfrac{1}{\sqrt{2}}$

Logarithmic Equations

A logarithmic equation is an equation that contains a logarithmic expression.

Example

Solve:

(a) $\log_4 x + \log_4 2 = 3$ 　　　(b) $\log_2 x + \log_2(x + 3) = 2$

(a) $\log_4 x + \log_4 2 = 3$

$\quad \log_4 2x = 3$ 　　　Write the expression as a single logarithm

$\quad\quad 2x = 4^3$ 　　　Write in index form

$\quad\quad 2x = 64$

$\quad\quad x = 32$

Alternatively we have

$$\log_4 x + \log_4 2 = 3$$

$$\log_4 2x = \log_4 64$$

$$2x = 64$$

$$x = 32$$

(b) $\log_2 x + \log_2(x + 3) = 2$

$$\log_2 x(x + 3) = 2$$

$$x(x + 3) = 4$$

$$x^2 + 3x - 4 = 0$$

$$(x + 4)(x - 1) = 0$$

$$x + 4 = 0 \quad \text{or} \quad x - 1 = 0$$

$$x = -4 \qquad\qquad x = 1$$

Possible solutions are $x = -4$ or $x = 1$. To check we substitute -4 and 1 in the original equation

When $x = 1$, we have

$$\log_2 1 + \log_2 4 = \log_2 4 = 2$$

and when $x = -4$, we have

$$\log_2 -4 + \log_2 -1 \neq 2$$

Since logarithms of negative numbers are not defined, the only solution for the original equation is $x = 1$.

Try this 17

Solve:

(a) $\log_2 x + \log_2 3 = 2$ (b) $\log_{10} x + \log_{10}(x - 3) = 1$

Exercise 12.3(c)

Solve:

1. $\log_{10} x + \log_{10} 4 = 2$

2. $\log_3 3x + \log_3 9 = 4$

3. $\log_2 x - \log_2 8 = 1$

4. $\log_5 5x - \log_5 3 = 2$

5. $\log_3 x + \log_3(2x + 3) = 2$

6. $\log_3(x + 3) + \log_3(x - 3) = 3$

7. $\log_3(x + 4) - \log_3 x = 2$

8. $\log_{10}(x + 9) = 1 + \log_{10}(2x - 1)$

9. $\log_3(x + 12) = \log_3 6 + \log_3(x - 3)$

10. $\log_{10}(2x + 5) - \log_{10} 5 = \log_{10}(x - 2)$

12.4 Tables of Common Logarithms

The logarithm to base 10 is called common logarithm. When a logarithm is written without a base, we assume the base to be 10.

The logarithm of a number to base 10 can be found from logarithm tables. The first table of common logarithms was compiled by the

English mathematician Henry Briggs. Logarithm tables have been replaced by electronic calculators and computers with logarithmic functions.

Finding logarithm of numbers between 1 and 10

The logarithm of a number has a decimal part called the mantissa, and an integer before the decimal point called the characteristic. For example, the common logarithm of 596 is 2.7752. The characteristic is 2, and the mantissa is 0.7752.

The mantissa is printed in the tables of logarithms. The table has three main columns. The first column gives the first two digits of the number. The second column gives the third digit and the difference column takes care of the fourth digit.

Note that all the logarithms are decimal fractions, but the decimal point is not printed. The numbers in the difference column have zeros in front of them, which are not printed.

We want to look up the logarithm of 5.62. Recall that the logarithm of a number is the index of a particular base. Now

$$1 = 10^0$$

$$10 = 10^1$$

5.62 lies between 1 and 10, so its logarithm must be between 0 and 1. To find the logarithm of 5.62 from a table

1. Move down the first column on the left until you reach 56.

2. Move right from 56 until you reach the column headed 2. You would find the figure 7497.

So the logarithm of 5.62 is 0.7497.

To obtain the logarithm of 5.627, find the logarithm of 5.62, as described above. Then add to it the number in the same row as 56 and under the column headed 7 in the difference column.

So the logarithm of 5.627 is $0.7497 + 0.0005 = 0.7502$

The working is set out full here, but this can be done mentally.

Try this 18

Find the logarithms of the following numbers from tables.

(a) 5.580   (b) 5.83   (c) 5.736   (d) 6.097

Logarithm of Numbers greater than 10

Study the table below carefully.

Number	Standard form	Number expressed as power of 10	Logarithm
2.75	2.75×10^0	$10^{0.4393}$	0.4393
27.5	2.75×10^1	$10^{1.4393}$	1.4393
275	2.75×10^2	$10^{2.4393}$	2.4393
2750	2.75×10^3	$10^{3.4393}$	3.4396
27500	2.75×10^4	$10^{4.4393}$	4.4393

Table 12.1

Notice that the mantissa is the same for the logarithms of the numbers in the first column, and the logarithms of these numbers differ only in their characteristic.

Finding the characteristic of the logarithms of numbers

You may obtain the characteristic of the common logarithms of numbers by writing the numbers in standard form, that is in the form, $a \times 10^n$. The index n is called the characteristic. However a practical rule for obtaining the characteristic is stated below:

The characteristic is one less than the number of digits before the decimal point.

For example, the characteristic of the logarithm of 64800 is (5 – 1) i.e. 4. Also, the characteristic of the logarithm of 853.07 is (3 – 1), i.e. 2

To find common logarithms of numbers

To obtain the common logarithm of a number, first you have to supply the characteristic and then read the mantissa from tables of logarithms, as described above.

Example

Use logarithm tables to find the logarithm of 468

First work out the characteristic, in this case it is 2. Then move down the left-hand column until you reach 46. Next move right until you reach column 8. You would find the figure 6702.

So the logarithm of 468 is 2.6702.

Try this 19

Use logarithm table to find the logarithms of:

(a) 36   (b) 2310   (c) 735.8   (d) 694200

Use of the Antilogarithm Tables

You can obtain the number that a given logarithm represents from Antilogarithm Tables. Often the antilogarithm table is on the next page of a logarithm table.

To find a number whose logarithm is known:

1. use the mantissa to find the number

2. use the characteristic to insert the decimal point

Example

Find the number whose logarithm is 1.6340

Move down the left-hand column until you find .63. Then move along the row until you reach the column headed 4. You would find the figures 4305.

The characteristic is 1 indicating that there would be two digits before the decimal point. Insert the decimal point after the first two digits.

So the number is 43.05

Note: add any difference

Try this 20

Use tables to write down the numbers whose logarithms are:

(a) 0.7321   (b) 1.8530   (c) 3.6645   (d) 2.3018

Exercise 12.4(a)

Use tables to find the logarithms of:

1. 4.60   2. 8.02   3. 6   4. 2.11   5. 1.02

6. 5.463   7. 5.082   8. 4.007   9. 2.995   10. 1.076

Read off the characteristics of the logarithms of the numbers:

11. 53   12. 751   13. 5839   14. 6.521

15. 24730   16. 912100   17. 3075.69   18. 10,700,000

19. 20823   20. 20.56

Use tables to write the logarithm of:

21. 28.0   22. 52.6   23. 4580   24. 714.5

25. 40.72   26. 63,700   27. 328300   28. 67,384

29. 10003   30. 731.26

State the number of digits before the decimal point in the numbers whose logarithms are:

31. 0.8534   32. 1.0120   33. 2.6240   34. 4.3172

35. 3.263   36. 5.8023   37. 6.0002   38. 0.2010

39. 5.2   40. 0.0032

Use tables to write down the numbers whose logarithms are:

41. 2.0360   42. 0.786   43. 3.0042   44. 4.921

45. 5.7962   46. 7.4870   47. 2.3   48. 4.0000

49. 1.787   50. 6.0354

Use of Logarithm Tables

We may use logarithm table in calculations involving multiplication and division.

Examples

Evaluate:

(a) 283×73.6 (b) 75.35^2 (c) $\sqrt[5]{483.5}$

(a) Look up the logarithms of the numbers. Then add the logarithms. Finally, use the table of antilogarithms to find the answer. We present the work in a table like the one shown below

No	Log
283	2.4518
73.6	1.8669
20830	4.3187

Thus, $283 \times 73.6 = 20830$

(b) Look up the logarithm of 75.35 and then multiply by 2. Finally, look up the antilogarithm of the result.

No	Log
75.35	1.8771
75.35^2	1.8771×2
5678	3.7542

Thus, $75.35^2 = 5678$

(c) Look up the logarithm of 483.5 and then divide by 5. Finally, look up the antilogarithm of the result.

No	Log
483.5	2.6843
$\sqrt[5]{483.5}$	$2.6843 \div 5$
3.443	0.5369

Thus, $\sqrt[5]{483.5} = 3.443$

Try this 21

Evaluate

(a) 127.8×10.94 (b) $136.2 \div 94.5$ (c) $\sqrt[3]{262.1}$ (d) 3.57^4

Evaluate $\sqrt[3]{\dfrac{817.3 \times 946.2}{625.2 \times 32.8}}$

The working is shown in the table below.

No	Log	
817.3	2.9124	
946.2	2.9760	
Numerator	5.8884	.... .5.8884
625.2	2.7960	
32.8	1.5159	
Denominator	4.3119	...........4.3119
		1.5765
	$1.5765 \div 3$	
3.354	0.5255	

Try this 22

Evaluate

$$\frac{183.7 \times 29.62}{472.3}$$

Exercise 12.4(b)

Find the value of the following

1. 35.2×76.4

2. 60.7×108

3. 73.01×4002

4. 5472×4306

5. $843 \div 91.5$

6. $71.6 \div 38.7$

7. $6274 \div 83.6$

8. $5000 \div 306.4$

9. $(72.49)^2$

10. $(107.4)^3$

11. $(5.27)^5$

12. $(9.106)^3$

13. $\sqrt[3]{278.5}$

14. $\sqrt{543.6}$

15. $\sqrt[4]{867.2}$

16. $\sqrt[5]{704.8}$

17. $\dfrac{52.04 \times 80.65}{97.53}$

18. $\dfrac{732.4}{1.642 \times 3.98}$

19. $\left(\dfrac{174.6 \times 10.8}{134.7}\right)^2$

20. $\sqrt[3]{\left(\dfrac{7134}{85.7 \times 2.69}\right)}$

21. $\sqrt{\dfrac{128.1 \times 13}{965.8}}$

To find logarithms of positive numbers less than 1

Number	Standard form	Number expressed as power of 10	Logarithm
2.07	2.07×10^0	$10^{0.3160}$	0.3160
0.207	2.07×10^{-1}	$10^{-1+0.3160}$	$\bar{1} + 0.3160$
0.0207	2.07×10^{-2}	$10^{-2+0.3160}$	$\bar{2} + 0.3160$
0.00207	2.07×10^{-3}	$10^{-3+0.3160}$	$\bar{3} + 0.3160$
0.000207	2.07×10^{-4}	$10^{-4+0.3160}$	$\bar{4} + 0.3160$

Table 12.2

Logarithms of positive numbers less than 1 are negative as can be seen from the Table 12.2.

Conventionally, they are written so that the mantissa is always positive. The logarithm of 0.0207 is taken as $\bar{2}.3160$ instead of -1.6840. The bar over the characteristic indicates that it is negative whilst the mantissa remains positive. $\bar{2}.3160$ is read as bar 2 point 3160.

The characteristic can be found by expressing the number in standard form. In practice we can apply the rule below:

The characteristic is numerically one greater than the number of zeros immediately following the decimal point.

The number 0.000312 has characteristic $\bar{4}$ since there are 3 zeros immediately following the decimal point. Thus, the logarithm of 0.000312 is $\bar{4}.4942$.

Various problems may arise when working with negative characteristic. The examples below illustrate the methods of approaching these problems. The working as shown in the examples may be done mentally. The difference between -6.5 and $\bar{6}.5$ must be noted.

$$-6.5 = -6 - 0.5 \text{ and } \bar{6}.5 = -6 + 0.5$$

Example

Find the value of the following leaving the decimal part positive

(a) $\bar{2}.8 + \bar{4}.7$ (b) $\bar{5}.2 - 2.7$ (c) $\bar{2}.4 \times 3$ (d) $\bar{1}.7 \div 3$

(a) Write $\bar{2}.8 + \bar{4}.7$ as shown below

$$
\begin{array}{r}
\bar{2} + 0.8 \\
\bar{4} + 0.7 \\
\hline
\bar{6} + 1.5 \\
\end{array}
$$

Thus $\bar{2}.8 + \bar{4}.7 = (\bar{6} + 1) + 0.5 = \bar{5}.5$

(b) We write $\bar{5}.2$ as $\bar{6} + 1.2$

$$
\begin{array}{r}
\bar{6} + 1.2 \\
2 + 0.7 \\
\hline
\bar{8} + 0.5 \\
\end{array}
$$

Thus $\bar{6}.2 - 2.7 = (\bar{6} - 2) + (1.2 - 0.7) = \bar{8}.5$

(c) The working is shown below

$$\begin{array}{r} \bar{2} + 0.4 \\ \times\, 3 \\ \hline \bar{6} + 1.2 \end{array}$$

Thus $\bar{2}.4 \times 3 = (\bar{6} + 1) + 0.2 = \bar{5}.2$

  (d) To ensure a positive decimal, -1 is expressed as $-3 + 2$. So $\bar{1}.7 = \bar{3} + 2.7$

$$3 \overline{\smash{)}\begin{array}{l} \bar{1} + 0.9 \\ \bar{3} + 2.7 \end{array}}$$

So $\bar{1}.7 \div 3 = \bar{1}.09$

Try this 23

Find the value of the following, leaving the decimal part positive

(a) $\bar{1}.7 + \bar{2}.5$ (b) $\bar{2}.3 - \bar{4}.7$ (c) $\bar{1}.8 \times 3$ (d) $\bar{6}.5 \div 5$

Exercise 12.4(c)

Find the characteristic of each of the following numbers

1. 0.321   2. 0.04632   3. 0.005.6

4. 0.1032   5. 0.000078   6. 0.000302

Write down the logarithms of the following numbers

7. 0.432   8. 0.0834   9. 0.0067

10. 0.209   11. 0.0003   12. 0.005216

Find the numbers whose logarithms are:

13. $\bar{1}.682$ 14. $\bar{2}.7645$ 15. $\bar{4}.8085$ 16. $\bar{3}.4133$ 17. $\bar{4}.0000$

18. $\bar{5}.9130$ 19. $\bar{2}.05$ 20. $\bar{3}.086$ 21. $\bar{6}.3016$ 22. $\bar{2}.0087$

Simplify each expression and write your answers in the bar notation

23. $\bar{2}.7 + 5.8$ 24. $\bar{1}.6 + \bar{2}.7$ 25. $\bar{4}.5 - \bar{2}.8$ 26. $\bar{2}.4 - 1.6$

27. $\bar{2}.8 \times 4$ 28. $\bar{1}.3 \times 5$ 29. $\bar{3}.2 \div 4$ 30. $\bar{4}.7 \div 3$

Example

Evaluate:

(a) 0.0863×0.278 (b) $0.3475 \div 0.0796$ (c) $(0.752)^2$ (d) $\sqrt[3]{0.6170}$

The method of solution is the same. Remember that in these cases the mantissa is always positive. Each solution is presented in a table as shown below.

(a)   (b)

No	Log
0.0863	$\bar{2}.9360$
0.278	$\bar{1}.4440$
0.02399	$\bar{2}.3800$

No	Log
0.3475	$\bar{1}.5409$
0.0796	$\bar{2}.9009$
4.365	0.6400

(c)   (d)

No	Log
0.752	$\bar{1}.8762$
0.752^2	$\bar{1}.8762 \times 2$
0.5654	$\bar{1}.7524$

No	Log
0.6170	$\bar{1}.7903$
$\sqrt[3]{0.6170}$	$\bar{1}.7903 \div 3$
0.8513	$\bar{1}.9301$

Try this 24

Evaluate:

(a) 0.0726×0.215 (b) $0.5327 \div 0.08406$ (c) $\sqrt{0.00658}$
(d) $(0.0816)^3$

Example

Evaluate:

(a) $\dfrac{0.09156}{0.214 \times 0.7506}$ (b) $\sqrt{\dfrac{0.00683}{0.517 \times 0.0408}}$

(a) The working is shown in the table below

No	Log
0.09156	$\bar{2}.9617$ ... ... $\bar{2}.9617$
0.214	$\bar{1}.3304$
0.7506	$\bar{1}.8754$
	$\bar{1}.2058$ ... ... $\bar{1}.2058$
	$\bar{1}.7559$
0.5701	$\bar{1}.7559$

(b) The working is shown in the table below

No	Log
0.00683	$\bar{3}.8344$ ... ... $\bar{3}.8344$
0.517	$\bar{1}.7135$
0.0408	$\bar{2}.6107$
	$\bar{2}.3242$ ... ... $\bar{2}.3242$
	$\bar{1}.5102$
	$\bar{1}.5102 \div 2$
0.569	$\bar{1}.7551$

Try this 25

Evaluate

$$\sqrt{\frac{0.3265 \times 0.08162}{0.007429}}$$

Exercise 12.4(d)

Find the values of the following

1. 0.0234×0.213 2. 0.00863×0.0924 3. $0.874 \div 0.632$

4. $0.0463 \div 0.00852$ 5. $(0.513)^4$ 6. $(0.0864)^2$

7. $\sqrt[4]{0.723}$ 8. $\sqrt[5]{0.009056}$ 9. $\sqrt{0.07234}$

10. $\dfrac{\sqrt{0.06473}}{0.1681 \times 0.5346}$ 11. $\left(\dfrac{0.06372 \times 0.1508}{0.08307}\right)^2$ 12. $\sqrt[3]{\dfrac{0.045}{1.246 \times 0.3146}}$

Review exercise 12

Simplify the following and write your answer in index form.

1. $3^7 \cdot 3^{-5}$ 2. $2^{-5} \cdot 2^4$ 3. $5 \cdot 5^{-3}$ 4. $5^8 \div 5^{12}$

5. $3^{-6} \div 3^{-8}$ 6. $3^{-1} \div 3$ 7. $(5^{-2})^{-1}$ 8. $(2^3)^2$

9. $(3^{-10})^{-\frac{3}{5}}$ 10. $\left(7^{\frac{4}{3}}\right)^{-\frac{3}{2}}$ 11. $\left(\frac{1}{3}\right)^0$ 12. $\left(\frac{1}{8}\right)^{-\frac{2}{3}}$

Without using a calculator evaluate:

13. $16^{\frac{1}{4}}$ 14. $216^{\frac{1}{3}}$ 15. $625^{\frac{1}{2}}$ 16. $243^{\frac{1}{5}}$

17. $8^{\frac{2}{3}}$ 18. $27^{\frac{2}{3}}$ 19. $81^{-\frac{3}{4}}$ 20. $64^{-\frac{2}{3}}$

Solve each of the following equations

21. $4^x = 32$ 22. $3^{2(x+1)} = 27$ 23. $27^x = 9$

24. $25^{3x+2} = 625$ 25. $9^x = \frac{1}{3}$ 26. $\left(\frac{1}{16}\right)^x = 32$

27. $36^x = 1$ 28. $25^{2-x} = \left(\frac{1}{625}\right)^x$

Write each of the following numbers in standard form

29. 597000   30. 640   31. 67.5   32. 7582

33. 0.65   34. 0.000804   35. 0.02178   36. 0. 00009

Evaluate each of the following and write your answer in standard form

37. 65000×0.083 38. 0.0093×0.52

39. $8600 \div 0.0043$ 40. $0.243 \div 0.00027$

41. $\dfrac{0.625 \times 0.0036}{0.0025 \times 0.045}$ 42. $\dfrac{0.00035 \times 0.56}{0.05 \times 0.14}$

43. $\dfrac{0.169 \times 0.0075 \times 3.6}{0.00039 \times 0.018}$ 44. $\dfrac{0.0045 \times 0.16 \times 0.081}{0.00018 \times 0.54}$

Write each of the following equations in logarithm form

45. $6^2 = 36$ 46. $8^{\frac{2}{3}} = 4$ 47. $16^0 = 1$ 48. $25^{-\frac{1}{2}} = \frac{1}{5}$

Write each of the following equations in index form

49. $log_6 216 = 3$ 50. $log_2 \frac{1}{32} = -5$

51. $log_{81} 9 = \frac{1}{2}$ 52. $log_{10} 10,000 = 4$

Write each of the following expressions as a single logarithm

53. $log_3 5 + log_3 7$ 54. $log_2 15 - log_2 3$

55. $1 + log_5 8 - log_5 16$ 56. $\frac{3}{2}log_4 36 - \frac{1}{3}log_4 27 - 2$

Without a calculator evaluate:

57. $log_2 128$ 58. $log_7 49$ 59. $\frac{3}{4}log_3 81$ 60. $\frac{log_4 64}{log_4 16}$

Given $log2 = 0.3010$ and $log3 = 0.4771$, evaluate

61. $log1.2$ 62. $log72$ 63. $log\left(\frac{1}{5}\right)$ 64. $log45$

Solve each of the following logarithmic equations

65. $log_6(x + 8) + log_6 3 = 2$ 66. $log_{10}x + log_{10}(x - 3) = 1$

67. $log_3(2x + 5) - log_3 x = 1$

68. $log_2(2x - 3) + 2 = log_2(x + 2)$

Use the logarithm tables to evaluate:

69. 38.74×25.12 70. $424.5 \div 82.7$ 71. 5.319^3

72. $\sqrt[5]{321.4}$ 73. 0.7264×0.02005 74. $0.03752 \div 0.4034$

75. $(0.739)^3$ 76. $\sqrt[4]{0.0658}$ 77. $\frac{12.8 \times 293.6}{47.56}$

78. $\frac{0.0762}{0.2035 \times 0.0629}$ 79. $\sqrt[3]{\frac{12.81 \times 1.3}{0.09658}}$ 80. $\left(\frac{998.1}{45.23 \times 12.64}\right)^5$

Chapter Test 12

Take this test as you would take a test in class. After you are done, check your work against the answers in the back of the book.

1. Simplify $(3^4)^{\frac{3}{4}} \times (3^{-6})^{\frac{1}{2}} \div 3^{-2}$ and write your answers in index form.

2. Work out the value of:

    (a) $8^{\frac{1}{3}} \times 27^{\frac{2}{3}}$ (b) $32^{\frac{3}{5}} \times 81^{\frac{1}{4}} \div 36^{\frac{1}{2}}$

3. Solve:

    (a) $\left(\frac{1}{2}\right)^x = \frac{1}{8}$ (b) $27^x = 9^{x+3}$

4. Write each of the following numbers in standard form

    (a) 0.00362   (b) 57.8

5. Work out $\frac{0.0625 \times 0.54}{0.225}$, and write your answer in standard form.

6. Without a calculator evaluate:

    (a) $log_5 625$ (b) $log_3 \frac{1}{234}$ (c) $\dfrac{\log 8 - \log 4 + \frac{1}{6}\log 64}{3 \log 2}$

7. Given $log_{10} 2 = 0.3010$ and $log_{10} 3 = 0.4771$, find

    (a) $log_{10} 60$ (b) $log_{10} 135$

8. Write each of the following expressions as a single logarithm

    (a) $log_3 8 + log_3 27 - log_3 36$ (b) $1 + log 18 - \frac{1}{2} log 225$

9. Solve each of the following equations

    (a) $log_2 x + log_2(x - 2) = 3$ (b) $2 log_4(x + 5) = 3$

10. Use tables of logarithms to evaluate

(a) $\dfrac{635.2}{71.65 \times 39.2}$

(b) $\sqrt[4]{\dfrac{18.37 \times 2.962}{0.4723}}$

11. Given   that $3.25 = 10^{0.5119}$, $5.06 = 10^{0.7042}$ and

$8.015 = 10^{0.9039}$ evaluate without using tables or a calculator

$\sqrt{\dfrac{32.5}{50.6}}$

12. Solve each of the following simultaneous equations

(a) $3^{x+y} = 243$

$2^{2x-y} = 16$

(b) $2^x \cdot 4^y = 16$

$9^x \div 3^y = 27$

13

Number Bases

We normally do our arithmetic in the base 10 system, but there are many number bases that we can work in. The base 10 system is also known as the decimal system.

The base 2 or binary system has become very important because it provides the basis on which computers are built.

In any base system, we use symbols to represent zero and each number less than the base. For example, the base 2 system uses two symbols, namely the numeral 0 and 1. In a base 5 system we use five numerals: 0, 1, 2, 3 and 4.

A number in a base system other than base 10 will be indicated by a subscript to the right of the number. Thus, 312_8 (or 312_{eight}) represents a number in base 8, read three-one-two base 8. The value of 312_8 is not the same as the value of 312_{10}. A base 10 number is written without a subscript.

13.1 Addition and Subtraction

We add or subtract in any number base system much the same way as we add or subtract in base 10. We add (or subtract) digit by digit starting from the right side.

Addition

Consider the decimal addition below

$$\begin{array}{cc} 2 & 9 \\ 4 & 3 \\ \hline \end{array}$$

To add these two numbers, we first consider the column on the right, called the 'ones column'. Now, 9 plus 3 gives 12. 12 is greater than 9, remember that base 10 operates with the nine digits: 0, 1, 2, 3, 4, 5, 6, 7, 8 and 9. 12 means we have 1 ten and 2 ones. We carry the 1 from the ones column to the next column, called the 'tens column' and leave the 2 in the ones column.

We move to the tens column and calculate $1 + (2 + 4)$, which gives 7. Since 7 is less than 9 we leave the 7 in the tens column.

Addition in all base systems works in the same way, except that only numerals within the base system we carry out the addition is used.

Example

Add 1101 and 101 in base 2

First, we rewrite the addition in a column form.

$$
\begin{array}{r}
1101 \\
\underline{101} \\
\end{array}
$$

We start the addition from the right column. Now, $1 + 1$, results in 10. Leave the 0 in the column and carry the 1 to the next column.

Move to the next column and calculate $1 + (0 + 0)$, which gives 1

Next move to the third column and calculate $1 + 1$, which results in 10. Leave 0 in the column and carry 1 to the fourth column

Finally, move to the fourth column and calculate $1 + 1$, which results in 10. Leave 0 in the column and carry the 1.

$$
\begin{array}{r}
1101 \\
\underline{101} \\
\underline{10010} \\
\end{array}
$$

Try this 1

(a) Add 123_4 and 12_4 (b) Add 7345_8 and 675_8

Exercise 13.1(a)

1. Do the following addition in base 2

(a) $110 + 11$ (b) $101 + 11$

(c) $111 + 101$ (d) $1011 + 1010$

(e) $1001 + 1101 + 110$ (f) $11001 + 101 + 11$

2. Do the following addition in base 3

(a) $12 + 21$ (b) $211 + 121$

 (c) $102 + 211$ (d) $1121 + 212$

 (e) $112 + 110 + 21$ (f) $2121 + 2001 + 122$

3. Do the following addition in base 4

(a) $32 + 21$ (b) $13 + 12$ (c) $321 + 213$

(d) $302 + 113$ (e) $112 + 321 + 211$ (f) $332 + 32 + 12$

4. Do the following addition in base 5

(a) $43 + 21$ (b) $32 + 13$

(c) $243 + 123$ (d) $311 + 1213$

(e) $2011 + 322 + 32$ (f) $1102 + 423 + 214$

5. Do the following addition in base 6

(a) $25 + 54$ (b) $32 + 45$ (c) $124 + 51$

(d) 5314 + 243   (e) 1152 + 212 + 41   (f) 4320 + 213 + 142

6. Do the following addition in base 7

(a) 56 + 34   (b) 42 + 15   (c) 436 + 234

(d) 565 + 413   (e) 126 + 254 + 342   (f) 205 + 164 + 563

7. Do the following addition in base 8

(a) 76 + 31   (b) 54 + 26   (c) 543 + 474

(d) 624 + 127   (e) 743 + 54 + 267   (f) 654 + 210 + 57

8. Do the following addition in base 9

(a) 25 + 32   (b) 74 + 53   (c) 675 + 123

(d) 542 + 237   (e) 322 + 157 + 24   (f) 654 + 1302 + 175

Addition in Base Twelve (Duodecimal)

The base twelve numeration system has the following twelve symbols:

0, 1, 2, 3, 4, 5, 6, 7, 8, 9, T, E

T and E represent ten and eleven respectively.

Example

Add 9E3 and 2T

$$
\begin{array}{rrr}
9 & E & 3 \\
+ \quad 2 & T \\
\hline
T & 2 & 1 \\
\hline
\end{array}
$$

Try this 2

Add 275 and 238 in base 12.

Exercise 13.1(b)

Do the following addition in base 12

1. $83 + 25$ 2. $1E3 + 57$ 3. $TE9 + 57$

4. $1E1 + 97T$ 5. $695 + 2E3 + 4T$ 6. $1TE3 + 527 + 11T$

Subtraction

Consider the following subtraction in base 10

```
   4 5
-  2 9
```

We start by subtracting 9 from 5. Since, 9, a larger number is subtracted from smaller number, 5 we borrow 1 from the tens column, that leaves 3 in the tens column. Then we calculate $(10 + 5) - 9$, which gives 6. We move to the tens column and calculate $3 - 2$, which gives 1.

```
  4 5
  2 9
  1 6
```

We work subtraction in all base systems in the same way, except that the 1 borrowed from a column is equal to the base.

Example

Subtract 132 from 321 in base 4.

$$
\begin{array}{r}
3\ 2\ 1 \\
1\ 3\ 2 \\
\hline
1\ 2\ 3 \\
\hline
\end{array}
$$

Try this 3

(a) Subtract 131_5 from 224_5 (b) Subtract 110_2 from 1101_2

Exercise 13.1(c)

1. Do the following subtraction in base 2

(a) $110 - 11$ (b) $1001 - 110$ (c) $10101 - 1010$

(d) $1000 - 111$ (e) $111011 - 11011$ (f) $10010 - 101$

2. Do the following subtraction in base 3

(a) $122 - 21$ (b) $211 - 22$ (c) $201 - 111$

(d) $1121 - 212$ (e) $200 - 111$ (f) $2011 - 122$

3. Do the following subtraction in base 4

(a) $233 - 21$ (b) $121 - 32$ (c) $321 - 132$

(d) $203 - 121$ (e) $1232 - 313$ (f) $3012 - 323$

4. Do the following subtraction in base 5

(a) $32 - 21$ (b) $432 - 143$ (c) $213 - 134$

(d) $1210 - 324$ (e) $4203 - 1321$ (f) $432 - 204$

5. Do the following subtraction in base 6

(a) $54 - 32$ (b) $131 - 52$ (c) $410 - 234$

(d) $2103 - 542$ (e) $3002 - 453$ (f) $1125 - 313$

6. Do the following subtraction in base 7

(a) $64 - 41$ (b) $53 - 36$ (c) $235 - 56$

(d) $421 - 254$ (e) $342 - 145$ (f) $1023 - 454$

7. Do the following subtraction in base 8

(a) $76 - 54$ (b) $53 - 37$ (c) $213 - 64$

(d) $342 - 145$ (e) $702 - 354$ (f) $1534 - 756$

8. Do the following subtraction in base 9

(a) $67 - 45$ (b) $82 - 67$ (c) $124 - 76$

(d) $506 - 278$ (e) $2134 - 753$ (f) $1673 - 734$

9. Do the following subtraction in base 12

(a) $78 - 52$ (b) $80 - 6T$ (c) $6T2 - 3E8$

(d) $E8T - 7E5$ (e) $E1T - 938$ (f) $19E3 - ET9$

13.2 Multiplication

Multiplication like addition and subtraction can be performed in other bases. We may do multiplication with the help of a multiplication table for the base desired.

Using the Base 5 Multiplication Table

Suppose we want to determine the product of $3_5 \times 4_5$. You may find it easier to multiply the values in base 10 and then change the product to base 5.

Multiplying 4×3 in base 10 gives 12, and converting 12 from base 10 to base 5 gives 22_5. Notice that 12 has two groups of five, and two units. Thus, $3_5 \times 4_5 = 22_5$. Similarly, in base 10, $2 \times 4 = 8$. Now, 8 is one group of five, and three units. Thus, $2_5 \times 4_5 = 13_5$ We calculate the rest of the values in Table 13.1 in the same way.

x	1	2	3	4
1	1	2	3	4
2	2	4	11	13
3	3	11	14	22
4	4	13	22	31

Table 13.1

To multiply in base 5, use the base 5 multiplication table to find the products. When the product consists of two digits, record the right digit and carry the left digit.

Example

Calculate $312_5 \times 23_5$

First, rewrite the multiplication in a column form.

$$
\begin{array}{r}
312_5 \\
\times\ 23_5 \\
\hline
1441 \\
1124\ \ \\
\hline
13231_5 \\
\end{array}
$$

1441_5 is obtained as follows:

From Table 13.1, $3_5 \times 2_5 = 11_5$. Record the 1 on the left and carry the 1 on the right.

Next, $(3_5 \times 1_5) + 1_5 = 3_5 + 1_5 = 4_5$. Record the 4

Finally, $(3_5 \times 3_5) = 14_5$. Record the 4 and carry the 1

Similarly, 1124_5 is obtained as follows:

$2_5 \times 2_5 = 4_5$. Record the 4

Next, $2_5 \times 1_5 = 2_5$. Record the 2

Finally, $2_5 \times 3_5 = 11_5$. Record the 1 and carry the 1

Try this 4

Multiply $204_5 \times 31_5$

Exercise 13.2(a)

1. Construct the base 3 multiplication table and use your table to calculate

    (a) 21×12 (b) 121×11 (c) 112×21 (d) 212×12

2. Construct the base 4 multiplication table and use your table to calculate

    (a) 23×3 (b) 321×21 (c) 112×23 (d) 1032×32

3. Construct the base 6 multiplication table and use your table to calculate

    (a) 24×5 (b) 432×12 (c) 451×23 (d) 3014×35

4. Construct a multiplication table in base 8 on the set $\{1, 2, 3, 4\}$ and use your table to calculate

(a) 42×3 (b) 234×21 (c) 431×32

5. Construct a multiplication table in base 12 on the set $\{2, 5, 7, 9\}$ and use your table to calculate:

(a) 25×7 (b) 57×29 (c) 292×72

Multiplying without a Table

Constructing a multiplication table is often tedious, especially when the base is large. To multiply in a given base without the use of table, multiply in base 10 and convert the product to the given base before you record them.

Example

Multiply $123_5 \times 42_5$

First, we multiply 123 by 2, which gives 301.
Next multiply 123 by 4, which gives 1102. You may put 0 as a placeholder and write 11020. Finally, add 301 and 11020, the sum of which is 11321.

$$
\begin{array}{r}
1\,2\,3 \\
4\,2 \\
\hline
3\,0\,1 \\
1\,1\,0\,2\,0 \\
\hline
1\,1\,3\,2\,1 \\
\hline
\end{array}
$$

We obtain 301 as follows:

Multiplying in base 10, $3 \times 2 = 6$. Converting 6 to base 5, we have 11_5.

Since the product consists of two digits, we record the right digit and carry the left digit

Now, the product 2×2 is 4 and then adding 1 gives 5. Converting 5 to base 5 we get 10_5

We record the 0, and then carry 1. Then calculate $(1 \times 2) + 1$, which gives 3.

Similarly 1102 was obtained as follows:

$(4 \times 3) = 12 = 22_5$

$(4 \times 2) + 2 = 8 + 2 = 10 = 20_5$

$(4 \times 1) + 2 = 4 + 2 = 6 = 11_5$

Try this 5

Multiply

(a) $413_5 \times 32_5$ (b) $304_6 \times 31_6$

Exercise 13.2(b)

1. Do the following multiplication in base 2

    (a) 1011×11 (b) 1001×101 (c) 11001×110

    (d) 11101×111 (e) 10110×101 (f) 1111×110

2. Do the following multiplication in base 3

    (a) 21×2 (b) 112×11 (c) 121×21

(d) 1012×112 (e) 2101×121 (f) 1122×212

3. Do the following multiplication in base 4

    (a) 32×3 (b) 123×34 (c) 211×32

    (d) 1302×132 (e) 321×103 (f) 1032×212

4. Do the following multiplication in base 5

    (a) 143×4 (b) 231×34 (c) 412×21

    (d) 3142×131 (e) 2014×23 (f) 1120×43

5. Do the following multiplication in base 6

    (a) 42×5 (b) 234×21 (c) 514×32

    (d) 1034×53 (e) 3120×23 (f) 1143×214

6. Do the following multiplication in base 7

    (a) 65×14 (b) 453×21 (c) 241×53

    (d) 3102×223 (e) 5643×121 (f) 1645×321

7. Do the following multiplication in base 8

    (a) 76×24 (b) 121×73 (c) 543×61

    (d) 2306×211 (e) 7124×32 (f) 3112×56

8. Do the following multiplication in base 9

    (a) 78×6 (b) 216×57 (c) 543×132

    (d) 8034×35 (e) 3416×42 (f) 1357×213

9. Do the following multiplication in base 12

    (a) 95×7 (b) 463×78 (c) $2T3 \times 52$

(d) $1E4 \times 65$ (e) $307 \times E61$ (f) $T1E \times 422$

13.3 Converting to Base Ten (Decimals)

A base system is based on the principle of place value. The principle of place value implies that the meaning of each digit in a numeral is determined by the position it occupies. This permits the writing of numerals for large numbers by the use of few symbols. For example, the 3 in 375 represents 3 hundreds, the 7 represents 7 tens and the 5 represents 5 ones. Going from right to the left, each place is associated with a number ten times as great as the one before. In the base 10 system, we identify the places by powers of 10. For bases other than 10, we identify the places by the power of the base. The numbers identifying the places of numbers in base 2, base 5 and base 8 are shown below

Column Headings	Number Base
$\cdots, 2^4, 2^3, 2^2, 2^1, 2^0$	2
$\cdots, 5^4, 5^3, 5^2, 5^1, 5^0$	5
$\cdots, 8^4, 8^3, 8^2, 8^1, 8^0$	8

To change a number in a base other than 10 to a base 10 number, we multiply each digit in the number by its respective positional value. Then find the sum of the products.

Examples

Convert (a) 1101_2 and (b) 312_5 to base 10

(a) The positional values are shown in the first row

$$
\begin{array}{cccc}
2^3 & 2^2 & 2^1 & 2^0 \\
1 & 1 & 0 & 1
\end{array}
$$

In the expended form, we have

$$1101_2 = 1 \times 2^3 + 1 \times 2^2 + 0 \times 2^1 + 1 \times 2^0$$

$$= 1 \times 8 + 1 \times 4 + 0 \times 2 + 1 \times 1$$

$$= 8 + 4 + 0 + 1$$

$$= 13_{10}$$

(b)

5^2	5^1	5^0
3	1	2

In the expanded form, we have

$$312_5 = 3 \times 5^2 + 1 \times 5^1 + 2 \times 5^0$$

$$= 3 \times 25 + 1 \times 5 + 2 \times 1$$

$$= 75 + 5 + 2$$

$$= 82_{10}$$

Try this 6

Convert 524_6 to base 10

Given $23_x = 13_{10}$, find x

First convert 23_x to a number in base 10

$$23_x = 2 \times x^1 + 3 \times x^0$$

$$= 2x + 3$$

Next equate $2x + 3$ and 13, and solve

$$2x + 3 = 13$$

$$2x = 10$$

$$x = 5$$

Thus, the base is 5

Try this 7

Given $156_x = 420_5$, find x

Exercise 13.3(a)

Convert the following to base 10

1. 1010_2 2. 11011_2 3. 101101_2 4. 121_3
5. 2120_3

6. 2021_3 7. 232_4 8. 1321_4 9. 1232_4
10. 243_5

11. 314_5 12. 10423_5 13. 324_6 14. 452_6
15. 2311_6

16. 504_7 17. 1245_7 18. 2063_7 19. 57_8
20. 465_8

21. 3127_8 22. 205_9 23. 4126_9 24. 538_{12}
25. $ET3_{12}$

Find x

26. $32_x = 17_{10}$ 27. $47_x = 39_{10}$ 28. $18_x = 32_5$
29. $15_x = 102_3$

30. $56_x = 45_9$ 31. $123_x = 38_{10}$ 32. $147_x = 124_{10}$
33. $136_x = 114_9$

34. $235_x = 95_{10}$ 35. $321_x = 57_{10}$ 36. $156_x = 132_9$

Converting from base 10 to any given base

We will use the method of continued division to change a decimal numeral to another base.

Example

Convert 134 to base 5

Dividing 134 by 5, gives a quotient of 26 and a remainder of 4. Write the quotient below the dividend and the remainder on the right as shown. Continue this process of division by 5 until the quotient is zero. The answer is read from the bottom number to the top number in the remainder column.

5	134	remainder
5	26	4
5	5	1
5	1	0
5	0	1

Thus $134 = 1014_5$

Try this 8

Convert 25 to base 2

Exercise 13.3(b)

1. Convert the following decimal numerals to base 2

   (a) 23   (b) 17   (c) 9   (d) 34   (e) 29

2. Convert the following decimal numerals to base 3

   (a) 13   (b) 27   (c) 32   (d) 37   (e) 40

3. Convert the following decimal numerals to base 4

   (a) 19   (b) 28   (c) 31   (d) 45   (e) 123

4. Convert the following decimal numerals to base 5

   (a) 18   (b) 132   (c) 49   (d) 121   (e) 107

5. Convert the following decimal numerals to base 6

   (a) 34   (b) 123   (c) 52   (d) 134   (e) 215

6. Convert the following decimal numerals to base 7

   (a) 39   (b) 65   (c) 146   (d) 324   (e) 413

7. Convert the following decimal numerals to base 8

   (a) 66   (b) 243   (c) 172   (d) 27   (e) 340

8. Convert the following decimal numerals to base 9

   (a) 20   (b) 55   (c) 145   (d) 307   (e) 516

9. Convert the following decimal numerals to base 12

   (a) 131   (b) 145   (c) 256   (d) 379   (e) 730

Converting from a given base to another base

To convert from a given base to another base we first convert to base 10 and then convert from base 10 to the required base as shown in the example below.

Example

Convert 121_4 to base 6

First we convert 121_4 to a number in base 10

$$121_4 = 1 \times 4^2 + 2 \times 4^1 + 1 \times 4^0$$

$$= 1 \times 16 + 2 \times 4 + 1 \times 1$$

$$= 16 + 8 + 1$$

$$= 25_{10}$$

Next convert 25_{10} to a number in base 6

$$
\begin{array}{r|l}
6 & 25 \\
\hline
6 & 4 \\
\hline
6 & 0 \\
\end{array}
\qquad
\begin{array}{c}
\text{remainder} \\
1 \\
4 \\
\end{array}
$$

Thus $121_4 = 41_6$

Try this 9

Convert 352_6 to base 5

Exercise 13.3(c)

Convert the numbers from the given base to the base stated

1. 111011_2 to base 5

2. 213_4 to base 6

3. 267_8 to base 7

4. 1305_6 to base 8

5. 314_5 to base 4

6. 121_3 to base 2

7. $2E3_{12}$ to base 9

8. 101101_2 to base 3

9. 1303_8 to base 12

10. 1265_7 to base 9

Review exercise 13

Work out

1. $1011_2 + 101_2$

2. $1101_2 + 11_2$

3. $211_3 + 122_3$

4. $1201_3 + 112_3$

5. $203_4 + 311_4$

6. $323_4 + 1032_4$

7. $143_5 + 321_5$

8. $132_5 + 1314_5$

9. $425_6 + 54_6$

10. $5112_6 + 544_6$

11. $656_7 + 34_7$

12. $364_7 + 432_7$

13. $476_8 + 32_8$

14. $345_8 + 474_8$

15. $725_9 + 165_9$

16. $245_9 + 738_9$

17. $ET4_{12} + 48_{12}$

18. $956_{12} + 74_{12}$

Work out

19. $1001_2 - 101_2$

20. $11010_2 - 1001_2$

21. $221_3 - 102_3$

22. $1211_3 - 221_3$

23. $323_4 - 32_4$

24. $3021_4 - 233_4$

25. $134_5 - 43_5$

26. $312_5 - 143_5$

27. $2101_6 - 345_6$

28. $2115_6 - 331_6$

29. $154_7 - 56_7$

30. $1402_7 - 516_7$

31. $423_8 - 56_8$ 32. $5143_8 - 2665_8$ 33. $567_9 - 388_9$

34. $1314_9 - 758_9$ 35. $2T51_{12} - E8_{12}$ 36. $1ET3_{12} - 8TE_{12}$

Work out

37. $1101_2 \times 101_2$ 38. $1001_2 \times 111_2$ 39. $121_3 \times 21_3$

40. $221_3 \times 12_3$ 41. $132_4 \times 31_4$ 42. $122_4 \times 23_4$

43. $234_5 \times 43_5$ 44. $1021_5 \times 32_5$ 45. $345_6 \times 14_6$

46. $124_6 \times 312_6$ 47. $512_7 \times 43_7$ 48. $345_7 \times 21_7$

49. $211_8 \times 37_8$ 50. $354_8 \times 16_8$ 51. $178_9 \times 23_9$

52. $315_9 \times 24_9$ 53. $E12_{12} \times 32_{12}$ 54. $206_{12} \times 13_{12}$

Convert to base 10

55. 11101_2 56. 1011_2 57. 122_3

58. 2112_3 59. 312_4 60. 1223_4

61. 432_5 62. 144_5 63. 432_6

64. 524_6 65. 261_7 66. 1352_7

67. 456_8 68. 1354_8 69. 215_9

70. 2143_9 71. EIT_{12} 72. $1T25_{12}$

Solve

73. $34_x = 19_{10}$ 74. $24_x = 22_{10}$ 75. $32_x = 43_5$

76. $132_x = 42_{10}$ 77. $321_x = 222_6$ 78. $233_x = 125_7$

Convert the decimal numerals to the base stated

79. 125 base 2   80. 37 base 2   81. 23 base 3

82. 67 base 3   83. 38 base 4   84. 123 base 4

85. 134 base 5   86. 87 base 5   87. 132 base 6

88. 203 base 6   89. 73 base 7   90. 135 base 7

91. 65 base 8   92. 134 base 8   93. 57 base 9

94. 154 base 9   95. 143 base 12   96. 562 base 12

Convert the number from the given base to the base started

97. 110111_2 base 5   98. 312_4 base 6   99. 726_8 base 7

100. 1234_6 base 8   101. 1413_5 base 9   102. 3102_9 base 12

103. $1ET_{12}$ base 6   104. 221_3 base 2   105. 1532_7 base 8

Chapter 13 Test

Take this test as you would take a test in a class. After you are done, check your work against the answers in the back of the book.

1. Work out:

   (a) $101101_2 + 10111_2$ (b) $1342_5 - 434_5$

   (c) $512_6 \times 34_6$

2. Work out:

   (a) $TIE_{12} + 85_{12}$ (b) $16E3_{12} - 708_{12}$

   (b) $2E4_{12} \times T2_{12}$

3. Find the missing number of the following subtraction in base 8

$$\begin{array}{ccc} * & * & * \\ -5 & 4 & 6 \\ \hline 1 & 6 & 7 \\ \hline \end{array}$$

4(a) Draw a multiplication table in base 9 on the set $\{2, 3, 4, 5\}$

(b) Use your table to calculate

(i) $24_9 \times 35_9$ (ii) $223_9 \times 54_9$

5. Convert 432_6 to base 10

6. Convert 543_{10} to base 7

7. Convert 627_8 to a base 5 numeral

8. Work out $132_4 + 213_4$ and write your answer in base 5

9. Work out $243_6 - 145_6$ and write your answer in base 8

10. Solve

(a) $23_x = 32_5$ (b) $142_x = 57_8$

14

Surds

The square root of a number, say 4 is the number which when multiplied by itself gives 4. Since, $2 \cdot 2 = 4$, the square root of 4 is 2. We denote the square root of 4 by $\sqrt{4}$. Also $\sqrt{9} = 3$, $\sqrt{16} = 4$ and $\sqrt{25} = 5$. We say that the numbers 4, 9, 16 and 25 are perfect squares because their square roots are exact whole numbers. The following table shows the most common square roots.

Square Roots

$\sqrt{1} = 1$	$\sqrt{36} = 6$	$\sqrt{121} = 11$
$\sqrt{4} = 2$	$\sqrt{49} = 7$	$\sqrt{144} = 12$
$\sqrt{9} = 3$	$\sqrt{64} = 8$	$\sqrt{169} = 13$
$\sqrt{16} = 4$	$\sqrt{81} = 9$	$\sqrt{196} = 14$
$\sqrt{25} = 5$	$\sqrt{100} = 10$	$\sqrt{225} = 15$

Most whole numbers are not perfect squares and do not have rational square roots. For example, $\sqrt{2} = 1.414213\cdots$. The decimal approximation of $\sqrt{2}$ can be given to as many decimal places as one required.

Numbers like $\sqrt{2}$, are often left in the square root form, and are called surds (or radicals). Note that surds are irrational numbers, and you can approximate them with a terminating decimal.

14.1 Simplest Surds

The surd $\sqrt{a}$, where a is a whole number is said to be in the simplest form if a has no perfect square factor. For instance, $\sqrt{10}$ is in the simplest form, since 10 has no perfect square factor.

However, $\sqrt{8}$ is not in the simplest form because 8 has a perfect square 4, as one of its factors.

Simplifying Surds

To simplify surds, we will need to apply one or more of the rules develop below. First, study the following statements.

1. $\left(\sqrt{4}\right)^2 = \sqrt{4} \cdot \sqrt{4} = 2 \cdot 2 = 4$

2. $\sqrt{6^2} = \sqrt{36} = 6$

3. $\sqrt{36} = 6 = 2 \cdot 3 = \sqrt{4} \cdot \sqrt{9}$

4. Now $\sqrt{\dfrac{9}{16}} = \dfrac{3}{4} = \dfrac{\sqrt{9}}{\sqrt{16}}$

These results suggest the following rules.
For any whole numbers a and b,

1. $\left(\sqrt{a}\right)^2 = a$

2. $\sqrt{a^2} = a$

3. $\sqrt{ab} = \sqrt{a} \cdot \sqrt{b}$

4. $\sqrt{\dfrac{a}{b}} = \dfrac{\sqrt{a}}{\sqrt{b}}$

A surd can be simplified as follows:

1. Write the whole number as product of two factors. One of the factors should be the largest perfect square.

2. Simplify the perfect square factor.

Example

Simplify:

(a) $\sqrt{12}$ (b) $\sqrt{27}$

(a) $\sqrt{12} = \sqrt{4 \cdot 3}$

$\qquad = \sqrt{4} \cdot \sqrt{3}$ Using rule 3

$\qquad = 2\sqrt{3}$ Simplify $\sqrt{4}$

Identifying a factor that is a perfect square may be sometimes difficult. In such cases, write the prime factors of the number, and pair each group of like factors. Note that each pair of like factors is a perfect square. You could write 12 as $2^2 \cdot 3$.

So, $\sqrt{12} = \sqrt{2^2 \cdot 3}$

$\qquad = \sqrt{2^2} \cdot \sqrt{3}$ Using rule 3

$\qquad = 2\sqrt{3}$ Using rule 2

(b) $\sqrt{27} = \sqrt{9 \cdot 3}$

$\qquad = \sqrt{9} \cdot \sqrt{3}$ Using rule 3

$\qquad = 3\sqrt{3}$ Simplifying $\sqrt{9}$

Alternatively, we have

$\sqrt{27} = \sqrt{3^2 \cdot 3}$

$\qquad = \sqrt{3^2} \cdot \sqrt{3}$

$\qquad = 3\sqrt{3}$

Try this 1

Simplify (a) $\sqrt{18}$ (b) $\sqrt{20}$

Exercise 14.1

Simplify the following

1. $\sqrt{24}$ 2. $\sqrt{50}$ 3. $\sqrt{8}$ 4. $\sqrt{32}$ 5. $\sqrt{48}$

6. $\sqrt{28}$ 7. $\sqrt{72}$ 8. $\sqrt{45}$ 9. $\sqrt{75}$ 10. $\sqrt{90}$

11. $\sqrt{80}$ 12. $\sqrt{108}$ 13. $\sqrt{128}$ 14. $\sqrt{147}$ 15. $\sqrt{180}$

16. $\sqrt{112}$ 17. $\sqrt{300}$ 18. $\sqrt{405}$ 19. $\sqrt{480}$ 20. $\sqrt{640}$

14.2 Operation of Surds

Addition and Subtraction

You can add and subtract surds in much the same way that you will add and subtract algebraic expressions. Note that you can only add or subtract like surds.

Example

Calculate:

(a) $5\sqrt{3} + 2\sqrt{3}$ (b) $7\sqrt{2} - 4\sqrt{2}$

(a) $5\sqrt{3} + 2\sqrt{3} = (5 + 2)\sqrt{3}$

$$= 7\sqrt{3}$$

(b) $7\sqrt{2} - 4\sqrt{2} = (7-4)\sqrt{2}$

$$= 3\sqrt{2}$$

The example shows that we can add or subtract like surd by multiplying the sum or difference of their coefficients by the common surd. In practice, the first step can be omitted.

Try this 2

Calculate:

(a) $3\sqrt{5} + 4\sqrt{5}$ (b) $5\sqrt{7} - 2\sqrt{7}$ (c) $7\sqrt{2} - 9\sqrt{2} + 2\sqrt{2}$

Sometimes some surds which look unlike can be simplified to produce like surds.

Example

Calculate $\sqrt{18} + \sqrt{50}$

Begin by writing $\sqrt{18}$ and $\sqrt{50}$ in the simplest form

$$\sqrt{18} + \sqrt{50} = 3\sqrt{2} + 5\sqrt{2}$$

$$= 8\sqrt{2}$$

Try this 3

Calculate: (a) $\sqrt{32} + \sqrt{8}$ (b) $\sqrt{27} - \sqrt{12}$

Exercise 14.2(a)

Calculate:

1. $4\sqrt{5} + 6\sqrt{5}$ 2. $2\sqrt{7} + 3\sqrt{7}$

3. $5\sqrt{3} + \sqrt{3}$

4. $8\sqrt{7} - 5\sqrt{7}$

5. $9\sqrt{3} - 7\sqrt{3}$

6. $9\sqrt{2} - 12\sqrt{2}$

7. $9\sqrt{2} - 20\sqrt{2} + 11\sqrt{2}$

8. $7\sqrt{3} + 2\sqrt{3} - 4\sqrt{3}$

9. $8\sqrt{5} - 12\sqrt{5} + 6\sqrt{5}$

10. $\sqrt{27} + \sqrt{48}$

11. $\sqrt{18} + \sqrt{8}$

12. $\sqrt{28} + \sqrt{63}$

13. $\sqrt{50} - \sqrt{32}$

14. $\sqrt{27} - \sqrt{12}$

15. $\sqrt{45} - \sqrt{20}$

16. $3\sqrt{2} - \sqrt{8} - \sqrt{50}$

17. $3\sqrt{8} + 5\sqrt{50} - 4\sqrt{32}$

18. $\sqrt{20} + 2\sqrt{5} - \sqrt{45}$

19. $5\sqrt{8} + 3\sqrt{18} - 4\sqrt{32}$

20. $\frac{3}{4}\sqrt{128} - \frac{2}{5}\sqrt{50}$

21. $\frac{2}{3}\sqrt{27} + \frac{3}{4}\sqrt{48} - \frac{1}{2}\sqrt{12}$

Multiplication of Surds

To multiply two or more surds we make use of the fact that $\sqrt{a} \cdot \sqrt{b} = \sqrt{a \cdot b}$.

Example

Calculate:

(a) $\sqrt{5} \times \sqrt{10}$ (b) $3\sqrt{5} \times 4\sqrt{3}$ (c) $\sqrt{8} \times \sqrt{24}$

(a) $\sqrt{5} \times \sqrt{10}$

   Multiply and then simplify

$$\sqrt{5} \times \sqrt{10} = \sqrt{50}$$

$$= 5\sqrt{2}$$

(b) $3\sqrt{5} \times 4\sqrt{2} = 3 \times \sqrt{5} \times 4 \times \sqrt{2}$

$$= 3 \times 4 \times \sqrt{5} \times \sqrt{2}$$

$$= 12 \times \sqrt{5 \times 2}$$

$$= 12\sqrt{10}$$

Note that to multiply the surds, you multiply the numbers outside the square root sign together, and multiply the numbers under the square root sign. Simplify the result if possible.

(c) Begin by writing the surd in the simplest form

$$\sqrt{8} \times \sqrt{24} = 2\sqrt{2} \times 2\sqrt{6}$$

$$= 4\sqrt{12}$$

$$= 4 \times 2\sqrt{3} \qquad \text{Simplify } \sqrt{12}$$

$$= 8\sqrt{3}$$

Try this 4

Calculate:

(a) $5\sqrt{3} \times 3\sqrt{2}$ (b) $\sqrt{12} \times \sqrt{20}$

Exercise 14.2(b)

Work out:

1. $\sqrt{12} \times \sqrt{27}$ 2. $\sqrt{20} \times \sqrt{32}$

3. $\sqrt{60} \times \sqrt{8}$ 4. $\sqrt{98} \times \sqrt{15}$

5. $\sqrt{15} \times \sqrt{27}$ 6. $\sqrt{50} \times \sqrt{30}$

7. $\sqrt{75} \times \sqrt{18}$ 8. $\sqrt{12} \times \sqrt{18}$

9. $\sqrt{27} \times \sqrt{72}$ 10. $\sqrt{30} \times \sqrt{2} \times \sqrt{27}$

11. $\sqrt{10} \times \sqrt{5} \times \sqrt{6}$ 12. $\sqrt{72} \times \sqrt{75} \times \sqrt{18}$

Rationalizing the Denominator

A fraction with a surd in the denominator can be expressed in the simplest form by removing the surd and then simplifying the fraction when possible. The process of removing the surd from the denominator is called rationalizing the denominator

To rationalize the denominator, we multiply both the numerator and denominator of the fraction by a number that will make the denominator a rational number.

Example

Rationalize the denominator of the fraction

(a) $\frac{4}{\sqrt{2}}$ (b) $\frac{3}{2\sqrt{3}}$

(a) Recall that, $\sqrt{a} \cdot \sqrt{a} = a$. We use this fact to rationalize the denominator.

$$\frac{4}{\sqrt{2}} = \frac{4}{\sqrt{2}} \times \frac{\sqrt{2}}{\sqrt{2}}$$

The fraction does not change because we have multiplied by 1

$$\frac{4}{\sqrt{2}} = \frac{4\sqrt{2}}{2} = 2\sqrt{2}$$

(b) $\frac{3}{2\sqrt{3}} = \frac{3}{2\sqrt{3}} \times \frac{\sqrt{3}}{\sqrt{3}}$

$$= \frac{3\sqrt{3}}{2 \times 3}$$

$$= \frac{1}{2}\sqrt{3}$$

Try this 5

Rationalize the denominator of:

(a) $\frac{4}{\sqrt{3}}$ (b) $\frac{10}{\sqrt{5}}$

Exercise 14.2(c)

Rationalize the denominator of:

1. $\frac{9}{\sqrt{3}}$ 2. $\frac{1}{\sqrt{2}}$ 3. $\frac{18}{\sqrt{12}}$ 4. $\frac{\sqrt{20}}{2\sqrt{5}}$ 5. $\frac{5\sqrt{6}}{\sqrt{15}}$

6. $\frac{14}{\sqrt{7}}$ 7. $\frac{8}{\sqrt{2}}$ 8. $\frac{3}{4\sqrt{2}}$ 9. $\frac{5}{\sqrt{10}}$ 10. $\frac{8}{\sqrt{32}}$

11. $\frac{\sqrt{12}}{\sqrt{3}}$ 12. $\frac{\sqrt{15}}{\sqrt{6}}$ 13. $\frac{\sqrt{18}}{\sqrt{27}}$ 14. $\sqrt{\frac{28}{3}}$ 15. $\sqrt{\frac{15}{8}}$

14.3 Multiplication of brackets containing surds

We can multiply expressions in brackets containing surds. Expand the expression in much the same way as in the expansion of algebraic expression. Note that the usual rules of algebra also hold when multiplying brackets containing surds.

Examples

Simplify:

(a) $3\sqrt{2}(2\sqrt{2} + \sqrt{3})$ (b) $(3\sqrt{2} - \sqrt{3})(2\sqrt{3} - 3\sqrt{2})$

(a) $3\sqrt{2}(2\sqrt{2} + \sqrt{3}) = 3\sqrt{2} \cdot 2\sqrt{2} + 3\sqrt{2} \cdot \sqrt{3}$

$$= 6 \cdot 2 + 3 \cdot \sqrt{6}$$

$$= 12 + 3\sqrt{6}$$

(b) $(3\sqrt{2} - \sqrt{3})(2\sqrt{3} - 3\sqrt{2})$

$$= 3\sqrt{2} \cdot 2\sqrt{3} - 3\sqrt{2} \cdot 3\sqrt{2} - \sqrt{3} \cdot 2\sqrt{3} + \sqrt{3} \cdot 3\sqrt{2}$$

$$= 6\sqrt{6} - 18 - 6 + 3\sqrt{6}$$

$$= 9\sqrt{6} - 24$$

Try this 6

Simplify:

(a) $2\sqrt{5}(3\sqrt{5} - \sqrt{2})$ (b) $(2\sqrt{3} - 3)(\sqrt{3} + 4)$

Simplify:

(a) $\left(2\sqrt{3} + 3\sqrt{2}\right)\left(2\sqrt{3} - 3\sqrt{2}\right)$ (b) $\left(3 + 2\sqrt{3}\right)^2$

(a) $\left(2\sqrt{3} + 3\sqrt{2}\right)\left(2\sqrt{3} - 3\sqrt{2}\right) = \left(2\sqrt{3}\right)^2 - \left(3\sqrt{2}\right)^2$

$$= 4 \cdot 3 - 9 \cdot 2$$

$$= 12 - 18$$

$$= -6$$

(b) $\left(3 + 2\sqrt{3}\right)^2 = 3^2 + 2 \cdot 3 \cdot 2\sqrt{3} + \left(2\sqrt{3}\right)^2$

$$= 9 + 12\sqrt{3} + 12$$

$$= 21 + 12\sqrt{3}$$

Try this 7

Simplify: (a) $\left(2\sqrt{2} - 3\right)\left(2\sqrt{2} + 3\right)$ (b) $\left(2\sqrt{3} - 1\right)^2$

Given $\sqrt{3} = 1.732$, find the value of $\sqrt{3}(2 + \sqrt{3})$

First multiply and then substitute 1.732 for $\sqrt{3}$

$\sqrt{3}(2 + \sqrt{3}) = 2\sqrt{3} + 3$

$$= 2(1.732) + 3$$

$$= 3.464 + 3$$

$$= 6.464$$

Try this 8

Given $\sqrt{2} = 1.414$, find the value of $4\sqrt{2} - \dfrac{3}{\sqrt{2}}$

Exercise 14.3(a)

Simplify:

1. $\sqrt{3}(\sqrt{3} + \sqrt{5})$

2. $\sqrt{2}(\sqrt{6} - \sqrt{2})$

3. $5\sqrt{2}(3\sqrt{2} - 2\sqrt{3})$

4. $2\sqrt{12}(\sqrt{3} + \sqrt{2})$

5. $(3 + \sqrt{2})(3 - \sqrt{2})$

6. $(\sqrt{12} - 3)(\sqrt{12} + 3)$

7. $(\sqrt{5} + 3\sqrt{2})(\sqrt{5} - 3\sqrt{2})$

8. $(2\sqrt{3} + 4\sqrt{2})(2\sqrt{3} - 4\sqrt{2})$

9. $(3 - \sqrt{2})^2$

10. $(\sqrt{3} + \sqrt{2})^2$

11. $(\sqrt{3} - 2\sqrt{2})^2$

12. $(\sqrt{6} - 2\sqrt{3})^2$

13. $(2 + 3\sqrt{2})^2$

14. $(3\sqrt{2} - 2\sqrt{3})^2$

15. $(3\sqrt{2} - \sqrt{6})(2\sqrt{3} + \sqrt{2})$

16. $(3\sqrt{2} + 2\sqrt{3})(5\sqrt{2} - \sqrt{3})$

17. $(2\sqrt{2} + \sqrt{3})(\sqrt{12} - \sqrt{8})$

18. $(1 - \sqrt{3})(2 + 3\sqrt{2})$

19. $2\sqrt{3}\left(4\sqrt{3} - \frac{8}{\sqrt{3}}\right)$

20. $3\sqrt{2}\left(2\sqrt{50} - \frac{20}{\sqrt{8}}\right)$

Given $\sqrt{2} = 1.414$ and $\sqrt{3} = 1.732$ find the value of:

21. $\sqrt{3}(4 + \sqrt{3})$

22. $\sqrt{2}(5\sqrt{2} - 3)$

23. $2\sqrt{3}\left(1 + \frac{6}{\sqrt{27}}\right)$

24. $(3 + \sqrt{2})^2$

25. $(3 - \sqrt{2})(1 + 2\sqrt{2})$

26. $(2 - \sqrt{3})^2$

27. $(2 - 3\sqrt{2})(\sqrt{2} - 3)$

28. $3\sqrt{2} - \frac{5}{\sqrt{2}}$

29. $2 + \frac{3}{\sqrt{12}}$

30. $(\sqrt{6} - \sqrt{2})(\sqrt{6} + 2\sqrt{2})$

Rationalizing fractions whose denominators are sums or difference of surds

Using the identity $(a - b)(a + b) = a^2 - b^2$, we can rationalize the denominators of fractions with denominators like $\sqrt{a} + b$ or $\sqrt{a} - \sqrt{b}$. For example, you can rationalize a fraction whose denominator is $\sqrt{3} + 2$, by multiplying both the numerator and denominator by $\sqrt{3} - 2$. The number $\sqrt{3} - 2$ is called the conjugate of $\sqrt{3} + 2$. Note that $\sqrt{3} + 2$ is the conjugate of $\sqrt{3} - 2$.

To rationalize the denominator involving such surds, multiply the numerator and the denominator by the conjugate of the surd in the denominator.

Example

Rationalize the denominator of $\dfrac{1}{2-\sqrt{3}}$

Multiply both the denominator and numerator by $2 + \sqrt{3}$

$$\frac{1}{2 - \sqrt{3}} = \frac{1}{2 - \sqrt{3}} \times \frac{2 + \sqrt{3}}{2 + \sqrt{3}}$$

$$= \frac{2+\sqrt{3}}{(2-\sqrt{3})(2+\sqrt{3})}$$

$$= \frac{2+\sqrt{3}}{2^2-(\sqrt{3})^2}$$

$$= \frac{2+\sqrt{3}}{4-3}$$

$$= 2 + \sqrt{3}$$

Try this 9

Rationalize the denominator of

$$\frac{3}{\sqrt{5}+2}$$

Exercise 14.3(b)

Rationalize the denominator of the following

1. $\dfrac{1}{\sqrt{2}-1}$ 　　2. $\dfrac{3}{2-\sqrt{3}}$ 　　3. $\dfrac{4}{\sqrt{5}-1}$

4. $\dfrac{2\sqrt{3}}{2-\sqrt{3}}$ 　　5. $\dfrac{4\sqrt{2}}{\sqrt{2}-\sqrt{3}}$ 　　6. $\dfrac{1}{1-2\sqrt{3}}$

7. $\dfrac{2-\sqrt{3}}{2+\sqrt{3}}$ 　　8. $\dfrac{\sqrt{3}-1}{2\sqrt{3}+1}$ 　　9. $\dfrac{3-2\sqrt{5}}{\sqrt{5}-2}$

10. $\dfrac{\sqrt{5}-2\sqrt{3}}{\sqrt{5}-\sqrt{3}}$

Review exercise 14

Simplify:

1. $\sqrt{60}$ 　　2. $\sqrt{45}$ 　　3. $\sqrt{18}$

4. $\sqrt{150}$ 　　5. $\sqrt{192}$ 　　6. $\sqrt{245}$

7. $\sqrt{216}$ 　　8. $\sqrt{243}$ 　　9. $\sqrt{384}$

10. $\sqrt{882}$

Simplify:

11. $2\sqrt{2} + 4\sqrt{2}$

12. $7\sqrt{3} + \sqrt{3}$

13. $5\sqrt{3} - 3\sqrt{3}$

14. $6\sqrt{5} - 7\sqrt{5}$

15. $3\sqrt{5} - 7\sqrt{5} + 9\sqrt{5}$

16. $6\sqrt{2} + 5\sqrt{2} - 12\sqrt{2}$

17. $4\sqrt{3} + \sqrt{12}$

18. $\sqrt{75} - \sqrt{48}$

19. $2\sqrt{20} + \sqrt{45}$

20. $2\sqrt{75} + 4\sqrt{12} - 2\sqrt{108}$

Simplify

21. $\sqrt{18} \times \sqrt{6}$

22. $\sqrt{8} \times \sqrt{10}$

23. $\sqrt{15} \times \sqrt{8}$

24. $\sqrt{20} \times \sqrt{27}$

25. $\sqrt{75} \times \sqrt{50}$

26. $\sqrt{12} \times \sqrt{45}$

27. $\sqrt{32} \times \sqrt{80}$

28. $\sqrt{72} \times \sqrt{27}$

29. $\sqrt{12} \times \sqrt{3} \times \sqrt{6}$

30. $2\sqrt{5} \times \sqrt{20} \times \sqrt{45}$

Rationalize the denominator of:

31. $\dfrac{4}{\sqrt{6}}$

32. $\dfrac{12}{\sqrt{10}}$

33. $\dfrac{8}{\sqrt{12}}$

34. $\dfrac{9}{\sqrt{27}}$

35. $\dfrac{15}{\sqrt{45}}$

36. $\dfrac{\sqrt{8}}{\sqrt{6}}$

37. $\dfrac{\sqrt{10}}{\sqrt{75}}$

38. $\dfrac{2\sqrt{8}}{\sqrt{72}}$

39. $\sqrt{\dfrac{4}{3}}$

40. $\sqrt{\dfrac{25}{8}}$

Simplify:

41. $\sqrt{2}\left(4\sqrt{2} + \sqrt{3}\right)$

42. $2\sqrt{3}\left(5\sqrt{2} - \sqrt{6}\right)$

43. $\sqrt{3}\left(2\sqrt{6}-\sqrt{3}\right)$ 44. $\left(\sqrt{3}+2\sqrt{2}\right)\left(2\sqrt{3}-3\sqrt{2}\right)$

45. $\left(\sqrt{5}-2\right)\left(2\sqrt{5}-2\right)$ 46. $\left(2\sqrt{7}-\sqrt{3}\right)\left(\sqrt{7}+2\sqrt{3}\right)$

47. $\left(2\sqrt{3}+3\right)\left(2\sqrt{3}-3\right)$ 48. $\left(5\sqrt{2}-7\right)\left(5\sqrt{2}+7\right)$

49. $\left(2\sqrt{5}+3\right)^2$ 50. $\left(3\sqrt{6}-\sqrt{3}\right)^2$

Given that $\sqrt{2} = 1.414$, $\sqrt{3} = 1.732$ and $\sqrt{5} = 2.236$ evaluate

51. $\sqrt{5}\left(3+\sqrt{5}\right)$ 52. $3\sqrt{2}\left(4-2\sqrt{6}\right)$

53. $\left(\sqrt{5}-\sqrt{3}\right)\left(\sqrt{15}+3\right)$ 54. $\left(4+\sqrt{5}\right)^2$

55. $\left(\sqrt{45}-2\right)^2$

Rationalize the denominator of:

56. $\dfrac{3}{2+\sqrt{5}}$ 57. $\dfrac{2}{3-2\sqrt{3}}$ 58. $\dfrac{\sqrt{3}}{2-\sqrt{3}}$

59. $\dfrac{\sqrt{5}}{\sqrt{5}+\sqrt{3}}$ 60. $\dfrac{\sqrt{10}+\sqrt{6}}{\sqrt{10}-\sqrt{6}}$

Chapter Test 14

Take this test as you would take a test in a class. After you are done, check your work against the answers in the back of the book.

1. Simplify:

    (a) $\sqrt{96}$ (b) $\sqrt{320}$

2. Work out:

    (a) $2\sqrt{5}+3\sqrt{5}-7\sqrt{5}$

(b) $2\sqrt{8} - 3\sqrt{50} + 5\sqrt{18}$

3. Work out:

(a) $\sqrt{32} \times \sqrt{27}$

(b) $\sqrt{3} \times \sqrt{5} \times \sqrt{30}$

4. Rationalize the denominator of:

(a) $\dfrac{9}{4\sqrt{6}}$

(b) $\dfrac{3\sqrt{7}}{\sqrt{21}}$

5. Rationalize the denominator of $\dfrac{2+\sqrt{3}}{\sqrt{3}}$, and write your answer in

the form $a + b\sqrt{3}$ where a and b are rational numbers.

6. Simplify:

(a) $\sqrt{30}\left(\sqrt{27} - \dfrac{1}{\sqrt{3}}\right)$

(b) $\left(\sqrt{3} + 2\sqrt{2}\right)\left(3\sqrt{5} - 5\sqrt{2}\right)$

7. Simplify:

(a) $\left(\sqrt{5} - \sqrt{3}\right)\left(\sqrt{5} + \sqrt{3}\right)$

(b) $\left(5\sqrt{3} - 2\right)^2$

8. Find p if:

(a) $\sqrt{2} - \dfrac{3}{\sqrt{2}} = p\sqrt{2}$

(b) $\left(p\sqrt{3} + \sqrt{2}\right)\left(p\sqrt{3} - \sqrt{2}\right) = 10$

9. Given that $\sqrt{2} = 1.414$ and $\sqrt{3} = 1.732$, evaluate

(a) $\sqrt{2}(5 - \sqrt{2})$ (b) $2\sqrt{3} + \dfrac{5}{\sqrt{3}}$

10. Rationalize the denominator of:

(a) $\dfrac{3\sqrt{2}}{2\sqrt{2}-3}$ (b) $\dfrac{\sqrt{8}+3}{\sqrt{18}+2}$

15

Sequences

A sequence is an ordered list of numbers. Two such sequences are the arithmetic progression and the geometric progression. Each number in a sequence is called a term of the sequence. A sequence can have a finite number of terms or an infinite number of terms. For example, the sequence

$$5, 7, 9, 11, 13, 15$$

is a finite sequence, whereas the sequence

$$7, 14, 28, 56, 112, \ldots$$

is an infinite sequence. Note that the three dots indicate that the sequence continues and has an infinite number of terms.

Series

The sum of the terms of a sequence is called a series. For example, $21 + 23 + 25 + 27$ is a finite series and $5 + 8 + 11 + 14 + 17 + \cdots$ is an infinite series.

15.1 Arithmetic Progression

A sequence is called an arithmetic progression when each term after the first is formed by adding (or subtracting) a constant number, called the common difference to the preceding term. For example, the sequence $3, 5, 7, 9, 11$ is an arithmetic progression obtained by adding 2 to each preceding term, and the sequence $5, 2, -1, -4, \ldots$ is an arithmetic progression obtained by subtracting 3 from each preceding term.

Example

Find the common difference and determine the next three terms of the sequence

(a) $5, 13, 21, 29, 37,\ ...$ (b) $27, 24, 21, 18, 15,\ ...$

(a) You can find the common difference by subtracting the first term of any two consecutive terms from the second term. Taking the first and second terms, we have

$$13 - 5 = 8$$

The common difference is 8

Notice that the difference between successive terms is 8.

The next three terms are $37 + 8 = 45,\ 45 + 8 = 53$ and $53 + 8 = 61$, i.e. 45, 53, 61.

(b) $27, 24, 21, 18, 15,\ ...$

The common difference is $24 - 27 = -3$

The next three terms are $15 - 3 = 12, 12 - 3 = 9$ and $9 - 3 = 6$, i.e. 12, 9, 6.

Try this 1

Find the common difference and determine the next three terms of the sequence $23, 18, 13, 8, \cdots$

Exercise 15.1(a)

1. Find the common difference of each of the following sequences

   (a) $5, 8, 11, 14, 17,\ ...$ (b) $-5, -1, 3, 7, 11,\ ...$

(c) $53, 47, 41, 35, 29, ...$

(d) $9, -3, -15, -27, -39, ...$

(e) $9, 13, 17, 21, 25, \ ...$

(f) $13, 11, 9, 7, 5, ...$

2. Write down the next three terms of the following sequences

(a) $9, 12, 15, 18, \ ...$

(b) $3, 9, 15, 21, \ ...$

(c) $10, 5, 0, -5, \ ...$

(d) $7, 13, 19, 25, \ ...$

(e) $7, -2, -11, -20, ...$

(f) $11, 15, 19, 23, \ ...$

The n th Term of an Arithmetic Progression

You may use u_1 to represent the first term of a sequence, u_2 the second term, u_3 the third term and so on to the n th term u_n.

Recall that successive terms of an arithmetic progression differ by the same number. Hence if the first term of an arithmetic progression is a and the common difference is d then the second term is $a + d$. The third term is obtained by adding d to the second, the fourth by adding d to the third term and so on. So

$$u_1 = a$$

$$u_2 = a + d$$

$$u_3 == a + 2d$$

$$u_4 == a + 3d$$

Following the pattern, the n th term u_n will have the form $u_n = a + (n - 1)d$, where n represent the number of terms of the sequence.

Examples

Calculate the tenth term of the sequence $13, 9, 5, 1, -3, \ldots$

Begin by finding the common difference.

$d = 9 - 13 = -4$

Here $a = 13, d = -4$ and $n = 10$

Using the formula $u_n = a + (n - 1)d$, we have

$u_{10} = 13 + (10 - 1) \times -4$

$u_{10} = 13 + 9(-4)$

$= -23$

The tenth term is -23

Try this 2

Calculate the twentieth term of the sequence $2, 5, 8, 11, 14, \ldots$

Find the number of terms in the sequence $8, 17, 26 \ldots, 143$

Begin by finding the common difference.

$d = 17 - 8 = 9$

Here $a = 8, d = 9$ and $u_n = 143$

Using the formula $u_n = a + (n - 1)d$, we have

$8 + (n - 1) \times 9 = 143$

$9(n - 1) = 135$

$n = 16$

Thus, the number of terms is 16.

Try this 3

Find the number of terms of the sequence. $12, 17, 22, 27, \dots, 82$.

Find the first three terms of the arithmetic progression having $u_6 = 20$ and $u_{10} = 32$

Using the definition of arithmetic progression and the fact that u_6 and u_{10} are four places apart, we have $u_{10} = 20 + 4d$, where d is the common difference.

So $20 + 4d = 32$

$$4d = 12$$

$$d = 3$$

If a is the first term of the sequence then

$$u_6 = a + 5(3) = a + 15$$

Hence $a + 15 = 20$

$$a = 5$$

This result can also be obtained in the following way:

$u_6 = a + 5d$ and $u_{10} = a + 9d$, so

$$a + 5d = 20 \qquad (1)$$

$$a + 9d = 32 \qquad (2)$$

Subtracting equation (1) from equation (2) gives

$$4d = 12$$

$$d = 3$$

Substituting 3 for d in equation (1), we have

$$a + 15 = 20$$

$$a = 5$$

So the first term is 5, the second term is $5 + 3 = 8$ and the third term is $8 + 5 = 13$

Try this 4

Find the first three terms of the arithmetic progression having $u_7 = -8$ and $u_{12} = -23$

Exercise 15.1(b)

1. Write down the terms indicated in each of the following sequences

   (a) $3, 11, 19, 27, \ldots$ 15 th   (b) $10, 7, 4, 1, \ldots$ 24 th

   (c) $2, 3\frac{1}{2}, 5, 6\frac{1}{2}, \ldots$ 18 th   (d) $61, 59, 57, 55, \ldots$ 30 th

   (e) $9, 8\frac{1}{2}, 8, 7\frac{1}{2}, \ldots$ 21 st   (f) $9, -3, -15, -27, \ldots$ 10 th

3. For each of the following sequences write an expression for the n th term and calculate the indicated term of the sequence

   (a) $7, 10, 13, 16, 19, \ldots$ 20 th term

   (b) $4, 8, 12, 16, 20, \ldots$ 25 th term

   (c) $5, 8, 11, 14, 17, \ldots$ 12 th term

   (d) $2, 9, 16, 23, 30, \ldots$ 21 st term

(e) $17, 15, 13, 11, 9, \ldots$ 10 th term

(f) $5, 7, 9, 11, 13, \ldots$ 16 th term

3. Find the number of terms in the following sequence

(a) $4, 6, 8, \ldots, 48$

(b) $49, 46, 43, \ldots, 7$

(c) $4, -7, -18, \ldots, -128$

(d) $2, 4\frac{1}{2}, 7, \ldots, 102$

(e) $11, 10, 9, \ldots, -20$

(f) $13, 11, 9, \ldots, -17$

4. Find the first three terms of the arithmetic progression having the terms indicated

(a) $u_4 = 14, \quad u_8 = 26$

(b) $u_5 = 25, \quad u_9 = 41$

(c) $u_6 = 3, \quad u_{11} = -7$

(d) $u_7 = 27, \quad u_{10} = 36$

(e) $u_8 = 49, \quad u_{10} = 73$

(f) $u_9 = 9, \quad u_{12} = 3$

Sum of Arithmetic Progression

If the first term of an arithmetic sequence is a, the common difference is d, and the n th term is u_n, then the sum, S_n, of the first n terms is given by

$$S_n = a + (a + d) + (a + 2d) + (a + 3d) + \cdots + u_n \qquad (1)$$

Writing S_n backward, we have

$$S_n = u_n + (u_n - d) + (u_n - 2d) + (u_n - 3d) + \cdots + a \qquad (2)$$

Adding (1) and (2), gives

$$2S_n = (a + u_n) + (a + u_n) + (a + u_n) + (a + u_n) + \cdots + (a + u_n)$$

$$= n(a + u_n)$$

Dividing each side by 2 yields

$$S_n = \frac{n}{2}(a + u_n) \tag{3}$$

But $u_n = a + (n - 1)d$, so we have

$$S_n = \frac{n}{2}[a + a + (n - 1)d]$$

$$= \frac{n}{2}[2a + (n - 1)d] \tag{4}$$

Examples

Calculate the sum of the arithmetic sequence $4 + 8 + 12 + \cdots + 36$

First find the number of terms of the sequence.

Here $a = 4$, $d = 4$ and $u_n = 36$

Using $u_n = a + (n - 1)d$, we have

$$4 + (n - 1) \times 4 = 36$$

$$n - 1 = 8$$

$$n = 9$$

Next, using $S_n = \frac{n}{2}(a + u_n)$, we have

$$S_9 = \frac{9}{2}(4 + 36)$$

$$= \frac{9}{2}(40)$$

$$= 180$$

Try this 5

Calculate the sum of the arithmetic sequence $5 + 10 + 15 + \cdots + 80$

Calculate the sum of the first twelve terms of the sequence $7 + 10 + 13 + \cdots$

Here $a = 7, d = 3$ and $n = 12$

Using $S_n = \frac{n}{2}[2a + (n - 1)d]$ gives

$$S_{12} = \frac{12}{2}[2(7) + (12 - 1) \times 3]$$

$$= 6(14 + 33)$$

$$= 6(47)$$

$$= 282$$

Try this 6

Calculate the sum of the first twenty terms of the arithmetic sequence $15 + 13 + 11 + \cdots$

Application of Arithmetic Sequence

Example

A theatre has 70 seats in the first row, 78 seats in the second row, 86 seats in the third row, and so on in the same increasing pattern. If the theatre has 20 rows of seats, how many seats are in the theatre?

The seating pattern forms the arithmetic progression $70, 78, 86,\ \ldots$

Using the formula for the sum, you can find the number of seats in the theatre

Here $a = 70, d = 8$ and $n = 20$

So, $S_{20} = \frac{20}{2}[2(70) + (20 - 1) \times 8]$

$\qquad = 10(140 + 152)$

$\qquad = 2920$

Thus, there are 2920 seats in the theatre

Try this 7

Kweku saved 5 Gp on the first day of June, 10 Gp on the second day, 15 Gp on the third day, and so on in the same increasing pattern. How much did he save in June?

Exercise 15.1(c)

1. Calculate the sum of each of the following arithmetic sequences

   (a) $7 + 9 + 11 + \cdots + 105$

   (b) $13 + 18 + 123 + \cdots + 88$

   (c) $3 + 5\frac{1}{2} + 8 + \cdots + 103$

   (d) $-7 - 4 - 1 + \cdots + 53$

   (e) $73 + 69 + 65 + \cdots - 51$

   (f) $12 + 8\frac{1}{2} + 5 + \cdots - 58$

2. Calculate the sum of each of the following arithmetic sequences as far as the term indicated

    (a) $9 + 15 + 21 + \cdots$ 12 th

    (b) $19 + 17 + 15 + \cdots$ 20 th

    (c) $8 + 9 + 10 + \cdots$ 50 th

    (d) $17 + 10 + 3 + \cdots$ 16 th

    (e) $3 + 4\frac{1}{2} + 6 + \cdots$ 12 th

    (f) $7 + 4\frac{1}{2} + 2 + \cdots$ 15 th

3. The first term of an arithmetic sequence is -1 and the fifteenth term is 27. Find the common difference and the sum of the first fifteen terms

4. The fourth term of an arithmetic sequence is 10 and the sixth term is 16. Find the first term, the common difference and the sum of the first twenty terms

5. A man's starting pay was GH¢2500, increasing at equal annual amount to GH¢8940 in 15 years. How much increase did he receive each year?

6. A theatre has 20 seats in the first row. Each successive row has two more seats than the previous row. How many seats are in the twelfth row?

7. Logs are stacked in a pile so that the top row has 15 logs and the bottom row has 21 logs. If the pile has 7 rows, how many logs are in the pile?

8. Kwesi purchased 20 books. The first book cost GH¢ 10, the second book cost GH¢15, the third book cost GH¢20, the fourth book cost GH¢25 and so on. How much did he pay for the 20 books?

9. Kofi saved 20 Gp on the first day of March, 25 Gp on the second day, 30 Gp on the third day, and so on. Find the total amount that he saved during the month.

10. Ama earns 25 Gp on the first day of the month, 50 Gp on the second day, 75 Gp on the third day, and so on. Find the total amount that she will earn during a 30 -day month.

15.2 Geometric Progression

A sequence is called a geometric progression when each term after the first is formed by multiplying (or dividing) the preceding term by a constant number, called the common ratio. For example, the sequence $1, 3, 9, 27, 81, \ldots$ is a geometric progression obtained by multiplying each preceding term by 3, and $32, 16, 8, 4, 2, \ldots$ is a geometric progression obtained by dividing each preceding term by 2.

Example

Find the common ratio and the next three terms of the sequence $3, 12, 48,$ ...

You can find the common ratio by dividing any term of the sequence by the preceding term.

Taking the first and the second terms, we have

$$\frac{12}{3} = 4$$

The common ratio is 4.

The next three terms are $48(4) = 192, 192(4) = 768$ and $768(4) = 3072$ i.e. $192, 768, 3072$

Try this 8

Find the common ratio and the next two terms of the sequence $36, 12, 4,$ ...

The u_n th Term of a Geometric Progression

If the first term of a geometric sequence is a, and the common ratio is r then the geometric progression has the form $a, ar, ar^2, ar^3,$ ...

$u_1 = a$

$u_2 = ar$

$u_3 = ar^2$

$u_4 = ar^3$

Following this pattern, the n th term will have the form $u_n = ar^{n-1}$

Examples

Find the eighth term of the geometric sequence $32, 16, 8, 4, \ldots$

Here $a = 32, r = \frac{1}{2}$ and $n = 8$

Using the formula $u_n = ar^{n-1}$, we have

$$u_8 = 32 \left(\frac{1}{2}\right)^7$$

$$= 32 \left(\frac{1}{128}\right)$$

$$= \frac{1}{4}$$

Try this 9

Find the tenth term of the geometric sequence $\frac{1}{2}, 1, 2, 4, 8, \ldots$

Given $u_4 = 6$ and $u_7 = 48$, of a geometric progression find the common ratio and the first term

The two terms are 3 places apart, so from the definition of a geometric sequence,

$$u_7 = (u_4)r^3$$

$$48 = 6r^3$$

$$8 = r^3$$

$$2 = r$$

This result can also be obtained in the following way

$$ar^3 = 6 \qquad (1)$$

$$ar^6 = 48 \qquad (2)$$

Dividing equation (2) by equation (1), gives

$$r^3 = 8$$

$$r = 2$$

Since $u_4 = ar^3$, we have

$$6 = a \times 2^3$$

$$6 = 8a$$

$$\frac{3}{4} = a$$

The common ratio is 2 and the first term is $\frac{3}{4}$

Try this 10

Given $u_3 = 18$ and $u_6 = 486$, of a geometric sequence find the common ratio and the first term

Exercise 15.2(a)

1. Find the common ratio of each sequence

(a) $9, 18, 36, 72, \ \ldots$

(b) $4, 12, 36, 108, \ \ldots$

(c) $-2, -0.2, -0.02, -0.002, \ldots$

(d) $\frac{1}{2}, \frac{1}{6}, \frac{1}{18}, \frac{1}{54}, \ \ldots$

(e) $-32, 16, -8, 4, \ \ldots$

(f) $\frac{3}{4}, 6, 48, 384, \ldots$

2. Find the next three terms of each of the following sequences

(a) $5, 10, 20, ...$ 　　　　　(b) $8, 4, 2, \ ...$

(c) $-6, 2, -\frac{2}{3}, \cdots$ 　　　　　(d) $1, 0.3, 0.09, \ ...$

(e) $1, \frac{1}{2}, \frac{1}{4}, \ ...$ 　　　　　(f) $10, 5, \frac{5}{2}, ...$

3. Write down the terms indicated in each of the following sequences

(a) $3, 6, 12, 24, \ ...$ 　　8 th 　　　(b) $6, 2, \frac{2}{3}, \frac{2}{9}, \ ...$ 　　　　7 th

(c) $8, 4, 2, 1, ...$ 　　　9 th 　　　(d) $4, -8, 16, -32, \ ...$ 　　10 th

(e) $3, 1, \frac{1}{9}, \frac{1}{27}, \ ...$ 　　6 th 　　　(f) $\frac{1}{32}, \frac{1}{16}, \frac{1}{8}, \frac{1}{4}, \ ...$ 　　　8 th

Sum of Geometric Progression

If a and r are the first term and the common ratio of a geometric sequence then the sum S_n of the first n terms is

$$S_n = a + ar + ar^2 + ar^3 + \cdots + ar^{n-1} \qquad (1)$$

Multiplying each side by r, gives

$$rS_n = ar + ar^2 + ar^3 + \cdots + ar^n \qquad (2)$$

Subtracting equation (2) from equation (1), gives

$$S_n - rS_n = a - ar^n$$

$$(1 - r)S_n = a(1 - r^n)$$

$$S_n = \frac{a(1-r^n)}{1-r} \qquad (3)$$

Multiplying both numerator and denominator by -1, we obtain

$$S_n = \frac{a(r^n-1)}{r-1} \qquad (4)$$

This form is more convenient if r is greater than 1.

Example

Calculate the sum of the first six terms of the geometric sequence $32 + 16 + 8 + 4 + \cdots$

Here, $a = 32, r = \frac{1}{2}$ and $n = 6$

Substituting in equation (3) gives

$$S_6 = \frac{32\left[1 - \left(\frac{1}{2}\right)^6\right]}{1 - \frac{1}{2}}$$

$$= 64\left[1 - \frac{1}{64}\right]$$

$$= 63$$

Try this 11

Find the sum of the first five terms of the geometric sequence $27, 9, 3, 1, \ldots$

Application of Geometric Sequence

Example

A man invested GH¢5000 at compound interest of 2% per annum. Find the value after 6 years.

The annual values form a geometric sequence with seven terms. Here the first term $a = 5000$ and the common ratio is $r = 1.02$.

Using the formula for the n th term of a geometric sequence we have:

The value after 6 years $= ar^n$

$$= 5000(1.02)^6$$

$$= 5630.81$$

The value after 6 years is GH¢5630.81

Try this 12

The value of a machine originally valued at GH¢6000 depreciates 12% per annum. Calculate its value after 5 years

Exercise 15.2(b)

1. Find the sum of each of the following sequences to the number

   of terms indicated:

   (a) $3 + 6 + 12 + 24 + \cdots$ 8 terms

   (b) $1 + 3 + 9 + 27 + \cdots$ 6 terms

   (c) $8 + 2 + \frac{1}{2} + \frac{1}{4} + \cdots$ 5 terms

   (d) $9 - 3 + 1 - \frac{1}{3} + \cdots$ 6 terms

   (e) $4 + 2 + 1 + \frac{1}{2} + \cdots$ 10 terms

   (f) $-3 + 6 - 12 + \cdots$ 8 terms

2. The first term of a geometric sequence is 2 and the seventh term

   is 1458. Find the common ratio and the sum of the first eight

   terms

3. The fifth and second terms of a geometric sequence are 27 and 8. Find the first term and the sum of the first five terms?

4. If the population of a town is 6000 and is increasing at 4% per annum. What will be the population in 5 years time?

5. The value of a car originally valued at GH¢18,500 depreciates 20% per annum. Calculate its value after 4 years

6. The value of a car originally valued at GH¢12,800 depreciates 15% per annum. If the car is sold when its value is GH¢1820 after how many years is the car sold?

7. A man invested GH¢3600 at compound interest of 5% per annum. Find the value after 8 years

8. If GH¢2500 is invested at compound interest at 6% per annum, find the time, correct to the nearest year, it takes to reach GH¢3984

Review exercise 15

1. Which of these sequences are arithmetic progressions?
   (a) $8, 10, 12, 14, \ ...$ (b) $7, 10, 15, 18, \ ...$

   (c) $10, 7, 4, 1, \ ...$ (d) $15, 17, 20, 24, ...$

   (e) $23, 20, 19, 14, \ ...$ (f) $-8, -10, -12, -14, - \cdots$

2. Which of these sequences are geometric progressions?
   (a) $7, 14, 28, 56, \ ...$ (b) $2, 6, 12, 24, \ ...$ (c) $16, 8, 4, 2, \ ...$

(d) $27, 9, 6, 1,$ ...   (e) $-2, 6, -18, 54,$ ...   (f) $8, 2, \frac{1}{2}, \frac{1}{8},$ ...

3. Write the next three terms of each of the following sequences

(a) $5, 9, 13,$ ...   (b) $12, 7, 2,$ ...   (c) $2, 5, 8,$ ...

(d) $7, 5, 3,$ ...   (e) $-12, -7, -2, ...$ (f) $11, 9, 7,$ ...

4. Write the next three terms of each of the following sequences

(a) $3, 6, 12,$ ...   (b) $25, 5, 1,$ ...   (c) $-4, 8, -16, ...$

(d) $16, -8, 4,$ ...   (e) $\frac{1}{32}, \frac{1}{8}, \frac{1}{2},$ ...   (f) $9, 3, 1,$ ...

5. Write down the term indicated in each of the following

   sequences

(a) $5, 13, 21, 29,$ ...   12 th   (b) $11, 8, 5, 2, ...$ 25 th

(c) $20, 17, 14, 11,$ ...   20 th   (d) $5, 10, 15, 20,$ ...   15 th

(e) $-10, -8, -6, -4, ...$ 18 th   (f) $7, 11, 15, 19 , ...$ 16 th

6. Write down the term indicated in each of the following

   sequences

(a) $-1, 3, -9,$ ... 7 th   (b) $18, 9, 4.5,$ .... 8 th

(c) $8, -4, 2,$ ... 9 th   (d) $\frac{1}{9}, \frac{1}{3}, 1, ...$ 6 th

(e) $2, 8, 32,$ ....   5 th   (f) $27, 9, 3,$ ... 6 th

7. Find the number of terms in each of the following sequences

(a) $5, 7, 9, ..., 49$ (b) $50, 47, 44, ..., 14$

(c) $5, -6, -17, ..., -127$ (d) $-8, -4, 0, ..., 48$

(e) $9, 8, 7, \dots, -21$ (f) $8, 13, 18, \dots, 103$

8. Find the number of terms in each of the following sequences

(a) $1, 2, 4, \ \dots, 128$ (b) $3, 1, \frac{1}{3}, \dots, \frac{1}{729}$ (c) $16, 8, 4, \ \dots, \frac{1}{64}$

(d) $\frac{1}{16}, \frac{1}{8}, \frac{1}{4}, \ \dots, 32$ (e) $\frac{1}{9}, \frac{1}{3}, 1, \ \dots, 243$ (f) $9, 6, 4, \ \dots, \frac{64}{81}$

9. Calculate the sum of each of the following sequences

(a) $3 + 5 + 7 + \cdots + 101$ (b) $8 + 13 + 18 + \cdots + 83$

(c) $12 + 14\frac{1}{2} + 17 + \cdots + 112$ (d) $-4 - 1 + 2 + \cdots + 56$

(e) $74 + 70 + 66 + \cdots - 50$ (f) $13 + 10 + 7 + \cdots - 44$

10. Calculate the sum of each of the following sequences as far as the term indicated

(a) $7 + 13 + 19 + \cdots$ 12 th   (b) $4 + 9 + 4 + \cdots$ 20 th

(c) $15 + 9 + 3 + \cdots$ 10 th   (d) $20 + 13 + 6 + \cdots$ 16 th

(e) $5 + 7\frac{1}{2} + 10 + \cdots$ 12 th   (f) $-12 - 4 + 4 + \cdots$ 15 th

11. Calculate the sum of each of the following sequences to the number of terms indicated

(a) $1 + 2 + 4 + \cdots$ 8 terms   (b) $27 + 9 + 3 + \cdots$ 6 terms

(c) $9 + 12 + 16 + \cdots$ 5 terms   (d) $-8 + 4 - 2 + \cdots$ 6 terms

(e) $6 + 9 + \frac{27}{2} + \cdots$ 7 terms   (f) $-3 + 6 - 12 + \cdots$ 8 terms

12. The third term of an arithmetic progression is 18 and the seventh term is 30. Find the common difference, the first term

and the sum of the first ten terms

13. The fourth term of an arithmetic progression is 7 and the common difference is -2. Find the first term and the sum of the first twenty terms

14. Find the sum of the even numbers between 99 and 199

15. The second term of a geometric progression is 27 and the fifth term is 1. Find the common ratio, the first term and the sum of the first six terms

16. The third term of a geometric progression is 8 and the fifth term is 2. Find two possible values of the common ratio and the second term in each case

17. A man accepted a job advertised at a starting salary of GH¢7830 with an annual increments of GH¢250

    (a) Find his salary in the tenth year

    (b) Calculate the total salary he will have received in the first 15 years

18. A theatre has 26 seats in the first row. Each successive row has two more seats than the previous row.

    (a) How many seats are in the twelfth row?

    (b) How many seats are in the theatre, if the theatre has twenty rows of seats?

19. If GH¢12,500 is invested at compound interest of 5% per annum find the value after 12 years

20. The value of a machine originally valued at GH¢15,800 depreciates 8% per annum. Calculates its value after 5 years

Chapter Test 15

Take this test as you would take a test in class. After you are done, check your work against the answers in the back of the book.

1. Write down the next three terms and the term indicated in each sequence

    (a) $20, 15, 10, \ldots \ 20$ th   (b) $40, 20, 10, \ldots \ 9$ th

2. For each of the following sequences write an expression for the n th term

    (a) $3, 5, 7, 9, 11, \ldots$ (b) $1, 4, 9, 16, 25, \ldots$

3. The n th term of a sequence is $2n^2 - 1$

    (a) Write down the first five terms of this sequence

    (b) Calculate the 10 th term of the sequence

4. Find the number of terms in each sequence

    (a) $7, 10, 13, \ldots, 55$ (b) $81, 27, 9, \ldots , \frac{1}{27}$

5. Calculate the sum of each sequence

    (a) $3 + 6 + 9 + \cdots + 99$ (b) $12 + 8 + 4 + \cdots \ 25$ th

6. Calculate the sum of each of the following sequences to number of terms indicated.

(a) $5, 10, 20, 40, \ldots$ 10 terms   (b) $12, 4, \dfrac{4}{3}, \ldots$ 6 terms

7. The fourth term of an arithmetic progression is 93 and the eighth term is 65. Find the first term and the common difference

8. The third term of a geometric progression is 12 and the sixth term is 96. Find the common ratio and the first term

9. Find the sum of the odd numbers between 10 and 50

10. A theatre has 80 seats in the first row, 86 seats in the second row, 92 seats in the third row, and so on in the same increasing pattern. If the theatre has twenty rows of seats, how many seats are in the theatre?

11. On commencing employment a man is paid a salary of GH¢8500 per annum and received annual increments of GH¢450. Find his salary in the 9 th year and calculate the total amount he will have received in the first 12 years

12. If GH¢450 is invested at compound interest of 6% per annum, find:

(a) the value after 12 years

(b) the time, correct to the nearest year, it takes to reach GH¢720

16

Ratio, Rate and Percentage

16.1 Ratios

We use ratios to make comparisons between two quantities. If Kwesi and Esi are 6 years and 2 years old respectively, we can compare their ages in two ways. We can say that Kwesi is 4 years older than Esi or we can say that Kwesi is three times as old as Esi. When we compare two quantities by stating how many times more one quantity is of the other we are using ratio. The ratio of the two ages is 6 to 2, written as 6 : 2.

A ratio can be written as a fraction with the numerator equal to the first quantity and the denominator equal to the second quantity. So the ratio 6 to 2 can be written as the fraction $\frac{6}{2}$.

The two quantities in a ratio must be the same, so, if the ratio involves units of measure, you must ensure that the units are the same.

Simplification of Ratios

You can simplify ratios the same way you simplify fractions. To simplify a ratio divide both numbers by their highest common factor. For example, the ratio 8 : 12 becomes 2 : 3, when you divide by 4. The ratio 2 : 3 is the simplest form of the ratio 8 : 12. We say that these two ratios are equivalent.

Example

Simplify each ratio

(a) 20 : 12   (b) 20 cm : 3 m   (c) $\frac{2}{3} : \frac{3}{4}$

(a) $20 : 12 = 5 : 3$ Divide each number by 4

(b) 250 cm : 3 m

Begin by expressing the ratio in the same unit. The 3 m must be changed to 300 cm

Thus, $250\ cm : 3\ m = 250\ cm : 300\ cm = 5 : 6$

Notice that the final ratio does not have units.

(c) $\frac{2}{3} : \frac{3}{4}$

First, find the least common multiply of 3 and 4, which is 12 and then multiply each number by 12

$\frac{2}{3} \times 12 : \frac{3}{4} \times 12 = 8 : 9$

Try this 1

Simplify each ratio

(a) 15 : 12   (b) 25 s : 1 min   (c) $\frac{2}{5} : \frac{3}{4}$

Exercise 16.1(a)

1. Write the following ratios in their simplest form:

(a) 16 : 20   (b) 14 : 18   (c) 32 : 24   (d) 1.8 : 1.5

2. Write the following ratios in their simplest form:

(a) $1\frac{1}{2} : 2$ (b) $7\frac{1}{2} : 12\frac{1}{2}$ (c) $\frac{5}{6} : \frac{3}{2}$ (d) $5\frac{1}{4} : 3\frac{1}{2}$

3. Write the following ratios in their simplest form:

(a) 15 cm : 25 mm   (b) 45 Gp : GH₵2

(c) 450 kg : 1 tonne   (d) $1\frac{1}{4}$ hr : 45 min

4. A class of 30 students has 18 boys. Find the ratio of girls to boys

5. A car travels 6 km in $\frac{1}{3}$ hr and a train travels 8 km in 40 min. Find the ratio of the speed of the car to that of the train

6. A football team has won 15 matches and lost 10. What is the ratio of the number of matches won to the number of games played?

7. A man sleeps $7\frac{1}{2}$ hours every day. Find the ratio of the time he is awake to a day

8. $\frac{3}{5}$ of the number of candidates who sat the WASSCE examination in a school passed in all the subjects. Find the ratio of the number of candidates who did not pass all the subject to the number of candidates

9. 25 students took a test in mathematics and physics. 15 took the test in mathematics and 20 took the test in physics. Find the ratio of the number of student who took both test to the number of students

10. An empty can weighs 2.5 gm. 37.5 gm of sugar is put into the can. Find the ratio of the weight of the sugar to the total weight of the sugar and can

Sometimes it is useful to write a ratio in the form $1:n$ or $n:1$, where n is any number.

Expressing a given ratio in the form $1 : n$

We need to divide the ratio by a number that will make the left-hand side of the ratio equal to 1.

Example

Express the ratio 2 : 6 in the form 1 : n

$2 : 6 = 1 : \dfrac{6}{2}$ Divide both numbers by 2

$ = 1 : 3$

Try this 2

Express the ratio 4 : 12 in the form 1 : n

Expressing a given ratio in the form $m : 1$

We need to divide the ratio by a number that will make the right-hand side of the ratio equal to 1.

Example

Express the ratio 9 : 6 in the form $m : 1$

$9 : 6 = \dfrac{9}{6} : 1$ Divide both numbers by 6

$ = \dfrac{3}{2} : 1$

Try this 3

Express 20 : 12 in the form $m : 1$

Exercise 16.1(b)

1. Express the ratio in the form $1 : n$

   (a) $12 : 36$ (b) $16 : 6$ (c) $4 : 18$

   (d) $20 : 12$ (e) $21 : 27$

2. Express the ratio in the form $m : 1$

   (a) $18 : 9$ (b) $10 : 4$ (c) $28 : 21$

   (d) $35 : 14$ (e) $20 : 25$

Proportion

A proportion is a statement that two given ratios are equal

Example

Given that $x : 20 = 3 : 4$, find x

Express each as a fraction and then solve for x

$$\frac{x}{20} = \frac{3}{4}$$

$$x = \frac{3}{4} \times 20$$

$$x = 15$$

Try this 4

Given $x : 15 = 27 : 45$

Use of Proportion

You can use   proportion to solve problems involving ratios. The examples below illustrate the use of proportion in solving everyday problem.

Direct proportion

Workers in a farm are paid according to how many oranges they picked. If Kweku picks twice as many oranges as Mensah, his wage will be twice. The wage paid to them is said to be in direct proportion to the number of oranges picked. In general, if one quantity increases as other increases the two quantities are said to be in direct proportion.

There are two methods for solving problems involving direct proportion as illustrated below.

Example

A school orders 3 cartons of milk for every 10 students. If there are 1200 students in the school, how many cartons of milk should be ordered?

The ratio method

Set up a proportion using the ratio of cartons to students and solve the equation for an unknown quantity in one of the ratios.

Let n be the unknown number of cartons. Then

$$\frac{n}{1200} = \frac{3}{10}$$

Multiplying each side by 1200, we have

$$n = \frac{3}{10} \times 1200$$

$n = 360$

The school should order 360 cartons of milk

The unitary method

10 students will be given 3 cartons

1 student will be given $\frac{3}{10}$ cartons

Therefore 1200 students will be given $\frac{3}{10} \times 1200$ i.e. 360 cartons

Try this 5

A shop offers 3 exercise books for GH¢ 4.50. How much would be charged for 15 exercise books

Inverse proportion

Sometimes an increase in one quantity causes a decrease in other quantity. For example, if more people do a job, it takes less time to complete the job. The time taken is said to be inversely proportional to the number of people doing the job.

Example

If 16 people can pick the oranges from the trees in 12 days, how long will it take 24 people?

16 people take 12 days

1 person will take 16×12 days

Therefore 24 people will take $\frac{16 \times 12}{24}$ i.e. 8 days

Try this 6

If 20 people plough a field in 6 hours, how many people working at the same rate would plough the same field in 4 hours?

Exercise 16.1(c)

1. Solve:

(a) $2 : 5 = a : 35$ (b) $5 : 4 = 40 : x$

(c) $x : 16 = 5 : 4$ (d) $9 : n = 36 : 38$

(e) $10 : 15 = m : 6$ (f) $5 : y = 25 : 30$

(g) $n : 100 = 4 : 5$ (h) $6 : 7 = n : 35$

(i) $6 : x = 12 : 34$ (j) $44 : 15 = 11 : x$

(k) $n : 13 = 20 : 52$ (l) $35 : m = 7 : 2$

2. If a car averages 35 kilometre per gallon, how many gallon should you expect to buy for a 900-kilometre trip?

3. The cost for water is GH¢ 1.42 per 750 gallons of water used. What is the water bill if 30,000 gallons are used?

4. A shop is offering 6 tins of milk for GH¢ 15. How much would be charged for 48 tins?

5. The cost for water is GH¢ 1.75 per 800 gallons of water used. How many gallons were used if the bill is GH¢ 21?

6. A shop offers 3 packs of tooth picks for GH¢ 0.75. How many packs were offered for GH¢ 6.25?

7. A gallon of paints cover 725 m^2. How much paint is needed to cover a house with a surface area of 5075 m^2?

8. A 30 kg bag of fertilizer will cover an area of 2500 m^2. How many kilograms are needed to cover an area of $28,000 \text{ m}^2$?

9. 3 bags of fertilizer each weighing 50 kg will cover an area of $4,500 \text{ m}^2$. How many bags will cover an area of $22,500 \text{ m}^2$?

10. A man who weighs 85 kg on Earth would weigh 14 kg on the moon. How much would his dog who weighs 15 kg weigh on the moon?

11. A machine manufactures 600 toys every 5 hours. How many hours does it take to manufacture 1800 toys?

12. A machine manufactures 500 toys every 3 hours. How many toys does it manufacture in 60 hours?

13. A hotel charges GH¢ 250 per week, including all meals. What is the charge for 28 days?

14. A hotel charges GH¢ 46.20 for 3 days, excluding meals. For how many days will GH¢ 227.20 be charged?

15. A firm makes a profit of GH¢ 750 when it sells 200 pieces of furniture. How much profit will it make if it sells 2500

pieces of furniture?

16. A machine prints five books in 12 minutes. How many will it print in 3 hours?

17. A farmer plants six orange trees in 25 minutes. If he continues to work at a constant rate, how long will it take him to plant 300 trees?

18. A bricklayer lays 1200 bricks in a 6-hour day. Assuming he continues to work at the same rate, calculate how many bricks he would expect to lay in a five -day week?

19. A bricklayer lays 1500 bricks in 8 hours. Assuming he continues to work at the same rate calculate how long ,to the nearest hour, it would take him to lay 9000 bricks?

20. A farmer used 25 litres of insecticide to spray 45 cocoa trees. How many litres of insecticide would be needed to spray 1350 trees?

21. A farmer used 30 litres of insecticide to spray 75 cocoa trees. How many trees could be sprayed with 400 litres of insecticide?

22. A train travelling at 120 km/h takes 5 hours for a journey. How long would it take a train travelling at 80 km/h?

23. A cyclist averages a speed of 24 km/h for 5 hours. At what average speed would he need to cycle to cover the same

distance in 3 hours?

24. A cyclist averages a speed of 18 km/h for 4 hours. How long will it take him to cover the same distance if the average speed is increased to 32 km/h?

25. A fishing pond is filled with water in 15 hours by three identical pumps. How much quicker would it be filled if five similar pumps were used instead?

26. A swimming pool takes 32 hours to be filled using three identical pumps. If the pool needs to be filled in 8 hours, how many of these pumps will be needed?

27. Four identical photocopying machines used simultaneously take 15 seconds to make 50 copies. How many of these copies would be made in 18 minutes?

Dividing a quantity in a given ratio

You can use ratio to divide a quantity. The method of solution is as follows: add the numbers of the ratios to find the total number of parts, next divide the quantity by the number of parts and finally multiply by the numbers of parts required

Example

Yaw, Kofi and Kojo share GH¢1080 in the ratio 4 : 2 : 3. How much does each receive?

The money is to be divided into $4 + 2 + 3$ i.e. 9 parts. One part will be equivalent to

GH¢ $1080 \div 9$ i.e.GH¢ 120

Yaw gets 4 of the 9 parts, so he will have GH¢ $120 \times 4 =$ GH¢ 480

Similarly, Kofi will have GH¢ $120 \times 2 =$ GH¢ 240

and Kojo will have GH¢ $120 \times 3 =$ GH¢ 360

Alternatively

Yaw's share is $\frac{4}{9} \times 1080 =$ GH¢ 480

Kofi's share is $\frac{2}{9} \times 1080 =$ GH¢ 240

Kojo's share is $\frac{3}{9} \times 1080 =$ GH¢ 360

You can check your work by adding their shares together to find if this add up to GH¢ 1,080.

Note $480 + 240 + 360 = 1080$

Try this 7

Share a prize of GH₵3,600 in the ratio $7 : 5 : 3$

Exercise 16.1(d)

1. Share GH¢ 2500 between two boys in the ratio $2 : 3$

2. Share GH¢ 3750 between Akosua and Esi in the ratio $3 : 2$

3. Share a prize of GH¢ 25,200 in the ratio $3 : 4 : 5$

4. Divide 810 km in the ratio 4 : 5

5. Divide 52 bags of rice in the ratio 7 : 6

6. Share GH¢ 31,000 between Ama and Kojo in the ratio 2 : 3. How much more did Kojo receive?

7. Mensa, Adjoa and Esi share GH¢ 9000 in the ratio 5 : 6 : 7 respectively. How much more does Esi get than Mensa?

8. Adjei and Ofori share an amount of money in the ratio 5 : 7. If Adjei's share is GH¢ 3,600, what is the total amount?

9. Acquah, Mensah and Araba shared an amount of money in the ratio 3 : 5 : 7 respectively. If Mensah received GH¢ 2,100 more than Acquah, what was the total amount shared?

10. The ratio of boys to girls in a class is 7 : 5. If the class has 36 students, what is the difference between the number of boys and the number of number of girls?

11. The cost GH¢ 560 of producing an article arises from the cost of electricity, labour and material in the ratio 2 : 3 : 5. Calculate the cost of electricity for producing 120 such articles?

12. A man saves $\frac{1}{4}$ of his salary and spent the rest on rent, transport and food in the ratio 3 : 4 : 5 respectively. If he spent GH¢ 200 on food, what was his salary?

Using Ratio to Solve Problems

Examples

$\frac{3}{7}$ of students in a class are girls. If there are 15 girls, how many boys are in the class?

Let the number of boys in the class be x. For every 7 students 3 are girls and 4 are boys.

So, $x : 15 = 4 : 3$

Writing each ratio as a fraction, we have

$$\frac{x}{15} = \frac{4}{3}$$

$$x = \frac{4}{3} \times 15$$

$$x = 20$$

Thus, there are 20 boys in the class

Try this 8

A man spent $\frac{3}{5}$ of his income on food and the rest on rent. If he spent GH¢ 240 on rent, how much did he spent on food

A prize of GH¢ 20,400 is shared among Kojo, Kofi and Yaw. Kofi received three times as Yaw and Kojo received one and a half as much again as Kofi. How much did each receive?

Take Yaw's share as 1 unit. Thus, Kofi's share is 3 units and Kojo's share is $3 + 1\frac{1}{2} = 4\frac{1}{2}$ units.

The prize is shared in the ratio $4\frac{1}{2} : 3 : 1$ i.e. $9: 6: 2$

Kojo's share is $\frac{9}{17} \times 20400 = $ GH¢ 10,800

Kofi's share is $\frac{6}{17} \times 20400 = $ GH¢ 7,200

Yaw's share is $\frac{2}{17} \times 20400 = $ GH¢ 2,400

Try this 9

A father shared GH¢ 27,180 among his three daughters, Esi, Akosua and Mansa. Akosua receive twice as much as Esi, and Mensa receive three times as Akosua. How much did each receive?

Exercise 16.1(e)

1. A boy spent $\frac{2}{5}$ of his pocket money. If he had GH¢ 21 left, how much did he spend?

2. A man spent $\frac{4}{7}$ of his income on food and the rest on rent. If he spent GH¢ 42 on rent, how much did he spend on food?

3. $\frac{3}{8}$ of a prize was donated to charity and the rest was invested in stocks. If GH¢ 22,800 was invested, how much was donated to charity?

4. For every 200 copies of a book printed, a publisher sold 170 copies. If 900 copies were left unsold, how many copies did he print?

5. An organiser of a concert donated 20 Gp in the cedi of his profit to charity. If his net profit was GH¢ 5000, how much did he donate?

6. A man added 60 Gp to each cedi his son saved. If the son had GH¢ 2,400, how much did he save?

7. Afua and Adjoa shared GH¢ 260. If Adjoa had half as much again as Afua, how much did each receive?

8. A mother shared GH¢ 1080 among her daughters Esi, Mensa and Abena. Mensa had twice as much as Abena and Esi had one and a half times as much as Mensa. How much did each receive?

9. A prize of GH¢ 1560 was shared among Kojo, Kofi and Adjoa. Kojo received twice as much as Kofi, and Kofi received three times as much as Adjoa. How much did Kofi receive?

10. A school admitted 520 students in 1992 to the Arts, Business and Science programs. The number of students admitted to the Arts program were $\frac{2}{3}$ of the number of students admitted to Business program and the number of students admitted to the Business program were two times the number admitted to the science program. How many students were admitted to each program?

11. Two numbers are in the ratio of 5 : 3. Their sum is 80. Find the two numbers.

12. The ratio of boys to girls in a class is 2 : 7. If there are 180 students in the class, how many girls are in the class?

Representative Fractions

The scale of a map is often given in the form of a ratio. For instance, the scale 10 cm to 1 km can be expressed as the ratio 1 : 10,000. When the scale of a map is given in the form 1 : n, it is called the representative fraction (RF) of the map.

Example

Find the R.F for the scale 2 cm to 5 km

$$\frac{2\ cm}{5\ km} = \frac{2}{500,000}$$

$$= \frac{1}{250,000}$$

Thus the R.F $= \frac{1}{250,000}$

Try this 10

Find the R.F for the scale 5 m to 1 km

Using Scales

The scale of a map shows how the distance on a map compares to the distance in real life. A scale of 1: 20,000 means that 1 cm on the map represents 20,000 cm on the ground. For example, if two villages are 8 cm apart on the map then the distance between the two villages is 8 × 20,000 i.e. 160,000 cm or 1.6 km.

Example

A map scale is 1 : 50,000. On the map the distance between two schools is 4 cm. What is the actual distance, in kilometres between the schools?

4 cm on the map represents:

4 × 50,000 = 200,000 cm

$$= 2 \text{ km}$$

Thus, the actual distance between the two schools is 2 km

Try this 11

The scale of a plan is 1 : 500. Find the length, in meters of a room which measures 2 cm on the plan.

Exercise 16.1(f)

1. The scale of a plan is 4 mm to 10 cm, find its R.F.

2. The scale of a map is 20 cm to 5 km, find its R.F.

3. The scale of a map is 1: 100,000. Find in kilometres the length of a road which is represented by a line 3.5 cm long on the map

4. The scale of a map is 1: 500,000. How far apart are two towns which are 2.4 cm apart on the map?

5. The scale of a map is 1 : 50,000. Find in cm, the length on the map of a road 2 km long

6. The scale of a plan is 1 : 500. Find the dimensions in metres of a room which measures 4 cm by 2.5 cm on the plan

7. A ground plan of a house is made on the scale 1 : 1,500. Find the length and width on the plan of a room 75 m by 60 m.

8. The scale of a map is 1: 200. Find the area on the ground represented by 2.5 cm^2 on the map

9. The scale of a map is 1: 50,000. Find the area of a farm represented by an area of 48 cm^2 on the map

10. The scale of a map is 1 : 250,000. If the actual area of a field is 25 km^2, what is the area on the map?

11. The area of a plot of land on a map with scale 1 : 50,000 is 16 cm^2. Find, in cm^2, the area of the plot of land on a map with scale 1 : 100,000

12. On a map, a square field has a perimeter of 24 cm. If the actual area of the field is 90,000 m^2, what does 1 cm on the map represent?

13. The scale of a model car is 1: 40. If the length of the real car is 5.8 m, what is the length of the modal car?

14. The scale of a model train is 1: 25. If the length of the model engine is 7.2 cm, what is the true length of the engine?

15. The scale of a model house is 1: 50. If the height of the model

house is 5.4 cm, what is the height of the actual building?

16.2 Rates

A rate is a ratio that compares two different quantities. If you walk 8 km in 2 hours, you walk at the rate of 4 kilometres per hour, written 4 km/ h. The fraction expressing this rate has units of distance in the numerator and units of time in the denominator. Similarly, if a man's weekly income is GH¢ 210, his rate of income is GH¢ 30 per day. Notice that rates have units.

When a rate remains the same, it is called a uniform rate. If the rate varies it is called non-uniform rate.

Examples

A labourer charges GH¢ 12.50 per hour for a job. How much will he receive for working 8 hours?

Amount receive = 12.50 × 8

$$= 100.00$$

Thus, the labourer receives GH¢ 100.00

Try this 12

A man who works at a factory earns GH¢ 15.60 per hour. How much will he earn in 6 hours?

In May 2011, one U. S. Dollar was worth GH¢ 1.51. Afua changed $ 65 to cedis. How much did she receive?

Amount received = 1.51 × 65

$$= 98.15$$

Thus, Afua received GH¢ 98.15

Try this 13

In May 2011, one British pound (£) was worth GH¢ 2.45, Kwesi changed £50 to cedis. How much did he receive?

Exercise 16.2(a)

1. Write the following rates in a simple form

   (a) 60 kilometre in 5 hours   (b) GH¢ 3O00 for 12 tickets

   (c) 1500 patients to 20 doctors   (d) 480 students in 12 classes

2. A hotel charges GH¢ 3000 for accommodations for 6 nights. How much does it charge per night?

3. A hotel charges GH¢ 35 per night for accommodation. Dzifa stayed 7 days in the hotel, how much did she pay?

4. A labourer charges GH¢ 12 per hour for a job. If he completed a job in 6 hours, how much did he receive for the job?

5. Water flows from a tap into an empty container. The volume of the container is 16 cm^3. If the container was filled in 12 seconds, at what rate is the tap flowing?

6. Water flows from a tap into an empty container at the rate of 3 cm^3 s^{-1}. If the container is filled in $1\frac{1}{2}$ minutes, what is the volume of the container?

7. The cost of repairing a fence wall is GH¢ 145 per square metre. If the total cost of repairs is GH¢ 3,625, what is the total area of the fence wall?

8. A car and a bus do 90 km and 45 km to the gallon respectively. The cost of 1 gallon of petrol is GH¢ 6.95. Calculate the total cost of petrol used by the car and bus together if they both travelled 270 km.

9. A 1680 cm^3 container was filled by two taps flowing together in 2 minutes. At what rate is the water flowing from the second tap if water flows from the other tap at the rate of 8 cm^3 s^{-1}?

10. A train passes a signal post completely in 1 second. Its speed is 180 km h^{-1}. What is the length of the train?

11. In May 2011, one U. S. Dollar was worth GH¢ 1.51. A tourist changed $ 120 to cedis. How much did the tourist receive?

12. In May 2011, one U. S. Dollar was worth GH¢ 1.51. Kofi changed GH¢ 3,020 to dollars. How many dollars did he receive?

13. In May 2011, one British pound (£) was worth GH¢ 2.44. Find, in pounds the value of a car which cost GH¢ 12,800.

14. In May 2011, petrol cost $ 3 per gallon in the U. S. A. Find the equivalent in cedis per litre in Ghana, if one dollar was worth

GH¢ 1.51   [Take 1 gallon = 3.8 litres]

15. In May 2011, one British pound (£) and one euro (€) were worth GH¢ 2.44 and GH¢2.15 respectively. A businessman changed £ 360 and € 750 to cedis, how many cedis did he receive?

16. A block-making machine produces 36 blocks in 8 minutes. How many blocks are produced in six hours?

17. A production line produces 800 litres of orange juice in 3 minutes. How long would it take to produce 36,000 litres?

18. A photocopying machine is capable of making 50 copies each minute. If six identical copiers are used simultaneously, how many copies would be made in 1 hour?

19. Water flows from a tap at the rate of 3.5 cm^3/s. How long would it take to fill a 2.52 litre container?

20. A tap issuing water at a rate of 1.5 litres per minute could fill a container in 4 minutes. If water leaks from the container at the rate of 1.2 litres per minute, how long would it take to fill the container?

Electricity and Water Tariffs

Example

In February 2002,the Electricity Corporation charged the following rates:

GH¢ 0.78 for the first 50 units or less

For each additional units at GH¢ 0.0242 per unit

Government special levy at GH¢ 0.0002 per unit

Street light levy at GH¢ 0.0001 per unit

If a household consumed 135 units of electricity, how much did this cost?

Amount paid for first 50 units = 0.78

Additional units consumed = 135 − 50 = 85

Amount paid for 85 additional units = 85 × 0.0242 = 2.057
= 2.06

Government special levy = 135 × 0.0002 = 0.027
= 0.03

Street light levy = 135 × 0.0001 = 0.0135
= 0.01

Total amount paid = 2.88

The household paid GH¢ 2.88

Try this 14

The rate of electricity consumed is GH¢ 0.15 per unit for the first 60 units and GH¢ 0.35 per units for each additional unit consumed. If a house consumed 210 units of electricity in a month, calculate the amount paid by the house?

Exercise 16.2(b)

1. A man paid for domestic electricity consumed each month at the following rates

   GH¢ 0.25 per unit for the first 50 units

   GH¢ 0.08 per unit for each additional units consumed

   If he consumed 80 units of electricity in a month, how much did he pay?

2. A man paid for domestic electricity consumed each month at the following rates

   GH¢ 0.40 per unit for the first 30 units

   GH¢ 0.80 per unit for the next 30 units and

   GH¢ 1.20 per unit for each additional units

   (a) If the man used 75 units of electricity in August, how much did he pay for the electricity?

   (b) If his electricity bill for September was GH¢ 68.40, find the increase in units of electricity consumed

3. The domestic electricity meter readings in June 1 and June 30 are 4641 and 4739 respectively. The Electricity Company charges GH¢ 0.12 per unit of electricity consumed. In addition the Electricity Company charges the following:

   Government special levy at GH¢ 0.05 per unit consumed

Street light levy at GH¢ 0.02 per unit consumed

What was the bill in June?

4. The Water Company charges the following rates in June 2000:

GH¢ 0.80 per 1,000 litres for the first 20,000 litres

GH¢ 1.20 for each additional 1000 litres consumed

Additional charges are

1% of total cost - fire fighting levy

2% of total cost – Rural water development levy

If a household consumed 104,000 litres of water in a month, how much should the household pay?

5. The water meter readings in 1000 litres at the beginning and end of a certain month were 11591 and 11751 respectively. The following rates were charged:

GH¢ 0.75 per 1000 litres for the first 20,000 litres

GH¢ 0.80 for each additional 1000 litres consumed

1% of total cost- Fire fighting levy

2% of total cost – Rural water development levy

What was the total bill?

Travel Graphs

A journey can be shown by drawing a graph of distance against time, called a travel graph.

Suppose Dzifa walked from home at a steady speed of 4 km h^{-1} for 3 hours, rested for an hour and then continued walking at a steady speed of 3 km h^{-1} for 2 hours. Her distances from home would be as shown in Table 16.1

Time (in hours)	0	1	2	3	4	5	6
Distance (in km)	0	4	8	12	12	15	18

Table 16.1

The travel graph for Dzifa's journey is shown below

Distance (in km)

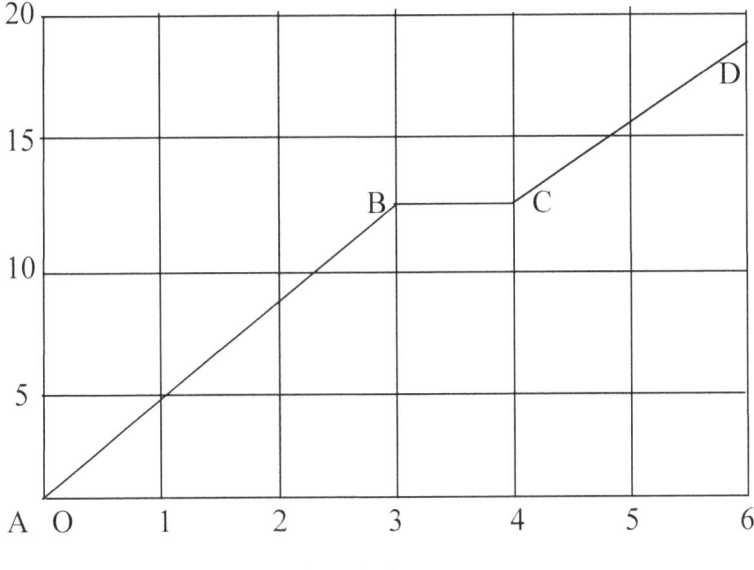

Time in hours

Figure 16.1

The gradient of a distance time graph measures the speed. In Figure 16.1, the line AB represents the walk at 4 km h^{-1}, the rest is shown by the horizontal line BC, and the line CD represents the walk at 3 km h^{-1}.

The gradient of a velocity- time graph represents the acceleration, and the area under a velocity – time graph is equal to the distance covered.

Example

A cyclist set out from home at 11-00 a.m. On the way his bike gets a puncture. He repairs it and cycle on for some time, and then return home. Figure 16.2 shows a graph of the distance in kilometre travelled against time.

(a) How long did it take to mend the puncture?

(b) What were his speeds in the three parts of the journey?

(c) How far had he gone when he returned home?

(d) When did he return home?

Distance (in km)

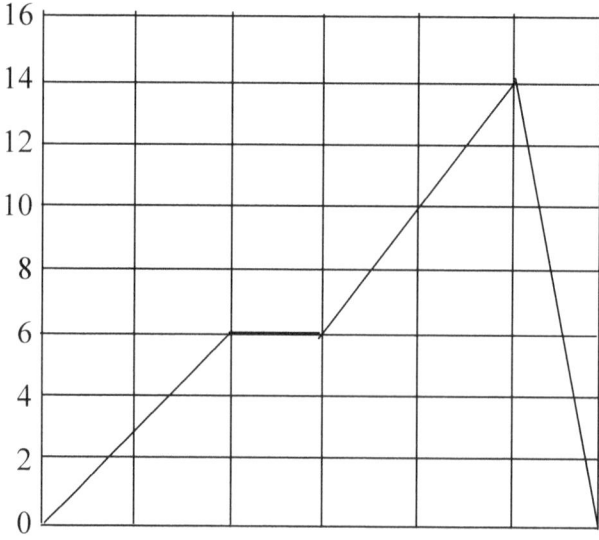

11.00 11.30 12.00 12.30 13. 00 13.30 14.00

Time

(a) He took 30 minutes to mend the puncture

(b) Speed from 11.00 a.m. to 12 noon. $= \dfrac{6\ km}{1\ hr} = 6\ \text{km h}^{-1}$

Speed from 12.30 p.m. to 1.30 p.m. $= \dfrac{8\ km}{1\ hr} = 8\ \text{km h}^{-1}$

Speed from 1.30 p.m. to 2.00 p.m. $= \dfrac{14\ km}{\frac{1}{2}\ hr} = 28\ \text{km h}^{-1}$

(c) He was 14 km away from home

(d) He returns home at 2.00 p.m.

Try this 15

The travel graph shows a cyclist's journey from his home

Distance(in km)

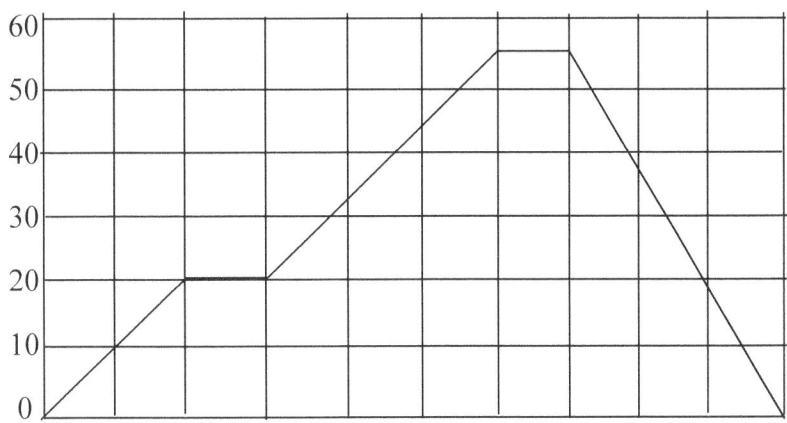

Time

(a) When did the cyclist leave home?

(b) How far away from home was he at 2 p.m.?

(c) At what times did he take a rest?

(d) Find his speed between

(i) 11 a.m. and 12 noon

(ii) 1 p.m. and 4 p.m.

(e) When did the cyclist return home?

Exercise 16.2(c)

1. A bus travels from Kumasi at a speed of 80 km h^{-1}. After 3 hours the driver stops and rest for 1 hour. He then continues at 60 km h^{-1} for 2 hours. Draw a travel graph showing the distance the bus has travelled

2. Esi walks three kilometre to a shop, taking 1 hour. She spends 15 minutes shopping, then walks back in 45 minutes. Draw a travel graph showing her distance away from home.

3. Kwesi walks 100 metres in 30 seconds. He rests for 10 seconds, and then walk a further 100 metres in 50 seconds. Draw a travel graph showing the distance he has travelled.

4. A salesman drives 120 kilometres at 60 km h^{-1} to meet a customer. He spends 1 hour with the customer, then drives home at 40 km h^{-1}. Draw a travel graph showing his journey.

5. Afua lives 15 kilometres north of Kojo. At the same time they set off to each other's House. Afua walks south at 3 km h^{-1} and Kojo walks North at 4.5 km h^{-1}. Draw graphs showing Afua's

and Kojo's distances from Kojo's house in terms of time.

(a) How long did it take them to meet?

(b) How far are they from Kojo's house when they met?

6.

Speed(km h^{-1})

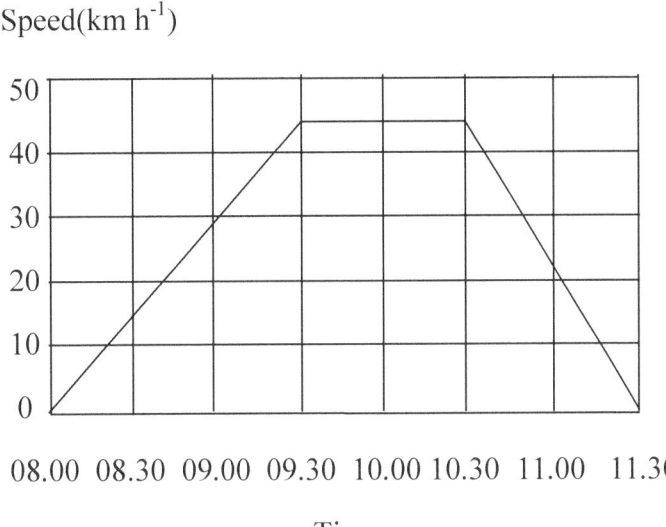

08.00 08.30 09.00 09.30 10.00 10.30 11.00   11.30

Time

The diagram shows the speed-time graph of a car's journey between two villages. The car starts from rest and accelerates uniformly to a maximum speed of 45 km/h.

(a) The car stopped at a rest stop on the way, how long did the car

stop?

(b) How long did it take the car to arrive at the second village?

Give your answer in hours and minutes

(c) Kwesi starts at the same time as the car and cycle at a constant

speed of 8 km/h along the car's route. How many kilometres

from the car is Kwesi at the time the car arrived at the second village?

7. The graph shows the speed-time graph of a bus journey between two towns

Speed (in ms^{-1})

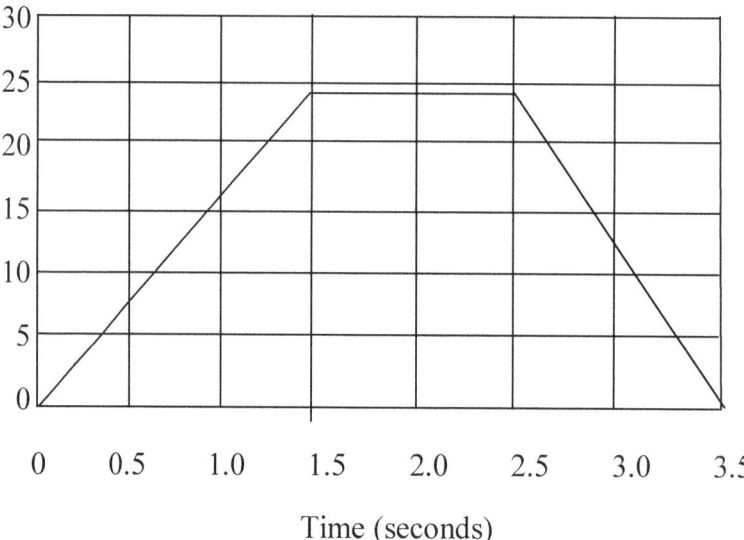

Time (seconds)

The bus starts from rest and accelerates uniformly to a maximum speed of 24 m/s.

(a) The bus stop on the way, how long did the bus stop?

(b) Find the acceleration of the bus.

(c) Calculate the distance travelled by the bus in $3\frac{1}{2}$ seconds

8. Kofi leaves his house at 12.00 to Afua's house 12 kilometre away. The graph shows Kofi's journey from his house to Afua's house.

Distance (in km)

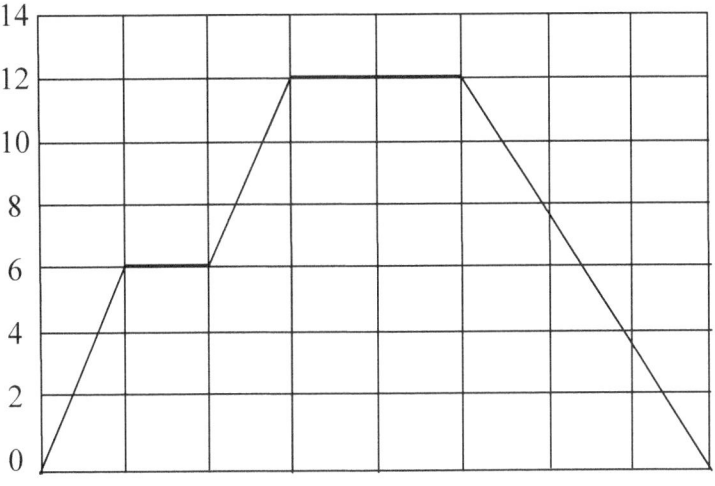

12.00 12.30 13.00 13.30 14.00 14.30 15.00 15.30 16.00

Time

(a) What happened at 12 45?

(b) Calculate Kofi's average speed from leaving the house to arriving at Afua's house. Give your answer in kilometres per hour.

(c) How long did he stay at Afua's house?

(d) When did Kofi return home?

16.3 Percentages

The term percent means per hundred. Percents are ratios whose second number is 100, and may be represented by a fraction whose denominator is 100. Percents are often written with the symbol %. For example, 5 percent is written as 5%.

Expressing fractions as percentages

Example

Express $\frac{3}{5}$ as a percentage

First change the fraction into an equivalent fraction with a denominator of 100.

$$\frac{3}{5} = \frac{3 \times 20}{5 \times 20} = \frac{60}{100}$$

Thus, $\frac{3}{5}$ is equivalent to 60 %

You can also use proportion to obtain the same result.

$$n : 100 = 3 : 5$$

Then $\dfrac{n}{100} = \dfrac{3}{5}$

$$n = \frac{3}{5} \times 100$$

$$= 60 \,\%$$

Try this 16

Express:

(a) $\frac{2}{3}$ (b) $\frac{7}{10}$

as a percentage

Expressing percentages as fractions

Example

Express 75 % as a fraction in its simplest form

$$75\% = \frac{75}{100}$$

$$= \frac{3 \times 25}{4 \times 25}$$

$$= \frac{3}{4}$$

Try this 17

Express:

(a) 12 %   (b) 65 %

as a fraction in its simplest form

Expressing decimal fractions as percentages

Express

Express 0.35 as a percent

$$0.35 = \frac{35}{100}$$

$$= 35\%$$

Try this 18

Express:

(a) 0.48   (b) 0.037

as a percentage

Expressing percentages as decimals

Example

Express 82 % as a decimal

$$82\% = \frac{82}{100}$$

$$= 0.82$$

Try this 19

Express:

(a) 32 %   (b) 19.4 %

as a decimal

Exercise 16.3 (a)

1. Express the following statements in percent form

(a) 20 students in every 100 students live in Tema

(b) A girl obtains 8 marks out of 10

(c) 3 students in every 50 students failed the examination

(d) The tax rate of a country is 3 Gp in every cedi

(e) 1 orange in every 4 oranges in a box is bad

2. Write these fractions as percentages:

(a) $\frac{3}{4}$ (b) $\frac{4}{5}$ (c) $\frac{8}{25}$ (d) $\frac{1}{6}$ (e) $\frac{6}{5}$ (f) $\frac{7}{2}$

3. Write these percentages as fractions:

(a) 85 %   (b) 15 %   (c) 6 %

(d) 250 %   (e) $37\frac{1}{2}$ %   (f) $4\frac{1}{2}$ %

4. Convert the following decimals to percentages

(a) 0.65   (b) 0.05   (c) 0.23

(d) 1.36   (e) 2.07   (f) 0.048

5. Convert the following percentages into decimals:

(a) 56 %   (b) 45 %   (c) $12\frac{1}{2}$ %

(d) 178 %   (e) 206 %   (f) $\frac{1}{5}$ %

Expressing one quantity as a percentages of another quantity

Example

What percent of 65 is 39?

$$\frac{39}{65} = 0.6 = 60\ \%$$

You can also use proportion to obtain the same result.

$$\frac{n}{100} = \frac{39}{65}$$

$$n = \frac{39}{65} \times 100 = 60\,\%$$

You can omit the first step. First write 39 as a fraction of 65 and then multiply by 100.

Try this 20

What percent of 1325 is 265?

Exercise 16.3 (b)

1. What percent of

(a) 55 is 11?   (b) 150 is 45?   (c) 96 is 48?

(d) 12.8 is 3.2?   (e) 32 is 16?   (f) 1 hour is 15 minute?

2. In an examination a student obtained 90 correct answers out of 120. Express this result as a percentage.

3. A man whose monthly income is GH¢ 240 spends GH¢ 30 on rent. What percentage of his income is spent on rent?

4. A man's monthly income increased from GH¢ 1500 to GH¢ 1800. Express the increase as a percentage of the original income

5. A man invested GH¢ 8500 and earned GH¢ 170 interest. Express his interest as a percentage of his investment?

6. In a class of 45 students, 20 are girls. Express the number of girls as a percentage of the number of boys

7. A man earned GH¢ 249 in June. If he saved GH¢ 83, express his savings as a percentage of his income.

8. A boy took a test in mathematics and science. There were 10 questions in each test and each question carry 1 mark. If he had 8 answers wrong express his mark as a percentage of the total mark.

9. An article cost GH¢ 1750 when new; after 1 year, its value cost GH¢ 1050. Express the decrease in value as a percentage of the original value

10. A tailor sells a shirt for GH¢ 80. The cost of material and labour was GH¢ 48. Express his profit as a percentage of his cost

Calculating the percentage of a given quantity

Example

Find 12 % of 480

$$12 \% \text{ of } 480 = \frac{12}{100} \times 480$$

$$= 57.6$$

Alternatively, write 12 % as the decimal 0.12 and multiply as shown below

$$12 \% \text{ of } 480 = 0.12 \times 480$$

$$= 57.6$$

Try this 21

Find 15 % of 1600

Exercise 16.3 (c)

1. Evaluate the following:

   (a) 25 % of 160   (b) 60 % of 1850   (c) $33\frac{1}{3}$ % of 2400

   (d) $\frac{3}{5}$ % of 10,000   (e) $32\frac{1}{2}$ % of 1080   (f) 45 % of 60

2. Ali's monthly income is GH¢ 800. He invested $8\frac{1}{2}$ % of this

   amount in stocks, how much did he invest?

3. A school admitted 400 students in 2010. If 64 % were boys, how

   many girls were admitted?

4. Mensah spends $12\frac{1}{2}$ % of his monthly income on transportation,

   25 % on rent and the rest on food. He earned GH¢ 600 in June

   2010, how much did he spend on food?

5. Addo received GH¢ 150 from his father. He spent 75 % of the

   amount and saved the rest, how much did he save?

6. Yayra took a test in mathematics consisting of 50 questions. She

   answered 62 % of the questions correctly, how many did she

   get wrong?

7. Edem and Afua shared GH¢ 1200. Edem received 65 % of the amount, how much did Afua receive?

8. A school offered 250 applicants programs in General Arts, Business and General Science. 35 % and 45 % of the applicants were offered programs in General Arts and Business respectively, how many were offered programs in General Science?

To calculate a quantity if a percentage of it is given

Example

150 is 75 % of what number?

If x is the number, then

$$x : 150 = 100 : 75$$

So, $\dfrac{x}{150} = \dfrac{100}{75}$

$$x = \dfrac{100}{75} \times 150$$

$$= 200$$

Alternatively, if x is the number then,

$$\dfrac{75}{100}x = 150$$

$$x = 150 \times \dfrac{100}{75}$$

$$= 200$$

Try this 22

405 is 45 % of what number?

Exercise 16.3 (d)

1. 25 % of a number is 32, find the number

2. 56 % of a number is 1120, find the number

3. $37\frac{1}{2}$ % of a number is 8400, find the number

4. $16\frac{2}{3}$ % of a number is 40, find the number

5. A man spends GH¢ 3,750 a year, and this is 75 % of his income. What is his income?

6. 45 % of the students in a class are girls. Find the number of students if there are 22 boys in the class.

7. Tawiah spent GH¢ 7420 on books, and this amount is $13\frac{1}{4}$ % of his income. What was his income?

8. A student paid $87\frac{1}{2}$ % of his fees. If GH¢ 150 remains to be paid, how much was the bill?

9. Kweku withdrew 35 % his savings. If he has GH¢ 1040 now in the account, what was the original amount in the account?

10. A man invested 30 % of his income in stocks and spent the rest. If he spent GH¢ 3,500, what was his income?

Percentage Changes

A change in the size or value of a quantity is often estimated by calculating what percent the change is of the original size or value. If a quantity increases from 60 to 75, then the percent increase is found by dividing the amount of increase, 15, by the original value, 60:

i.e. $\dfrac{75-60}{60} = \dfrac{15}{60} = 0.25 = 25\,\%$

The increase in value is 25 %.

The calculation can be written as:

$$\dfrac{15}{60} \times 100 = 25\,\%$$

Examples

Increase 640 by 45 %

The number increase by:

$0.45 \times 640 = 288$

Thus, the new value is

$640 + 288 = 928$

Note that the original value represents 100 %, therefore an increase of 45 % means the original value increase by 145 %.

New value $= 145\,\%$ of 640

$$= 1.45 \times 640$$

$$= 928$$

Try this 23

Increase 1300 by 12 %

Decrease 50 by 16 %

The percent decrease is 16 %, therefore

Decrease in value = 16 % of 50

$$= 0.16 \times 50$$

$$= 8$$

New value = 50 − 8

$$= 42$$

A decrease of 16 % means the new value is 84 % of the original value

New value = 84 % of 50

$$= 0.84 \times 50$$

$$= 42$$

Try this 24

Decrease 4520 by 25 %

Exercise 16.3 (e)

1. Esi's salary increased from GH¢ 200 per month to GH¢ 234 per month. What was the percentage increase in her salary?

2. The price of an item increased from GH¢ 75 to GH¢ 80, what was the percentage increase in price?

3. The value of a stock increases from GH¢ 1.25 per share to GH¢ 2.00 per share, what is the percentage increase in the value of the stock?

4. If Opoku's expenditure decreases from GH¢ 750 to GH¢ 600, what was the percentage decrease in his expenditure?

5. If the value of a stock decreases from GH¢ 3.00 to GH¢ 2.25, what is the percentage decrease in the value of the stock?

6. A shop reduces the price of an item from GH¢ 125 to GH¢ 95, what is the percentage decreased in the price of the item?

7. Increase the following by the given percentages:

    (a) 8000 by 35 %   (b) 416 by 125 %

    (c) 260 by $17\frac{1}{2}$ %   (d) 3140 by 12 %

8. Decrease the following by the given percentages

    (a) 750 by 40 %   (b) 2160 by $37\frac{1}{2}$ %

    (c) 450 by 36 %   (d) 1080 by $7\frac{1}{2}$ %

9. A particular stock is valued at GH¢ 3.50 per share. If the value is increased by 20 %, what will be the value of the stock per share after the increase?

10. The price of a gallon of petrol was GH¢ 112.50 in 2008. If the price was reduced by $15\frac{1}{2}$ % in 2009, what was the price of a gallon of petrol after the reduction?

11. A school had a population of 160 students in 1992. The number of students increased by 25 % the following year, how many students were in the school in 1993?

12. An investment in a mutual fund increased by 12 % in a single day. If the value of the investment before the increase was GH¢ 1300, what was the value after the increase?

13. A man whose salary is GH¢ 380 per month receives 15 % increase, what is his salary after the increase?

14. The price of a computer is reduced by $7\frac{1}{2}$ % in a sale. If the computer was priced at GH¢ 4800, what is its price in the sale?

15. Dzifa's pocket money was reduced by 3 %. If her pocket money was GH¢ 15 before the decrease, what is her pocket money after the decrease ?

Successive Percent Changes

If a quantity is increased by 5 % and then the result is increased by 10 %, the total increase is not 15 % of the original quantity. Note that the base of each successive percentage change is the result of the preceding percentage change. This is illustrated in the following example.

Examples

A number is increased by 10 % and the result is decreased by 15 %, what is the cumulative percent change?

If x is the number then an increase of 10% means that the new value after the increase is

$1.10x$

This value is further increased by 15% so the final value is

$(1.15)(1.10x) = 1.265x$

The percentage equivalent of 1.265 is 126.5 %.

Thus, the original number will be increased by 26.5 %.

So the cumulative percent change is a 26.5 %

Try this 25

A number is decreased by 8 %, and the result is increased by 6 %, what is the cumulative percent charge?

Increase 2500 by 50 % and then decrease the result by 10 %

The result of the 50 % increase can be found by multiplying 2500 by 150 %

$2500 \times 1.50 = 3750$

The base for the second percentage change is 3750. The second percentage change is

$3750 \times 0.90 = 3375$

Thus, the result when 2500 is increased by 50 % and the result decrease by 10 % is 3375.

This result can also be obtained as follows:

$2500 \times 1.50 \times 0.90 = 3375$

Try this 26

Decrease 500 by 10 % and then increase the result by 20 %

Exercise 16.3 (f)

1. Increase 800 by 20 % and then increase the result by 20 %

2. Decrease 2500 by 30 % and then decrease the result by 10 %

3. Increase 4000 by 5 % and then decrease the result by 2 %

4. Decrease 4800 by $12\frac{1}{2}$ % and then increase the result by 20 %

5. A smart phone is priced at GH¢ 1,200. During a sale the price is reduced by 20 %. If the price is reduced further by 5 %, calculate the price it is finally sold for?

6. The price of a computer is increase by 12 %. During a sale the price after the increase is reduced by 25 %. If the original price is GH¢ 4,500, calculate the price it is sold for during the sale.

7. A manufacturer takes 5000 shirts to be sold at a fair. In the first day he sells 8 % of the shirts. In the second day he sells 10 % of those that were left. How many shirts has he sold in total?

8. A particular stock valued at GH¢ 4.50 per share decreases by 5 % one week and by a further 8 % the following week

(a) What was the overall percentage decrease for the two weeks?

(b) What is the value of the stock in the second week?

Review exercise 16

1. Simplify

(a) 32 : 8   (b) 9 : 27   (c) 2.5 : 3.0   (d) $2\frac{1}{2} : 3\frac{1}{4}$

2. Simplify

(a) 40 min. : 1 hr   (b) 45 Gp : GH¢ 1   (c) 1 day : 8 hr

(d) 1 km : 500 m   (e) $1\frac{3}{4} hr$: 45 min   (f) 2 kg : 150 g

3. Express the following ratios in the form 1 : n

(a) 8 : 12   (b) 27 : 81   (c) 25 : 200   (d) 16 : 24

4. Express the following ratios in the form n : 1

(a) 18 : 27   (b) 50 : 40   (c) 3.2 : 2.4   (d) 16 : 24

5. Solve

  (a) $8 : 5 = x : 25$ (b) $x : 18 = 3 : 2$ (c) $27 : 18 = 9 : x$

  (d) $1.5 : x = 5 : 28$ (e) $15 : 8 = x : 24$ (f) $x : 38 = 2 : 19$

6. A team won 35 of their first 54 games. If the team continues to win at the same rate, how many of their 216 games can they expect to win this season?

7. A car travel a distance of 120 kilometres in 1.5 hours. At that rate how long would it take the car to travel a distance of 600 kilometres?

8. A man picked 17 kilograms of oranges in 2.5 hours. He is paid 22 Gp per kilogram. At that rate how much can he earn in an 8-hour day?

9. A car travelled 210 kilometres on 15 gallons of petrol. At this rate how far should it travel on a full tank of 20 gallons?

10. Afua earned GH¢ 4.20 in $2\frac{1}{2}$ hours picking oranges for a farmer. At that rate how much can he earn in an 8- hour day?

11. If 80 metres of wire weigh 72 kilograms, what will 250 metres of the same wire weigh?

12. Akosua typed seven pages of manuscript in 50 minutes. How long should it take her at this rate to type a manuscript of 245 pages? Give your answer in hours and minutes.

13. Divide:

   (a) 640 in the ratio 3 : 5   (b) 1248 in the ratio 5 : 7

   (c) 1080 in the ratio 2 : 7   (d) 4530 in the ratio 8 : 7

14. A father share GH¢ 1950 among his two children in the ratio 2 : 3. How much did each receive?

15. Esi and Ama share GH¢ 72 in the ratio 7 : 5. How much does Esi receive?

16. Akosua and Kwame share GH¢ 180 in the ratio 7 : 5. How much more does Akosua get?

17. GH¢ 63 was divided between Esi, Dela and Afua so that Esi had twice as much as Dela, and Dela had one-third as much as Afua. How much did each receive?

18. GH¢ 210 was divided between Ali, Dzifa and Kwesi so that Ali had one and a half times as much as Dzifa, and Kwesi had three times as much as Ali. How much did each receive?

19. Find the representative fraction (R. F.) for the following scales

   (a) 4 m : 1 km   (b) 5 cm : 1 m   (c) 25 cm : 1 km

   (d) 15 mm : 3 m   (e) 2 cm : 5 km

20. The distance between two town is 16 km. A map shows it as 8 cm. What is the scale of the map? Give your answer as a ratio.

21. A map scale is 1 : 20,000. On a map the distance between two villages is 6 cm. What is the actual distance between the villages?

22. A map is drawn on a scale of 1 : 1,000. The area of a field as shown on the map is 5 cm². What is the actual area of the field in square metres?

23. On a map whose scale is 1 cm to $\frac{1}{2}$ kilometre the area of an island is $1\frac{1}{2}$ m². What is the area of the island?

24. A man at a factory is paid hourly. On Friday he earned GH¢ 72 working 6 hours, how much did he earn an hour?

25. A man earns GH¢ 15 per hour. How much does he receive for working 8 hours a day?

26. A hotel charges a customer GH¢ 12 per night. How much does a man pay for staying 5 nights at the hotel?

27. A factory worker is paid GH¢ 15 per hour. How much does he receive if he works 36 hours and then 12 hours at time and a half?

28. A plumber charges GH¢10 ''call-out'' charge, then GH¢12 for every hour after that. Find how, much he would charge for the work taking (a) 1 hour (b) 2 hours (c) 3 hours

29. In July 2011, one dollar was worth GH¢ 1.49. A tourist changed $170 to cedis. How much did he receive?

30. In July 2011, one dollar was worth GH¢ 1.49. A trader changed GH¢ 1250 to dollars. How much did he receive?

31. In July 2011, one British pound was worth GH¢ 2.38. Work out the value, in cedis, of a computer which cost £150

32. The rate of electricity consumed is GH¢ 0.45 per unit for the first 50 units and GH¢ 1.20 per unit for each additional unit used. If a household consumed 120 units of electricity in a month, calculate the amount paid by the household

33. A man set off from home at 8.00 am and walked at a speed of 5 km h^{-1} for 3 hours. He stopped for a meal. After 1 hour he continued to walk at a speed of 3 km h^{-1} for 2 hours, then walked back reaching home at 5.00 pm. Draw a travel graph showing his journey.

(a) How far away from home was he at 10 am?

(b) How far away from home was he at 3 pm?

(c) How far away from home was he when he turned back?

(d) How long did it take him to walk back home?

(e) At what speed did he walk home?

34. Express the following fractions as percentages

   (a) $\frac{3}{8}$ (b) $\frac{19}{25}$ (c) $\frac{5}{8}$ (d) $\frac{17}{20}$

35. Express the following percentages as fractions in their simplest

   form

   (a) 15 %   (b) 72.5 %   (c) 150 %   (d) $112\frac{1}{2}$ %

36. Express the following decimals as percentage

   (a) 0.21   (b) 0.735   (c) 1.36   (d) 0.054

37. What is:

   (a) 25 % of 1256   (b) $12\frac{1}{2}$ % of GH¢ 800

   (c) $33\frac{1}{3}$ % of 7215   (d) 75 % of GH¢ 1040

38. What percent of:

   (a) 150 is 30   (b) 225 is 75

   (c) 2 hours is 45 minutes   (d) GH¢ 2000 is GH¢ 150

39. Find the number if:

   (a) 25 % of the number is 210

   (b) 60 % of the number is 48

   (c) $12\frac{1}{2}$ % of the number is 720

   (d) $15\frac{3}{4}$ % of the number is 5670

40. Increase:

    (a) 200 by 15 %   (b) 1500 by 45 %

    (c) 620 by $12\frac{1}{2}$ %   (d) 1080 by $23\frac{1}{3}$ %

41. Decrease:

    (a) 720 by 25 %   (b) 540 by $14\frac{1}{2}$ %

    (c) 1320 by $13\frac{1}{3}$ %   (d) 2150 by 60 %

42. A woman is given a pay increase of 15 %. If she earned GH¢ 1200 before the increase, how much does she earn after the increase?

43. An item is originally priced as GH¢ 150, but during a sale all prices are reduced by 20 %. How much does the item cost now?

44. The weekly food bill of a typical family used to be GH¢ 60. How much is it after inflation of 5 %?

45. A wage increase raises Esi's wages from GH¢ 250 to GH¢ 282. What is the percentage increase?

46. At a concert 250 out of an audience of 1000 were adults. What percentage were children?

47. An item is priced at GH¢ 150 after 20 % reduction during a sale. How much did the item cost before the sales?

48. A women earned GH¢1500 after her pay was increase by 20 %.

How much does she earn before the increase?

49. A man invested GH¢ 200 for two years. In the first year the interest rate was 5 %. In the second year it was 4 %. How much money was in the account after 2 years?

50. In 2009, the cost of an item rose 15 % in June, and by a further 20 % in July. What was the overall percentage increase for the two months?

Chapter Test 16

Take this test as you would take a test in class. After you are done, check your work against the answers in the back of the book

1. Write these ratios in their simplest form

   (a) 8 : 24   (b) 2.5 : 3.5

   (c) 15 cm : 0.5 m   (d) 1 hour : 15 minutes

2. (a) Out of 30 pupils in a class, 16 are boys. Find in its simplest form the ratio of boys to girls

   (b) Kwame, Kwesi and Ama share GH¢ 4200 in the ratio 3 : 4 : 5. How much more does Ama receive than Kwame?

3. A mother shares GH¢ 8500 among his children Esi, Adwoa and Kwame. Esi had three times as much as Adwoa and Kwame had one and a half more than Esi. How much did each receive?

4. (a) Find the RF for the scale 25 m to 5 km

   (b) A map has a scale 1 : 2000. What area, in square kilometres is represented by a piece of ground shown as 3 cm by 2 cm on the map?

5. Workers at a factory were paid hourly. On Thursday, a man earned GH¢ 45 working $2\frac{1}{2}$ hours at the factory

   (a) What is the rate of pay per hour?

   (b) How much will he receive if he works 36 hours in a week?

6. In July 2011, Kofi changed 150 dollars to cedis. The exchange rate was $1 = GH¢1.49. How much did he receive?

7. A man buys a car for GH¢ 18,500 in July 2011. The exchange rate was $1 = GH¢1.49. Work out the value of the car in dollars, correct to the nearest dollar.

8. Dzifa's employer has promised her a 20 % pay increase. She presently earns GH¢ 5 per hour. How much will she earn in 5 hours after the increase?

9. (a) On a test of 80 items, a student got 62 correct, what percent were wrong?

   (b) A worker's wages increases from GH¢ 75 to GH¢ 90. What is the percentage increase?

10. (a) The weekly food bill of a family is GH¢ 60.48 after an increase of 8 %. How much was the bill before the increase?

(b) A man buys an item for GH¢ 80. During the first year its value falls by 20 %, and then by 10 % in the second year. Calculate its value at the end of the second year.

17

Application of Percentage

We often use percentages in our daily life. For instance, changes in prices, profit and loss can be expressed as percentages

17.1 Commission, Discount and Depreciation

Commission

Many sales people earn an amount of money, called a commission, by selling a product. The commission is usually calculated as a percentage of total sales.

$Commission = Amount\ of\ sales \times Rate\ of\ commission$

Examples

A salesgirl earns 15 % commission on her total sales. If she sold GH¢ 6,000 worth of items in a month, what was her commission for that mouth?

$Commission = 6.000 \times \frac{15}{100}$

$$= 900$$

The commission is GH¢ 900

Try this 1

A salesman earns a 5 % commission on his total sales. If he sold GH¢ 1200 worth of products in a month, what was his commission?

A newspaper agent earns a commission of 25 % on his total sales. If his commission in a week was GH¢ 3,000, what was the amount of sales he made?

25 % of total sales = GH¢ 3000

$$100\text{ \% of total sales} = 3000 \times \frac{100}{25}$$

$$= 12{,}000$$

The amount of sales is GH¢ 12,000

Alternatively, if GH¢ x is the total sales for the week then

$$\frac{25}{100}x = 3000$$

$$= 3000 \times \frac{1000}{25}$$

$$= 12{,}000$$

The amount of sales is GH¢ 12,000

Try this 2

A salesman receives a commission of 12 % on his total sales. If his commission in a month was GH¢ 1836, what amount of sales did he make that month?

Exercise 17.1(a)

1. A salesman receives $2\frac{1}{2}$ % commission on his total sales. In a week he made GH¢ 80,000 sales, how much was his commission?

2. A salesgirl earns a 20 % commission on his total sales. If in one week his total sales was GH¢ 18,000, how much was her commission?

3. A book salesman earned GH¢ 150 commission selling a set of books for GH¢ 2000. Find the rate of commission

4. A salesman for a furniture company made total sales of GH¢ 75,000 in a week. If he earned GH¢ 3000 commission, what was the rate of commission?

5. A car salesman earns 5 % commission on sales. If he sells a car for GH¢ 35,700, how much commission will he earn?

6. A computer salesman sold 5 computers at GH¢1800 each. If he was paid a 3 % commission on each sale, how much did he receive?

7. A man sold his house through a real estate agent for GH¢ 27,000. If the real estimate agent gets a $4\frac{1}{2}$ % commission, how much money will the man receive?

8. A man sold his house through a real estate agent for GH¢ 30,000. How much money will he have after he pay his real estate agent a 5 % commission?

9. A salesman earns 15 % commission on his total sales. If his commission in a week was GH¢1800, what was his total sale?

10. Suppose a car salesman commission rate is 12 %. If the commission on one of the cars is GH¢ 1,740, what is the purchase price of the car?

11. Yaa is paid GH¢ 500 a week plus 12.5 % commission on sales. Calculate her weekly sales if she earned GH¢ 1,200?

12. Kwame a real estate agent earns 3 % commission on the first GH¢ 100,000 of the sale of a house and 5 % on any amount over GH¢ 100,000. Determine his pay if he sold a property worth GH¢ 450,000.

13. A real estate agent received a 6 % commission on the selling price of a house. If his commission was GH¢ 14,910, what was the selling price of the house?

14. A salesgirl commission rate is 8 %. If the commission on one item sold is GH¢ 240, what is the purchase price of the item?

15. A man works at a store where he is paid a monthly salary of GH¢ 500 plus a $5\frac{1}{2}$ % commission on his total sales. One month he sold goods worth GH¢ 3940. How much did he earn that month?

16 A salesman earns a monthly salary of GH¢ 750 plus a 8 % commission on all sales above GH¢ 5000. One month he had total sales of GH¢ 10,875. How much did he earn in all that

month?

17. A salesman earns a monthly salary of GH¢ 1,200 plus a commission on his total sales. If his total income in a month he made GH¢ 9375 sales, is GH¢ 1950, find the rate of commission.

18 A car salesman earns a 12 % commission on sales. If he sells a car for GH¢ 36,800, what is his net income if he spent GH¢ 120 on advertisement?

Discount

A shop may offer a discount on some goods, by reducing their prices. Usually, the discounts are calculated as a percentage of the marked price of the item. For instance, if an item marked at GH¢ 150 is sold for GH¢ 135, the discount is GH¢ 15, This is 10 % of the marked price.

$Discount = Original\ price \times Rate\ of\ discount$

Examples

A discount of 20 % is offered on a watch marked at GH¢ 350. How much was paid for the watch?

First, find the discount

$Discount = 350 \times \frac{20}{100}$

$= GH¢\ 70$

Next, subtract the discount from the original price

Amount paid $= 350 - 70$

$\qquad = 280$

The sale price is GH¢ 280

You can obtain the same result as follows:

The original price represents $100\,\%$, since a discount of $20\,\%$ is offered, the amount paid is $80\,\%$ of the original price.

$Amount\ paid = 80\,\%\ of\ GH¢\ 350$

$\qquad = 0.8 \times 350$

$\qquad = GH¢\ 280$

Try this 3

A discount of $12\,\%$ is offered on an item marked at GH¢ 650. What is the sale price?

A store offers a discount of $16\,\%$ on a dress. If Ama paid GH¢ 252 for a dress, what was the marked price?

Since a discount of $16\,\%$ is offered, the sale price is $84\,\%$ of the marked price

So $84\,\%$ of the marked price $= GH¢\ 252$

$\therefore\ 100\,\%$ of the marked price $= 252 \times \dfrac{100}{84}$

$\qquad\qquad\qquad = 300$

The marked price is GH¢ 300

Alternatively, if GH¢ x is the marked price of the dress then

$\dfrac{84}{100}x = 252$

$$x = 252 \times \frac{100}{84}$$

$$= 300$$

The marked price is GH¢ 300

Try this 4

A shop offers a discount of 15 % on an item. If Kwame paid GH¢ 108 for the item, what was the marked price?

Exercise 17:1(b)

1. A discount of 12 % is offered on a GH¢ 900 TV set. What is the discount?

2. A shop offers a 15 % discount on a GH¢ 40 dress. What is the discount?

3. A store offers 10 % discount on a shirt marked at GH¢ 15. What is the discount?

4. A discount of 15 % is offered on a GH¢ 750 cooker. What is the sale price?

5. A department store offers a 5 % discount on all items. Find the sale price of an item marked at GH¢ 6500

6. In a shop, a shirt is marked GH¢ 60. If the shop offers a discount of 15 % on the shirt, what is the sale price of the shirt?

7. A shop offers a discount of 10 % on all items. If Ofori paid GH¢ 7200 for an item, calculate the original price of the item?

8. A shop offers 20 % discount on the marked price of all items. If a man paid GH¢ 3200 for an item, what was the original price of the item?

9. A shop offers a discount of 35 Gp in the cedis on the marked price of all items. What is the sale price of an item marked at GH¢ 8,500

10. A trader offers 5 % discount on all bills. If a customer's bill is GH¢ 36,000 of which 20 %, is gross profit, find his next profit

Depreciation

A car costs GH¢ 15,000 when new. Due to age and use its value at the end of one year was GH¢ 13,200. When an item decreases in value over a period of time, it is said to depreciate. In this case the car depreciates by 12 % of its value. Depreciation is usually given as a percentage of the value of the item at the beginning of the period.

$Depreciation = Original\ value \times Rate\ of\ depreciation$

Example

A machine bought for GH¢ 3,000 depreciates by 10 % of its value at the beginning of that year. What is the value of the machine at the end of the first year?

Depreciation $= 3000 \times \frac{10}{100}$

$$= 300$$

The value of the machine after first year $= 3000 - 300$

$$= 2700$$

The value of the machine is GH¢ 2700

Alternatively, the value of the machine at the end of the first year is 90 % of the original value.

So, the value of the machine $= 3000 \times \frac{90}{100}$

$$= GH¢ \ 2700$$

Try this 5

A machine which cost GH¢ 750 when new, depreciates by 20 % of its value at the beginning of that year, what is the value at the end the first year?

Exercise 17.1(c)

1. A machine cost GH¢ 900 when new. Each year its value depreciates by 15 % of its value at the beginning of that year, what is its value at the beginning of the second year?

2. A man buys a machine for GH¢ 1,850. If the value of the machine depreciates by 10 % each year, what is its value at the beginning of the second year?

3. A car cost GH¢ 22,500 when new. Each year its value depreciates by 15 % of its value at beginning of that year. What is the value of the car at the beginning of the third year?

4. A machine losses 20 % of its value at the beginning of each year. If it cost GH¢ 1,200 when new, what is its value at the beginning of the fifth year?

5. A man buys a car for GH¢ 16,500. The value of the car depreciates by 20 % in the first year, and then by 10 % in the second year. What is its value at the end of the second year?

6. The value of a car depreciates by 25 % each year. If the value of the car at the beginning of the second year is GH¢ 15,000, what is the original value of the car?

7. The value of a car depreciated by 20 % each year. After two years its value was GH¢ 32,000. What was its value when new?

8. The value of a machine depreciates by 10 % each year. If the value of the machine at the beginning of the fourth year is GH¢ 36,450, what is the original value of the machine?

9. A machine that cost GH¢ 80,000 when new had a resale value of GH¢ 14,000 after 12 years, what is the annual depreciation?

10. A man bought a car at GH¢ 15,000 and sold it for GH¢ 12,000 after 2 years. What is the annual rate of depreciation?

17.2 Simple Interest and Compound Interest

Often we need to buy an item but do not have the cash to do so immediately. We may choose to borrow the money from a bank or other lending institution. The money the bank is willing to lend to you is called the amount of credit extended or the principal of the loan. The bank will usually charge interest on the principal. Interest can be defined as money added by the bank to the principal.

A bank may borrow from an individual in the form of a saving account. The bank pays interest on saving accounts. How much interest the bank pays depends on the interest rate (usually expressed as a percent per year), the length of time before the debt is repaid, and how the interest is calculated.

Simple Interest

Simple interest is computed on the principal for the entire period it is borrowed. If a principal p is borrowed at a simple interest rate of r % per year for a period of t years, the interest charge i is $i = prt$

Time is expressed in the same period as the rate. For example, if the rate is 5 % per month, the time must be expressed in months.

Amount

The amount A owed at the end of t years is the sum of the principal borrowed and the interest charge

$A = p + i$

$\quad = p(1 + rt)$

Examples

A man borrowed GH¢ 5000 for 3 years at 4 % per year simple interest. How much interest does he pay?

Substitute 5000 for p, 3 for t and 4 % for r in the formula $i = prt$

Interest $= 5000 \times \dfrac{4}{100} \times 3$

$\qquad = 600$

The interest is GH¢ 600

Try this 6

A man borrows GH¢ 1200 for 5 years at 10 % per year simple interest. How much interest does he pay?

Ama borrows GH¢ 5400 for 6 years at 5 % per year simple interest. Calculate the total amount she must pay?

Interest $= 5400 \times \dfrac{5}{100} \times 6$

$\qquad = $ GH¢ 1620

Amount $= 5400 + 1620$

$\qquad = 7020$

Total amount paid is GH¢ 7020

Try this 7

Kofi borrows GH¢ 6450 for 5 years at 8 % per year simple interest. Calculate the total amount he must pay?

Exercise 17.2(a)

1. Find the simple interest on:

   (a) GH¢ 2500 for 5 years at 6 % per year

(b) GH¢ 30,000 for 4 years at 5 % per year

(c) GH¢ 7500 for 6 years at 8 % per year

(d) GH¢ 6,450 for 3 years at $2\frac{1}{2}$ % per year

2. What will:

(a) GH¢ 24,000 amount in 8 years at 3 % per year simple interest

(b) GH¢ 7,200 amount in 5 years at 2 % per year simple interest

(c) GH¢ 650 amount in 12 years at 5 % per year simple interest

(d) GH¢ 9400 amount in 6 years at $2\frac{1}{2}$ % per year simple interest

3. What principal will earn:

(a) GH¢ 270 interest in $2\frac{1}{2}$ years at 10 % per year simple interest?

(b) GH¢ 8900 interest in 5 years at 6 % per year simple interest?

(c) GH¢ 1050 interest in 3 years at 8 % per year simple interest?

(d) GH¢ 1,785 interest in 3 years at 7 % per year simple interest?

4. Calculate the rate of interest per year if:

(a) GH¢ 6300 earns GH¢ 1890 interest in 5 years

(b) GH¢ 75,200 earns GH¢ 7896 interest in 3 years

(c) GH¢ 6300 earns GH¢ 1890 interest in 8 years

(d) GH¢ 56,800 earns GH¢ 13,632 interest in 4 years

5. In what time will:

(a) GH¢ 1200 amount to GH¢ 1800 at 6 % per year simple interest?

(b) GH¢ 6000 amount to GH¢ 6900 at 5% per year simple interest?

(c) GH¢ 6400 amount to GH¢ 8,960 at 8% per year simple interest?

(d) GH¢ 7,200 amount to GH¢ 8172 at $4\frac{1}{2}$% per year simple interest?

6. At what rate of interest per year simple interest will:

(a) GH¢ 7600 amount to GH¢ 8740 in 3 years

(b) GH¢ 4,560 amount to GH¢ 5244 in 6 years

(c) GH¢ 36,900 amount to GH¢ 41,205 in 5 years

(d) GH¢ 15,200 amount to GH¢ 18,848 in 3 years

Compound Interest

Simple interest is calculated once for the period of a loan. Most banks pay compound interest on savings. The interest is paid periodically on the existing account balance, which includes both the original principal and previous interest payments. This form of interest results in significantly higher earnings over a long period of time. Interest may be calculated annually, semi annually, quarterly, monthly or daily.

Examples

Find the compound interest on GH¢ 15,000 for 3 years at 10 % per annum

Begin by computing the interest for the first year using the simple interest formula. Add this interest to the principal to find the amount at the end of the first year. Using, this amount as principal for the second year, calculate the interest and add this interest to the new principal. Repeat the process for the third year, this time using the amount at the end of the second year as principal. The difference between the amount at the end of the third year and the original principal is called the compound interest. The solution may be worked as shown below:

	GH¢
Principal for first year	= 15, 000
Interest earned in the first year = $15,000 \times 0.1$	= 1,500
Amount at the end of first year	= 16,500
Interest earned in the second year = $16,500 \times 0.1$	= 1,650
Amount at the end of second year	= 18,150
Interest earned in the third year = $18,150 \times 0.1$	= 1,815
Amount at the end of third year	= 19,965

The compound interest = $19,965 - 15,000 = 4,965$

The compound interest earned is GH¢ 4,965

Try this 8

Find the compound interest on GH¢ 2,800 for 4 years at 5 % per annum

Kwame borrows GH¢ 50,000 to renovate his house. He borrows the money at 8% compound interest payable half yearly. If he repays the amount in full after $1\frac{1}{2}$ years, what interest will he pay?

The rate 8 % per annum is equivalent to 4 % per half-year

	GH¢
Principal for first half-year	= 50,000.00
Interest earned in the first half-year	=2,000.00
Amount at the end of first half-year	= 52,000.00
Interest earned in the second half-year	= 2,080.00
Amount at the end of second half-year	= 54,080.00
Interest earned in the third half-year	= 2,163.20
Amount at the end of third half-year	= 56,243.20

The compound interest = $56,243.20 - 50,000 = 6,243.20$

The interest is GH¢ 6,243.20

Try this 9

Calculate compound interest on GH¢ 1,250 for 1 year at 12 % per annum, payable quarterly.

Using the Compound Interest Formula

The compound interest may be calculated by using the formula

$$A = P(1 + r)^n$$

where A = amount

P = principal

r = rate of interest per a period

n = no times interest is calculated

Note that the interest cannot be calculated directly

Example

Calculate the compound interest on GH¢ 2,500 for 2 years at 6 % per annum payable half-yearly

Here $P = 2,500$, $r = 3\%$ and $n = 4$. So

$A = 2,500(1 + 0.03)^4$

$= 2,813.77$

∴ Compound interest $= 2,813.77 - 2,500.00$

$= 313.77$

The compound interest is GH¢ 313.77

Try this 10

Calculate the compound interest on GH¢ 1200 for 1 year at 8 % per annum payable quarterly.

Exercise 17.2(b)

1. Find the amount if

   (a) GH¢ 4,300 is invested for 3 years at 5 % per annum compound interest

(b) GH¢ 90,000 is invested for 2 years at $4\frac{1}{2}$% per annum compound interest

(c) GH¢ 31,250 is invested for $1\frac{1}{2}$ years at 4% per annum, interest payable half-yearly

(d) GH¢ 4,500 is invested for 2 years at 12% per annum, interest payable quarterly

2. Calculate the compound interest on

(a) GH¢ 8,000 for 2 years at 5% per annum

(b) GH¢ 12,000 for 3 years at 10% per annum

(c) GH¢ 10,000 for 2 years at 4% per annum, interest payable half-yearly

(d) GH¢ 25,000 for 1 year 6 month at 8% per annum, payable quarterly

3. A man invests GH¢ 2000 in the bank at 2% per year and adds GH¢ 200 to the sum at the end of each year. How much money is in his account after four years?

4. A man borrows GH¢ 5000 at 4% payable half-yearly, and arranges to pay back GH¢ 500 at the end of each half-year. How much is he owing after 2 years?

5. Mensah borrows GH¢ 75,000 to invest in a business. He borrows the money at 5 % compound interest and repays it in full after 3 years. What interest will he pay?

6. Esi borrows GH¢ 12,500 to buy a set of furniture. She borrows the money at 15 % compound interest payable half yearly. If she repays it in full after 2 years, what interest will she pay?

17.3 Hire Purchase

A shop may offer a buyer immediate purchase of an item for which he cannot afford to pay the price as a lump-sum. The buyer usually pays some deposit and then pay off the balance in regular weekly or monthly payments over an agreed period of time. A system by which a buyer pays for an item by making regular payments over a period is called hire purchase.

The cost of an item sold on hire purchase is called the hire purchase price. The hire purchase price is the sum of all the weekly or monthly payments and the deposit.

You pay more than the cash price, when you buy on hire purchase. A small amount called the interest is charged on the amount owned after the deposit is paid. The interest is the hire price minus the cash price. The interest is normally included in the monthly instalments.

Examples

An item is sold on hire purchase for GH¢ 200 deposit and 8 monthly payments of GH¢ 50, what is the hire purchase price?

Hire purchase price

= deposit + (monthly payment × number of payment)

= 200 + (8 × 50)

= 200 + 400

= 600

The hire purchase price is GH¢ 600

Try this 11

An item is sold for GH¢ 800 deposit and 5 monthly payment of GH¢ 60, what is the hire purchase price?

An item is bought on hire purchase for a GH¢ 50 deposit and 6 monthly payments of GH¢ 25. If the cash price of the item is GH¢ 185, find the interest charged.

Hire purchase price = $50 + 25 \times 6$

$$= 50 + 150$$

$$= 200$$

$Interest = hire\ purchase\ price - cash\ price$

Interest = $200 - 185$

$$= 15$$

The interest is GH¢ 15

Try this 12

An item which cost GH¢ 950 is bought on hire purchase for a GH¢ 600 deposit and 5 monthly payments of GH¢ 80, what is the interest charged?

Rate of Interest

Example

A man buys an item for which the cash price is GH¢ 2495 on hire purchase. He paid GH¢ 495 deposit and GH¢ 450 a month for 5 months. What is the rate of interest charged?

Begin by finding the hire purchase price and the interest.

$$\text{Hire purchase price} = 495 + 450 \times 5$$

$$= 495 + 2250$$

$$= \text{GH¢ } 2745$$

$$\text{Interest} = 2745 - 2495$$

$$= \text{GH¢ } 250$$

The interest you pay when you buy on hire purchase is considered as interest on a loan. Notice that the principal for the loan is the amount the buyer still owes after he makes the deposit. The monthly principal is found by dividing the amount owned by the number of weekly or monthly payments.

In this case the buyer pays a deposit of GH¢ 495, so he owes the ship GH¢ 2000. Because the item will be paid for in 5 months, he must pay GH¢ 400 each month. The GH¢ 400 is the monthly principal.

He pays 1 month interest on the GH¢ 400, he returns at the end of the first month, two months interest on the GH¢ 400, he returns at the end of the second month, three months interest on the GH¢ 400, he returns at the end of the third month, and so on.

If R is the rate of interest, then the interest on the loan is

$$400 \times R \times \frac{1}{12} + 400 \times R \times \frac{2}{12} + 400 \times R \times \frac{3}{12} + 400 \times R \times \frac{4}{12} + 400 \times R \times \frac{5}{12}$$

$$= 400 \times R \times \left(\frac{1+2+3+4+5}{12} \right)$$

$$= 400 \times R \times \frac{15}{12}$$

This interest will amount to GH¢ 250. Using the simple interest formula $i = prt$, we have

$$250 = 400 \times R \times \frac{15}{12}$$

$$R = \frac{250 \times 12}{400 \times 15}$$

$$= 0.5$$

Therefore the rate of interest is 50 %.

Notice that the sum of the interest for 1 month, 2 months, 3 months, 4 months and 5 months is the same as the buyer paying interest on GH¢ 400 for 15 months.

Try this 13

A man buys an article for which the cash price is GH¢ 450 on hire purchase. He paid GH¢ 80 deposit and GH¢ 65 for 6 months. What is the rate of interest charged?

Using the Standard Formula

A standard formula for finding the rate of interest charged on a hire purchase is

$$R = \frac{24I}{P(n+1)}$$

where R = rate

   I = interest

   P = amount owing after the deposit

   n = number of payment

Example

A DVD player costing GH¢ 600 cash is bought on hire purchase for GH¢ 300 deposit and 5 monthly payments of GH¢ 70. What is the rate of interest?

Hire purchase price $= 300 + 70 \times 5$

$$= 300 + 350$$

$$= GH¢\ 650$$

$$Interest = 650 - 600$$

$$= GH¢\ 50$$

Now, $I = 50$, $P = 600 - 300 = 300$ and $n = 5$ so

$$R = \frac{24(50)}{300(5+1)}$$

$$= 0.667$$

The rate of interest is 66.7 %

Try this 14

A television set costing GH¢ 800 cash is bought on hire purchase for GH¢ 300 deposit and 9 monthly payments of GH¢ 60. What is the rate of interest?

Exercise 17.3

1. The cash price of a bicycle is GH¢ 1500. It is available on hire purchase by paying a deposit of two fifth of the cash price followed by 6 monthly payments of GH¢ 200. Find the hire purchase price

2. A DVD player cost GH¢ 250 cash. Ama bought the DVD player on hire purchase by making a deposit of 32 % of the cash price followed by 8 monthly payments of GH¢ 25, how much did she pay in total?

3. The cash price of an item was GH¢ 480. A man paid 25 % of the price as deposit. He then paid GH¢ 80 a month for 6 months. How much did he pay in total?

4. A car is bought on hire purchase, for a GH¢ 8,000 deposit and 6 monthly payments. If the total payment is GH¢ 15,600, find the monthly payment.

5. The cash price of a computer is GH¢ 1500. Kweku bought the computer on hire purchase making 40 % deposit followed by 5

equal monthly payments. If the total payments is GH¢ 1800, find the amount he pay each month.

6. Esi bought a television priced at GH¢ 450 on hire purchase by making GH¢ 120 deposit and 8 monthly payments of GH¢ 45. How much would she saved by paying cash?

7. An item is offered for sale on hire purchase for GH¢ 300 deposit and 8 monthly payments of GH¢ 75. If the cash price of the item is GH¢ 850, what is the interest charged?

8. A set of furniture costing GH¢ 1,280 cash is sold on a hire purchase for GH¢ 720 deposit and 4 monthly payments of GH¢ 160 each. What is the interest charged?

9. The cash price of an item was GH¢ 720. A man paid 20 % of the price as deposit. He then paid GH¢ 45 a month for 12 months. How much more did he pay for the item?

10. A man buys an article which cost GH¢ 1,200 cash on hire purchase. He paid GH¢ 500 deposit and a monthly payments of GH¢ 150. If the ship charged GH¢ 200 interest on the article, how many monthly payments did he make?

11. A computer which cost GH¢ 3290 cash is offered for sale on hire purchase. A man pays GH¢690 deposit and 12 monthly payments of GH¢ 240 each, what is the rate of interest charged?

12. An item priced at GH¢ 4725 is bought on hire purchase for GH¢ 725 deposit and 8 monthly payments of GH¢ 550. What is the rate of interest?

13. The cash price of a computer is GH¢ 2500. Ali bought the computer on hire purchase making a deposit of GH¢ 750 followed by 8 equal monthly payments. If the interest rate is 25.9 % per annum, find his monthly payments.

14. Araba bought a fridge priced at GH¢ 1200 on hire purchase. She paid 10 % deposit followed by 6 monthly payments. If interest is calculated at 16 % per annum, what would be her monthly payments?

15. An item is priced at GH¢ 5400 but can be bought with 12 % deposit and 7 monthly payments. If interest is calculated at 25 % per annum, find the total price of the item on hire purchase

17.4 Partnership

A business own jointly by two or more persons is called a partnership. The partners draw up agreement called the partnership agreement. The partnership agreement states among other things, how much capital each partner has contributed and how profits and losses will be shared. The most common methods of sharing profits are:

1. The profits are shared equally among all the partners

2. The profits are divided among the partners in proportion to the capital invested

3. Each partner receives interest on his investment. After the interest is deducted from the total profit, the remainder is shared equally among the partners or in proportion to the capital invested

Example

Adjei and Ofori invested GH¢ 6000 and GH¢ 4000 respectively in a business. They agreed to share the profit in the ratio of their investment. How much did each partner receive at the end of the year in which they made a profit of GH¢ 30,000?

The profit is shared in the ratio 6000 : 4000 = 3 : 2

$\therefore$ Adjei receives $\frac{3}{5} \times 30{,}000 =$ GH¢ 18,000

Ofori receives $\frac{2}{5} \times 30{,}000 =$ GH¢ 12,000

Try this 15

Esi and Kojo invested GH¢ 2,500 and GH¢ 3,000 respectively in a business. They agreed to share any profit in the ratio of their investment. How much did each partner receive at the end of the year in which they made a profit of GH¢ 2,640?

Exercise 17.4

1. Kofi and Ama invested GH¢ 2,700 and GH¢ 4,500 respectively in a business. They agreed to share each year's profit in the ratio of their investment. If the first year's profit was GH¢ 1,680, what was each partner's share?

2. A business is formed by two partners. One invested GH¢ 2,500 and the other GH¢ 1,500. They agreed to share the profit in the ratio of their investments. How much did each receive if they made a profit of GH¢ 6,400 in the second year?

3. A small company, formed by Dzifa and Esi, makes a profit of GH¢ 15,600. They divided the profit between them in the ratio of their investment. If Dzifa invested GH¢ 3,000 and Esi GH¢ 3,500, calculate the amount of the profit they will each receive.

4. A business was formed by Dela and Kwesi, Dela invested GH¢ 3,500; but Kwesi invested GH¢ 2,500. They agreed that each will receive 5 % interest on his investment and that this interest will be deducted from the profit. The remainder will be divided equally between them. If the business makes a profit of GH¢ 15,600, how much will each of them receive?

5. Adjei, Mensah and Araba invested GH¢ 5000, GH¢ 3000 and GH¢ 2000 respectively in a business. They agreed that each will receive 5 % interest on their investment and that this will be deducted from the profit. The remainder will be divided among them in the ratio of their investment. The profit in the fourth year was GH¢ 25,000, how much did each receive?

6. Yaw, Kofi and Kojo contributed GH¢ 10,500, GH¢ 4,500 and GH¢ 7,500 respectively to form a company. They agreed to divide the profit the company makes among them in the ratio of their contribution. If Yaw's share of the profit the company made in the year 2012 is GH¢ 2,800, find the total profit.

7. Two partners invested GH¢ 20,000 and GH¢ 12,000 respectively in a business. They agreed to divide 20 % of the profit equally between them and the remainder in the ratio of the capital invested. The company made a profit of GH¢ 26,000 in the first year, how much did each receive?

8. Afua and Adjoa invested GH¢ 30,000 and GH¢ 20,000 in a business. They agreed to share the profit as follows: Afua was paid $2\frac{1}{2}$ % of the total profit for her services as the manageress. The remainder of the profit was shared between them in the ratio of their contribution to the capital. If in the year 2010, Afua's share of the total profit was GH¢ 45,600, how much did Adjoa receive?

17.5 Taxes

The money required by the government to maintain its schools, hospitals, roads and other services is obtained mostly from taxes. A proportion of our income and the money we spend are taken by the government in taxes.

Value Added Tax (VAT)

VAT is a tax you pay when you buy goods or services from a VAT registered business. However, it is not all goods and services which attract VAT; items like food and books are exempted. The rate of tax varies. The rate of VAT in Ghana in the year 2012 was $12\frac{1}{2}$ %.

Calculating VAT

You can find the amount of VAT charged by multiplying the purchase price by the VAT rate. The price you pay for an item you buy from VAT registered shop includes the VAT.

Example

A television set is priced at GH¢ 950. If VAT at $12\frac{1}{2}$ % is added on to the price, calculate the total price of the television set.

$$VAT = 950 \times \frac{25}{200}$$

$$= GH\cent \ 118.75$$

Total price $= 950 + 118.75$

$$= GH\cent \ 1,068.75$$

Try this 16

A shop charges $12\frac{1}{2}\%$ VAT on a radio. If the purchase price of the radio is GH¢ 450, calculate the total price of the radio?

Exercise 17.5(a)

1. A shop charges $12\frac{1}{2}\%$ VAT on an item. If the purchase price of the item is GH¢ 756, what is the VAT on the item?

2. A shop charges $12\frac{1}{2}\%$ VAT on a gas stove. If the purchase price of the gas stove is GH¢ 735, what is the VAT on the gas stove?

3. A shop charges $12\frac{1}{2}\%$ VAT on a furniture. If the purchase price of the furniture is GH¢ 2,835, what is the total cost of the furniture?

4. A decorator charges a price of GH¢ 1,950 for a job. If he adds VAT at $12\frac{1}{2}\%$ to the price, calculate the total bill for the job.

5 .Cement costs GH¢ 15.60. What is the total cost after VAT at $12\frac{1}{2}\%$ is added?

6. The total price of an item is GH¢ 585. If the total price includes a VAT of $12\frac{1}{2}$%, calculate the purchase price of the item?

7. The bill of a decorator is GH¢ 1,068.75 including a VAT of $12\frac{1}{2}$%. How much does the decorator charge?

8. The total price of an item is GH¢ 201.25, including a VAT of $12\frac{1}{2}$%. How much is the purchase price of the item?

9. A shop charges $12\frac{1}{2}$% VAT on all items. If the VAT on a radio is GH¢ 7.50, what is the purchase price of the radio?

10. A shop charges $12\frac{1}{2}$% VAT on a refrigerator. If the VAT on the refrigerator is GH¢ 209.38, what is the purchase price of the refrigerator?

Sales Tax

Sales tax is a tax levied on goods and services purchased and are normally a certain percentage of the cost price. The sales tax is added to the buyer's cost.

Example

The sales tax on an item is 8 % of the purchase price. If the purchase price of the item is GH¢ 850, what is the total price of the item?

Sales tax $= 850 \times \frac{8}{100}$

$$= 68$$

Total price $= 850 + 68$

$$= 918$$

The total price is GH¢ 918

Try this 17

The sales tax on a refrigerator is 6 % of the purchase price. If the price of the refrigerator is GH¢ 1,250, what is the total price of the refrigerator?

Exercise 17.5(b)

1. The sales tax on an item is 6 % of the purchase price. If the price of the item is GH¢550, how much sales tax must be paid?

2. Kojo bought a computer priced at GH¢ 1,625. If he had to pay a sales tax of 5 %, calculate the amount of sales tax he paid.

3. The sales tax on an item is 3 %. If the price of the item is GH¢ 725, what is the total price?

4. Ama bought a radio priced at GH¢ 387. If she had to pay a sales tax of 5 %, calculate the total cost of the radio.

5. The total price of an item is GH¢ 159. If the total price includes a sales tax of 6 %, what is the purchase price?

6. The total price of an item is GH¢ 1,312.50. If the total price includes a sales tax of 5 %, what is the purchase price?

7. The sales tax rate on an item in a shop is 4 %. If the sales tax on the item is GH¢ 5.44, what is the purchase price, and what is the total price of the item?

8. The sales tax rate on all items in a shop is 3 %. If the sales tax on a radio is GH¢ 4.36, what is the purchase price, and what is the total price of the radio?

9. The price of a radio including sales tax is GH¢ 415.50. If the purchase price of the radio is GH¢ 396, what is the sales tax rate?

10. The price of an item including sales tax is GH¢ 201.82. If the purchase price of the item is GH¢ 197, what is the sales tax rate?

11. The marked price of a mobile phone is GH¢ 650. If Ofori gets a discount of 15 % and pays 5 % sales tax, how much did he pay for the mobile phone?

12. A shop offers a discount of 25 % on a computer. The sales tax on the computer is 5 %. If Akosua paid GH¢ 945, what is the marked price of the computer?

Income Tax

Every person whose income exceeds a certain amount pay a fraction of it as tax. The income tax is usually a percentage of the taxable income. The taxable income is the income minus the tax-free allowances. The rate of income tax varies.

Examples

A man has an income of GH¢ 1,500. His tax-free allowance is GH¢ 450, but he pays tax at 25 % on his taxable income up to GH¢ 600 and 40 % after that. How much income tax does he pay?

The taxable income is the amount left after the tax free allowance is deducted.

Taxable income = 1,500 − 450

$$= GH¢ \ 1,050$$

Tax on GH¢ 600 = $600 \times \frac{25}{100}$

$$= GH¢ \ 150$$

The remainder of the taxable income = 1,050 − 600

$$= GH¢ \ 450$$

Tax on GH¢450 = $450 \times \frac{40}{100}$

$$= GH¢ \ 180$$

Total tax = 150 + 180

$$= \ 330$$

The income tax is GH¢ 330

Try this 18

A man earns an income of GH¢ 3,580. His tax-free allowance is GH¢ 1,200, but he pays tax at 25 % on his taxable income up to GH¢ 3000 and 40 % after that. How much income tax does he pay?

A man has a wife and 6 children and his total income in the year 2007 was GH¢ 8,750. He was allowed the following free of tax:

Personal	GH¢ 1,200
Wife	GH¢ 300
Each child	GH¢ 250 for a maximum of 4
Dependent relations	GH¢ 400
Insurance	GH¢ 250

The rest was taxed as follows:

The first GH¢ 2000 at 10 %

The next GH¢ 2000 at 15 %

The next GH¢ 2000 at 25 %

The next GH¢ 2000 at 30 %

Calculate

(a) his tax-free allowance

(b) his taxable income

(c) his monthly tax

(d) his net monthly income

(a) Allowance for children = 250 × 4 = GH¢ 1,000

Tax-free allowance $= 1,200 + 300 + 1,000 + 400 + 250$

$$= GH\cancel{c}\ 3,150$$

(b) Taxable income $= 8,750 - 3,150 = GH\cancel{c}\ 5,600$

(c) It sometimes helps to set out the solution in a table

Taxable income	Amount taxed	Tax (in GH$\cancel{c}$)
5,600	2,000	$2,000 \times 0.10 = 200$
3,600	2,000	$2,000 \times 0.15 = 300$
1,600	1,600	$1,600 \times 0,25 = 400$
		Total tax $= GH\cancel{c}\ 900$

$$\text{Monthly tax} = \frac{900}{12}$$

$$= 75$$

The monthly tax is GH$\cancel{c}$ 75

(d) Net income is what is left after deductions such as tax, insurance and pension contribution are taken from the gross income.

$$\text{Net monthly income} = \frac{8,750 - 900}{12}$$

$$= 654.17$$

The net monthly income is GH$\cancel{c}$ 654.17

Try this 19

A man has a wife and 3 children and his total income in 2007 was GH$\cancel{c}$ 12,500. He was allowed the following free of tax

Personal	GH¢ 1,200
Wife	GH¢ 300
Each child	GH¢ 250 for a maximum of 4
Dependent relations	GH¢ 400
Insurance	GH¢ 250

The rest was taxed as follows:

The first GH¢ 2000 at 10 %

The next GH¢ 2000 at 15 %

The next GH¢ 2000 at 25 %

The next GH¢ 2000 at 30 %

On the remainder at 40 %

Calculate

(a) his tax-free allowance

(b) his taxable income

(c) his monthly tax

(d) his net monthly income

Exercise 17.5(c)

1. A man earns an income of GH¢ 2,540 in a year. He pays no income tax on the first GH¢ 750, but he pays tax at 20 % on the rest. How much income tax does he pay?

2. A man earns an income of GH¢ 18,000. His tax-free allowance is GH¢ 3,000, but he pays tax at 25 % on the first GH¢ 6000 and 30 % on the rest. How much income tax does he pay?

3. A man earns an income of GH¢ 40,000. He pays no income tax on $\frac{3}{8}$ of his income, but he pays tax at $2\frac{1}{2}$ % on the first GH¢ 17,500 and 5 % on the remainder. How much income tax does he pay?

4. Tax is paid at 20 Gp in the cedi of taxable income. Kwame's tax free income is GH¢ 2,700, and this is 15 % of his total income. What was his net income?

5. A man's annual income is GH¢ 30,000. His tax free allowance is $\frac{7}{12}$ of his income, but he pays tax at 25 % on his taxable income. If his pension contribution is $5\frac{1}{2}$% of his income, calculate his net income.

6. In a certain country annual income tax is calculated as follows:

Taxable pay	Rate of tax
On the first GH¢ 200	4 %
On the next GH¢ 400	8 %
On the next GH¢ 600	12 %
On the remainder	15 %

If a man's taxable income is GH¢ 1,450, how much income tax does he pay?

7. The annual income tax payable is assessed at the following rates:

Taxable Pay	Rates of tax
On the first GH¢ 1,000	5 %

On the next GH¢ 2,000	$7\frac{1}{2}\%$
On the next GH¢ 2000	10 %
On the remainder	12 %

If a man's annual income in the year 2008 was GH¢ 10,000, find

(a) his annual tax

(b) his net annual income

if no tax is paid on $\frac{1}{4}$ of his income

8. The annual income tax payable by a man in a certain year was assessed at the following rates:

	Rate of tax
The first GH¢ 200	Nil
The next GH¢ 300	5 %
The next GH¢ 400	$7\frac{1}{2}\%$
The next GH¢ 500	12 %
The next GH¢ 600	15 %
The next GH¢ 800	20 %

If the man earned GH¢ 1,750 that year, calculate

(a) his income tax

(b) his net income

9. A man has a wife and 3 children and his total income in the year 2007 was GH¢ 12,500. He was allowed the following free of tax

Personal	GH¢ 1,200
Wife	GH¢ 600
Each child	GH¢ 250 for a maximum of 4
Dependent relations	GH¢ 550

The rest was taxed as follows:

The first GH¢ 2000 at 10 %

The next GH¢ 3000 at 15 %

The next GH¢ 4000 at $22\frac{1}{2}$ %

The next GH¢ 5000 at 30 %

Calculate

(a) his taxable income

(b) his monthly tax

(c) his net monthly income

10. A man's annual salary is GH¢ 10,750. He pays no tax on $\frac{3}{5}$ of his income, but he pays tax on the first GH¢ 1,000 at 5 %, the next GH¢ 2,000 at $7\frac{1}{2}$ %, the next GH¢ 3,000 at 10 % and the remainder at 12 %

(a) How much tax does he pay?

(b) If the rate of tax assessment is the same as above, what was his

income in a year that he paid GH¢ 608 as tax?

Review exercise 17

1. A salesgirl earns 12 % commission on her total sales. If her total sales in a month was GH¢ 7,500, how much commission did she earn?

2. A newspaper agent earns a commission of 20 %. If he makes a total sales of GH¢ 5000 in a week, how much commission did he earn?

3. A salesman for a car company made total sales of GH¢ 150,760 in a month. If he earned GH¢ 22,614 commission, what was the rate of commission?

4. A salesman earns 12 % commission on his total sales. If he earns GH¢ 991.20 in a month, how much was his total sales?

5. A salesman earns a monthly salary of GH¢ 650 plus a 5 % commission on his total sales. If he made total sales of GH¢ 21,620, how much did he earn that month?

6. A shop offers 15 % discount on a refrigerator marked GH¢ 1,860. What is the discount?

7. A discount of 25 % is offered on an item marked at GH¢ 750, what is the sale price?

8. A shop offers a discount of 12 % on an item. If a man paid GH¢ 528 for the item, what is the marked price?

9. A machine bought for GH¢ 5,600 depreciates by 8 % of its value at the beginning of that year. What is the value of the machine at the end of the first year?

10. A car cost GH¢ 15,800 when new. Each year its value depreciates by 9 % of its value at the beginning of that year. What is its value at the end of the second year?

11. The value of a car depreciates by 20 % each year. If the value of the car at the end of the first year is GH¢ 20,560, what is the original value of the car?

12. A man borrows GH¢ 3,600 for 2 years at 15 % per year simple interest. How much interest does he pay?

13. A woman borrows GH¢ 8,420 for 3 years at 5 % per year simple interest. Calculate the total amount she must pay?

14. Find the compound interest on GH¢ 12,000 for 4 years at 5 % per year?

15. Calculate the compound interest on GH¢ 25,000 for 2 years at 6 % per year, payable half-yearly.

16. An item is sold for GH¢ 350 deposit and 6 monthly payments of GH¢ 20, what is the hire purchase price?

17. A car is sold for GH¢ 15,000 deposit and 8 monthly payments of GH¢ 2,500, what is the hire purchase price?

18. A radio is sold on hire purchase for a GH¢ 120 deposit and 5 monthly payments of GH¢ 35. If the cost price of the radio is GH¢ 280, find the interest charged.

19. An item costing GH¢ 6,000 is bought on hire purchase for GH¢ 3,000 deposit and 5 monthly payments of GH¢ 700. What is the rate of interest?

20. Two men invested GH¢ 8,000 and GH¢ 12,000 in a business. They agreed to share the profit in the ratio of their investment. How much did each man receive in a year in which they made a profit of GH¢ 6,520.

21. Dzifa and Kwame invested GH¢ 6,000 and GH¢ 8,000 respectively in a business. They agreed to share the profit as follows: Kwame was paid 5 % of the total profit for managing the

business. The remainder of the profit was shared between them in the ratio of their investment. If they made a profit of GH¢ 3,675 in a year, how much did each receive?

22. An interior decorator charges GH¢ 1,540 plus VAT to decorate a room. If the VAT rate is $12\frac{1}{2}$%, how much VAT does he add to the bill?

23. A man bought an item costing GH¢ 190 in a shop that charges $12\frac{1}{2}$% VAT. How much does he pay for the item?

24. The marked price of a watch is GH¢ 436.50, including a VAT of $12\frac{1}{2}$ %. How much does the watch cost?

25. The sales tax on a computer is 6 % of the purchase price. If the computer cost GH¢ 1,890, what is the total price of the computer?

26. The total price of a gas cooker is GH¢ 540. If the total price includes a sales tax of 8 %, what is the purchase price?

27. The sales tax on a DVD recorder is GH¢ 12.25. If the sales tax rate is 5 %, what is the total price of the DVD recorder?

28. A man earns GH¢ 13,500 in a year. His tax-free allowance is GH¢ 5,000, but he pays tax at 15 %, on the first GH¢ 2,000, 20 % on the next GH¢ 3,000 and 30 % on the rest. How much income tax does he pay?

29. In a certain country annual income tax is calculated as follows:

Taxable pay	Rate of tax
On the first GH¢ 200	4 %
On the next GH¢ 400	8 %
On the next GH¢ 600	12 %

On the remainder   15 %

A man has a tax-free allowance of GH¢ 1,560 and an income of GH¢ 3,160, how much income tax does he pay?

30. A man has a wife and 6 children and his total income in the year 2007 was GH¢ 10,850. He was allowed the following free of tax

Personal	GH¢ 1,500
Wife	GH¢ 350
Each child	GH¢ 250 for a maximum of 4
Dependent relations	GH¢ 500
Insurance	GH¢ 300

The rest was taxed as follows:

The first GH¢ 2000 at 10 %

The next GH¢ 2000 at 15 %

The next GH¢ 2000 at 25 %

The next GH¢ 2000 at 30 %

Calculate

(a) his monthly tax

(b) his net monthly income

Chapter Test 17

Take this test as you would take a test in class. After you are done, check your work against the answers in the back of the book

1. A man is paid GH¢ 150 per month and a commission of 5 % on sales over GH¢ 2,000. How much will he earn in a month, if his sales amount to GH¢ 8,265

2. A shop gives a discount of 20 % for an article marked at GH¢ 3,600. How much will Kofi pay for the article?

3. Mr Addo buys a car for GH¢ 18,500. The value of the car depreciates by 20 % each year, what is the value of the car after 2 years?

4. A car is sold for GH¢ 9,500 deposit and 8 monthly payments of GH¢ 2,500. If the cash price of the car is GH¢ 26,000, find the interest charged.

5. Ama saved GH¢ 7,200 for 4 years at 15 % per year simple interest. How much interest does pay?

6. Kwame borrowed GH¢ 5,000 at 12 % per year compound interest. How much does he pay in 3 years?

7. A man received a bill of GH¢ 157.41 from a shop. The bill incorrectly included sales tax of 6 %. How much should he pay?

8. A shop charges $12\frac{1}{2}$ % VAT on a gas cooker. If the gas cooker is marked at GH¢ 279, find the selling price of the gas cooker.

9. Esi and Kwesi invested GH¢ 15,000 and GH¢ 18,000 respectively in a business. They agreed to donate 15 % of each year's profit to an orphanage. The rest was shared in the ratio of

their investment. If they made a profit of GH¢ 20,619.50 in 2010, how much did each receive?

10. In a certain country annual income tax is calculated as follows:

Taxable pay	Rate of tax
On the first GH¢ 200	4 %
On the next GH¢ 300	8 %
On the next GH¢ 500	12 %
On the remainder	15 %

A man's tax free allowance is GH¢ 2,704, and this is 65 % of his total income. What tax does he pay?

18

Plane Geometry

In this chapter you will study shapes, like lines, triangles and circles, that can be drawn on a flat surface called a plane.

Three basic geometric figures are points, lines and planes. The term point, line and plane are considered undefined terms so we will only explained the terms, using examples and descriptions.

Planes

A plane can be thought of as a flat surface such as this page of your book, a floor or a tabletop. Unlike these surfaces, a plane has no thickness and extends in all directions without end. Planes are often represented using four-sided figures.

Points

A point is an exact position on a plane surface. We indicate the position of a point by placing a dot. A point has no size. A point is usually named by using a capital letter. All geometric figures consist of points.

Lines

The geometric figure shown in Figure 18.1 is called a line. A line is understood to be a straight line that extends in both directions without end and has no width.

Figure 18.1

Lines are named by any two points on the line. In Figure 18.1 line AB passes through the points A and B, and goes off in both directions indefinitely, and is perfectly straight. A line could also be named by a single small letter such as x. This method is sometimes used when the line does not have two points on it to define it.

The part of a line that contains two points and all the points between them is called a line segment. In Figure 18.1, the line segment AB, usually denoted by $\overline{AB}$ consists of points A and B and all points on the line between A and B. The two points A and B are called endpoints.

18.1 Angles

When two straight lines meet at a common end point, they form an angle as shown in Figure 18.2

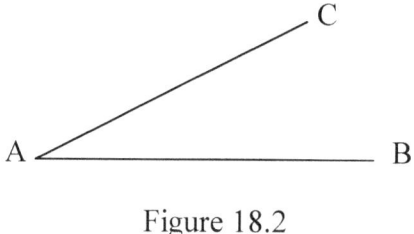

Figure 18.2

The line AC and AB are called sides of the angle, and the end point A is called the vertex (plural vertices). The symbol for angle is $\angle$.

Angles are named by indicating the lines which meet to form the angle. The angle shown in Figure 18.2 is named as $\angle CAB$ (or $\angle BAC$). Notice that the name of the vertex is in the middle.

When no other angle shares the same vertex, the angle can be named by just the vertex. So in Figure 18.2 the angle could also be called simply $\angle A$.

Occasionally, an angle is named by placing a small letter in the angle, as shown in Figure 18.3

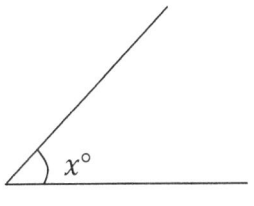

Figure 18.3

Angles are usually measured in degrees. There are 360 degrees in one full rotation. The symbol for a degree is a little circle, " ° ". For example 60° means 60 degrees.

Try this 1

Name the angle in each diagram by indicating the lines which meet to form the angle

(a)   (b)

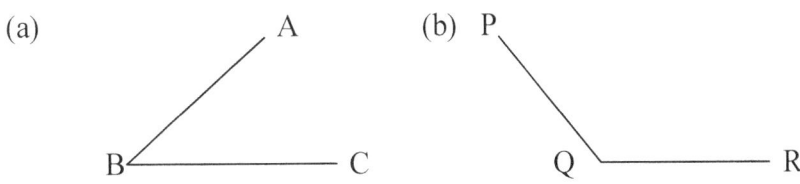

(c)

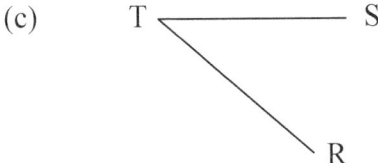

Types of Angles

Right Angles

Angle whose measure is 90° is called a right angle.

Straight Angles

A straight angle has its sides lying along a straight line. The straight angle is 180°.

Acute Angles

An angle which lies between 0⁰ and 90⁰ is called an acute angle

Obtuse Angles

An angle which lies between 90⁰ and 180⁰ is called an obtuse angle

Reflex Angles

An angle which lies between 180⁰ and 360⁰ is called a reflex angle

Acute angle   Right angle   Straight angle

Obtuse angle   Reflex angle   Reflex angle

Perpendicular Lines

When the angle between two straight lines is a right angle, either line is said to be perpendicular to the other. In Figure 18.4 the line AN is perpendicular to the line BC, written as $AN \perp BC$.

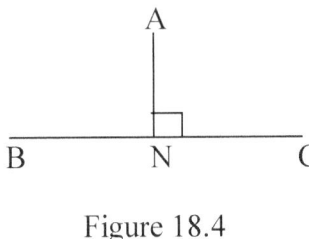

Figure 18.4

Adjacent Angles

Two angles are adjacent angles if they share a vertex and have a common side and lie on the opposite side of this line.

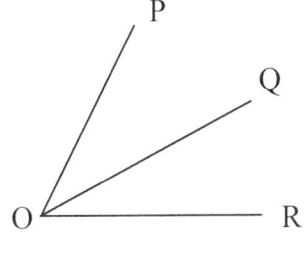

Figure 18.5

In Figure 18.5, $\angle POQ$ and $\angle QOR$ are adjacent angles. Note that $\angle POR$ and $\angle QOR$ are not adjacent angles.

Vertically opposite Angles

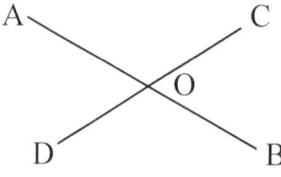

Figure18. 6

When two lines intersect at a point, they form four angles, as indicated in Figure 18.6. Each angle has a vertex at point O, called the point of intersection of the two lines.

The pair of opposite angles, angle AOC and BOD, and angle AOD and BOC are called vertically opposite angles.

Complementary Angles

Two angles whose sum is a right angle (90^0) are called complementary angles. For example, 30^0 and 60^0 are complementary angles. Similarly, 16^0 and 74^0 are complementary angles.

Supplementary Angles

Two angles whose sum is a straight angle(180^0) are said to be supplementary. For example, 120^0 and 60^0 are supplementary angles. Similarly, 32^0 and 148^0 are supplementary angles.

Try this 2

State whether the following pairs of angles are complementary, supplementary or neither.

(a) 75^0, 15^0 (b) 86^0, 94^0 (c) 79^0, 32^0 (d) 65^0, 25^0

(e) 142^0, 38^0 (f) 137^0, 53^0 (g) 110^0, 70^0 (h) 48^0, 42^0

Measuring Angles

We measure angles in degrees using a protractor. The normal protractor measures 0° to 180°. A protractor normally has two scales, an inner scale and an outer scale, reading in opposite directions. Each scale is divided into 180 equal parts. The origin of the protractor is the centre of the base line.

To measure an angle, place the origin of the protractor at the vertex of the angle to be measured, and align the base line with one arm of the angle. Read on the scale which has its zero mark on the arm you place the base line on, and increases in the direction of the other arm of the angle.

Drawing Angles

You can use a protractor to draw angles. Suppose you are required to draw $\angle BAC = 110^\circ$. Draw the line AB. Place the origin of the protractor on A, and align the base line with the line AB. Using the appropriate scale mark the point C where the protractor reads 110°. Remove the protractor and then join C to A.

Exercise 18.1(a)

1. Name the angles marked with a single letter, in terms of the lines which meet to form the angles

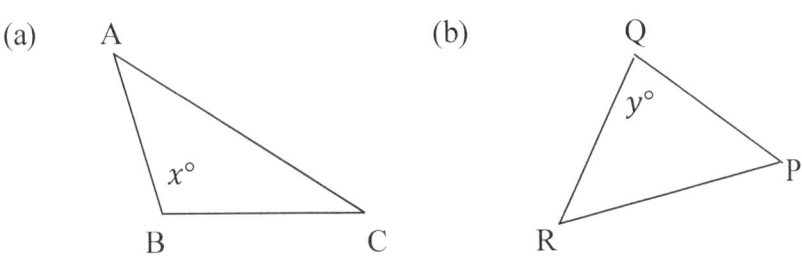

(a)   (b)

(c)   Q   S   (d)   F

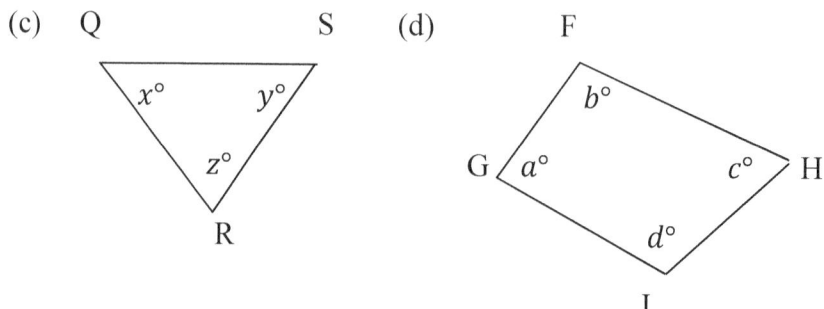

2. Use a protractor to draw the following angles:

(a) 35⁰   (b) 70⁰   (c) 54⁰   (d) 68⁰   (e) 85⁰

(f) 97⁰   (g) 105⁰   (h) 139⁰   (i) 165⁰   (j) 155⁰

3. Measure each of the following angles:

(a)   (b)

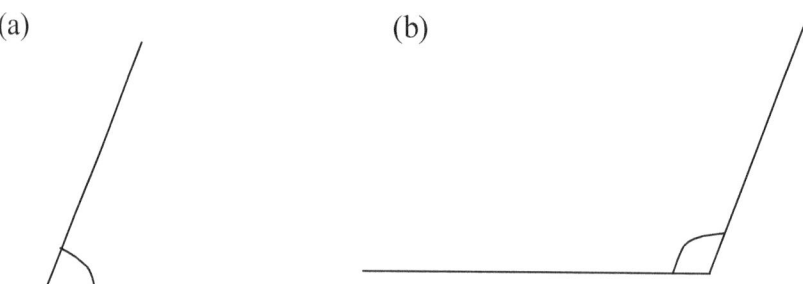

(c)   (d)

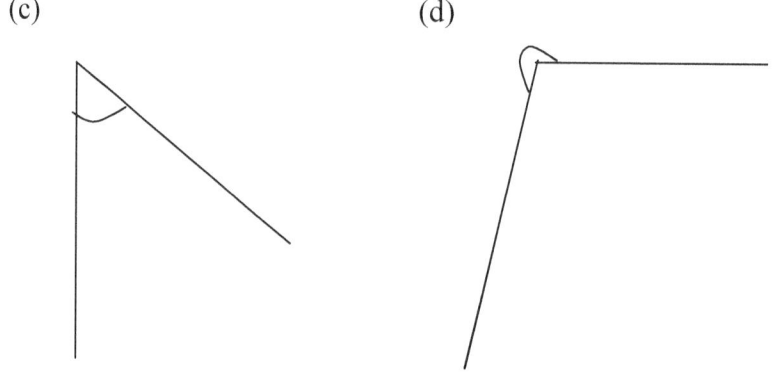

Angle Properties of Lines

1. Vertically opposite angles are equal

2. The sum of all angles around a given point is 360°

3. The sum of angles with a common vertex on a line is 180°

In Figure 18.7, $x° + y° + z° = 180°$

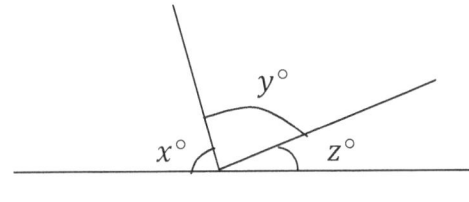

Figure 18.7

Examples

In the figure, find the value of x

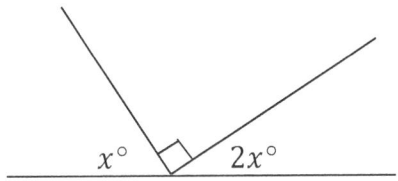

The sum of the angles is 180°.

Therefore $x + 90 + 2x = 180$

$$3x = 90$$

$$x = 30$$

Try this 3

In the figure, find the value of x

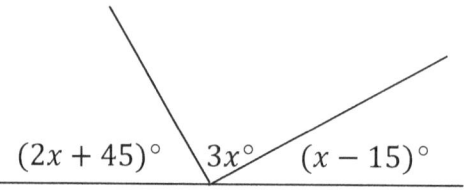

$(2x + 45)°$ $3x°$ $(x - 15)°$

Find the value of x in the figure below

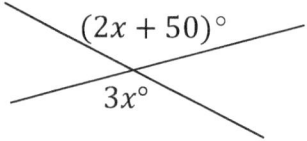

$(2x + 50)°$

$3x°$

Vertically opposite angles are equal

Therefore $3x = 2x + 50$

$x = 50$

Try this 4

Find the value of x in the figure below:

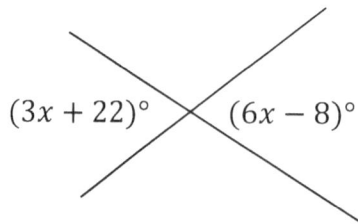

$(3x + 22)°$ $(6x - 8)°$

Exercise 18.1(b)

Find x and y in the following figures

1.

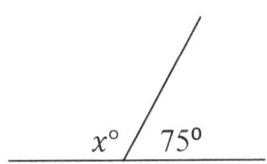

2.

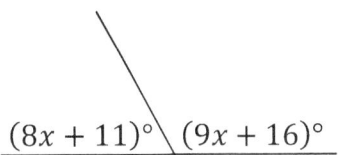

3.

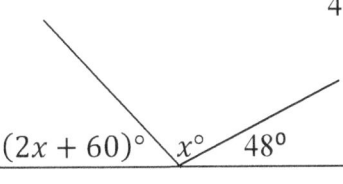

4.

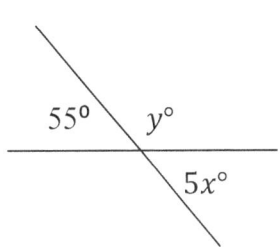

5.

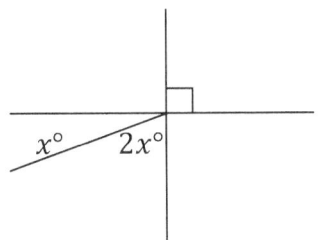

6.

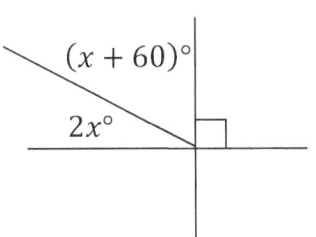

7.

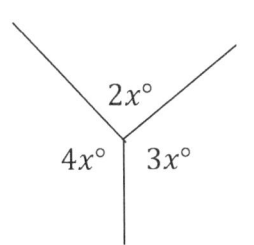

8.

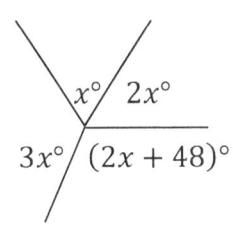

9.   10.

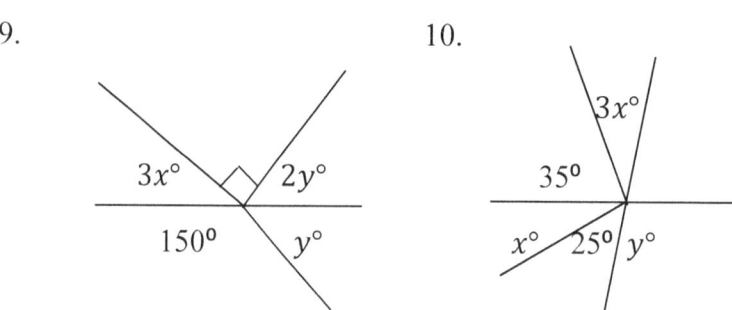

11.   12.

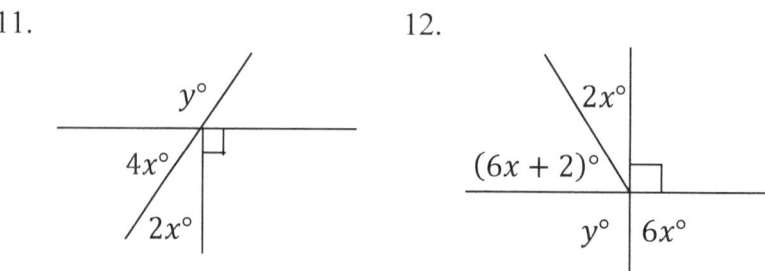

Parallel Lines

Two lines in the same plane that are the same distance apart and do not meet no matter how far they are extended are said to be parallel. Figure 18.8 shows two lines ℓ_1 and ℓ_2, that are parallel, denoted by $\ell_1 /\!/ \ell_2$. To show that lines are parallel, we draw small arrows mark on them. In Figure 18.8, note the arrows on the lines ℓ_1 and ℓ_2. If the diagram has another set of parallel lines they would have two arrows each, and so on.

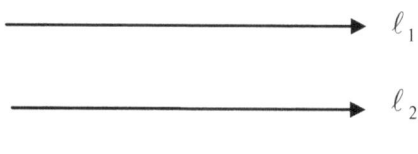

Figure 18.8

Angles formed by Parallel Lines and a Transversal

A line that intersects two or more lines in a plane is called a transversal.

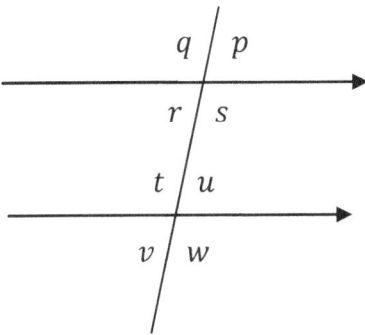

Figure 18.9

The transversal and the two parallel lines form eight angles. The four angles between the given lines are called interior angles; the four angles outside the given lines are called exterior angles. If two angles are on opposite sides of the transversal, they are called alternate angles.

r, s, t and u are interior angles

p, q, v and w are exterior angles

The pairs of angles r and u, and s and t are called alternate angles

The pairs of angles p and u, q and t, r and v, and s and w are called corresponding angles.

Angle Properties of Parallel Lines

When two lines cut by a transversal are parallel lines then:

1. the alternate angles are equal

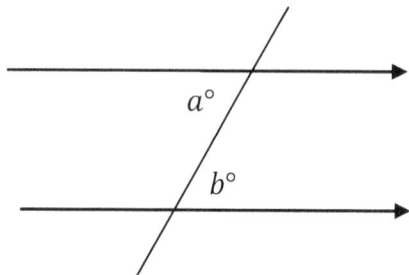

In the diagram above $a° = b°$

2. the corresponding angles are equal

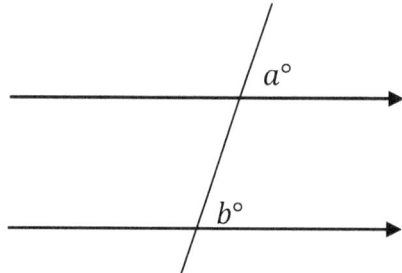

In the diagram above $a° = b°$

3. the sum of the interior angles on the same side of the transversal is 180⁰

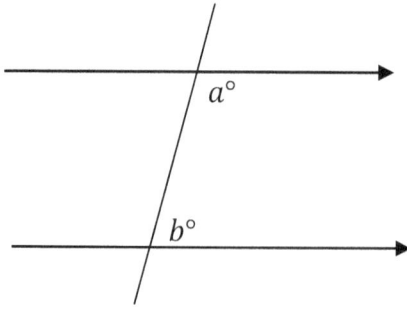

In the diagram above $a° + b° = 180°$

Notice that at the point where the transversal intersect the parallel lines there are two pairs of vertical opposite angles

Example

In the figure, two parallel lines are cut by a transversal. Find the value of x and y.

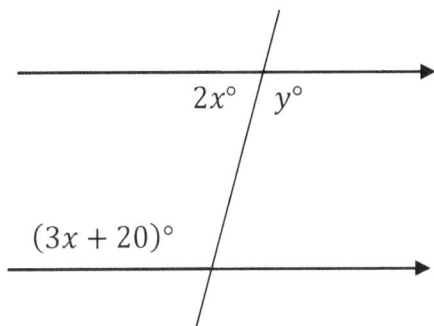

The sum of the two interior opposite angles is 180°, so

$$3x + 20 + 2x = 180$$

$$5x = 160$$

$$x = 32$$

Because y° and $(3x + 20)^\circ$ are alternate angles, $y = 3x + 20$

Substituting 32 for x, we have

$$y = 3(32) + 20 = 116$$

This result can also be obtained as follows:

The sum of the angles at a point on a line is 180°.

Therefore $y + 2(32) = 180$

$$y = 180 - 64$$

$$y = 116$$

Try this 5

Find the value of x and y

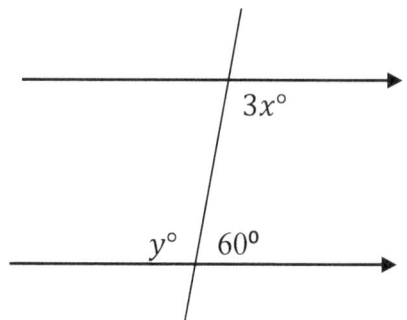

Exercise 18.1(c)

1. Find the value of angles marked in each diagram

(a)   (b)

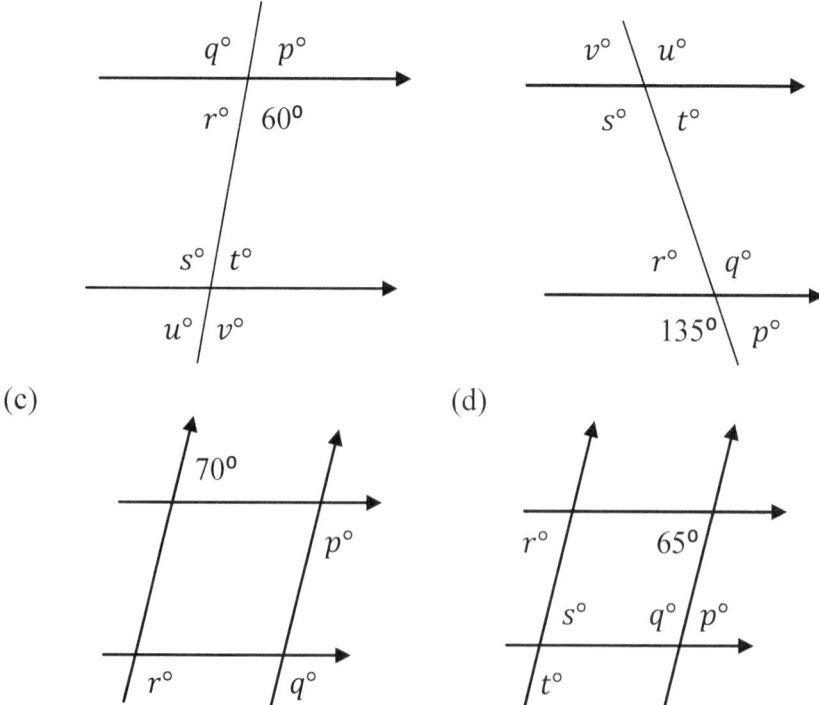

2. Find the value of x and y

(a)

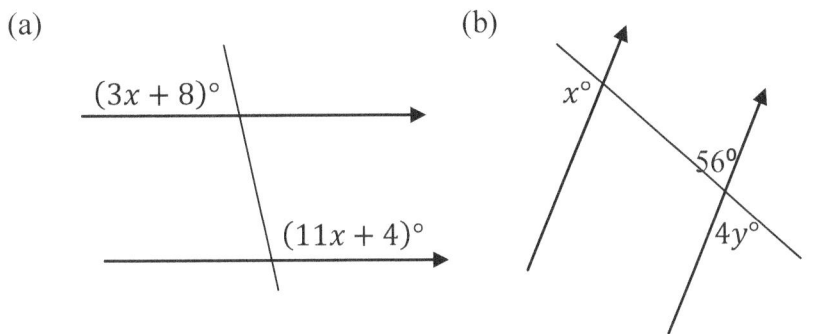

$(3x + 8)°$

$(11x + 4)°$

(b)

$x°$

$56°$

$4y°$

(c)

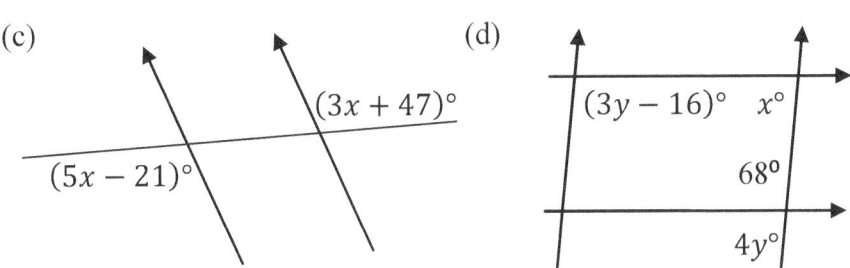

$(3x + 47)°$

$(5x - 21)°$

(d)

$(3y - 16)°$ $x°$

$68°$

$4y°$

(e)

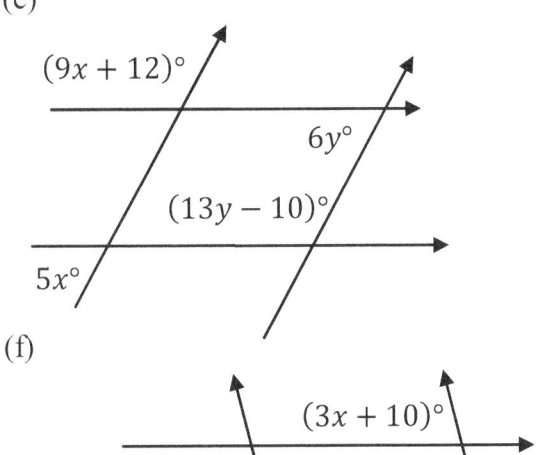

$(9x + 12)°$

$6y°$

$(13y - 10)°$

$5x°$

(f)

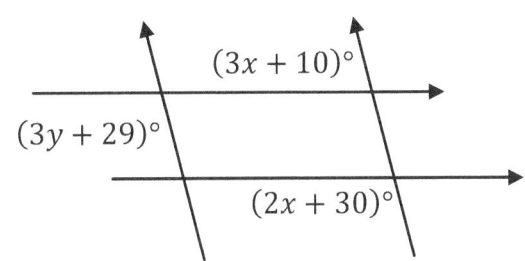

$(3x + 10)°$

$(3y + 29)°$

$(2x + 30)°$

3. Find the angle marked in each diagram

(a)   (b)

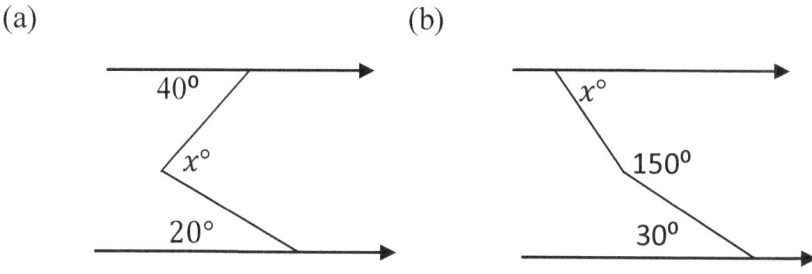

(c)   (d)

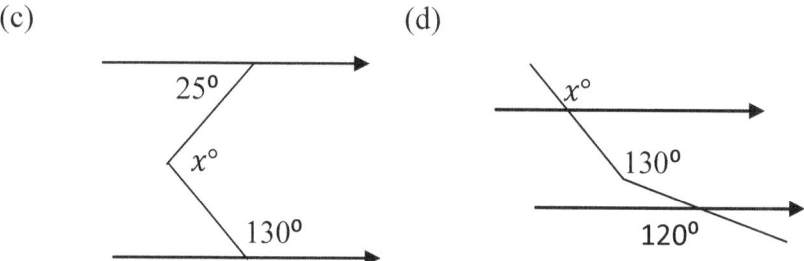

18.2 Triangles

A closed plane figure bounded by three line segments is called a triangle. Every triangle has three sides and three interior angles.

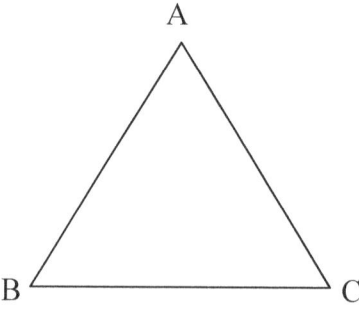

Figure 18.10

In Figure 18.10, AB, BC and CA are the sides of the triangle ABC. The points A, B and C are called vertices (singular vertex). The side opposite a vertex is called the base. The line segment drawn from any vertex perpendicular to the opposite side is called the height (or altitude). The symbol for a triangle is Δ.

Types of triangles

There are six main types of triangles. They are named by the size of their angles or the length of their sides.

Scalene triangles

Scalene triangles have sides of different length and angles of different sizes.

The shortest side of a scalene triangle is the side opposite the smallest angle, and the longest side is the side opposite the largest angle.

Isosceles triangles

Isosceles triangles have two sides of equal length. The angles opposite the equal sides are the same size.

Equilateral triangles

Equilateral triangles have three sides of equal length and the three angles are the same size. Each angle is 60°.

Right-Angled Triangle

Right-angled triangles contain one right angle. The sum of the other two angles is 90°. The side of a right-angled triangle opposite the right angle is called the hypotenuse. A little square on the corner of a triangle tells us it is a right-angled triangle.

Acute –angled Triangles

Acute-angled triangles have three acute angles. All the angles are less than 90⁰

Obtuse –angled Triangles

Obtuse-angled triangles have one angle greater than 90⁰

Angle properties of triangles

Sum of interior angles of a triangle

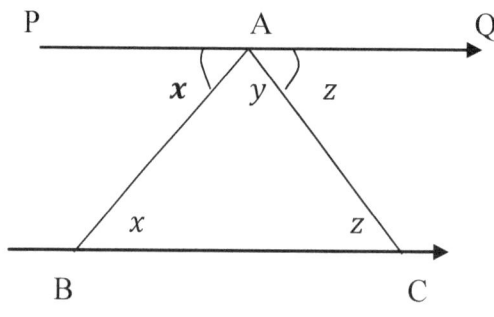

Figure 18.11

In Figure 18.11, the line PQ is drawn parallel to BC. $\angle ABC$ and $\angle BAP$, and $\angle ACB$ and $\angle CAQ$ are two pairs of alternate angles. So, $\angle ABC = \angle BAP = x°$ and $\angle ACB = \angle CAQ = z°$. Since angle PAQ is a straight angle, $x° + y° + z° = 180°$. Thus, the sum of the three angles in any triangle is 180⁰.

Angle relation between interior angles and exterior angles

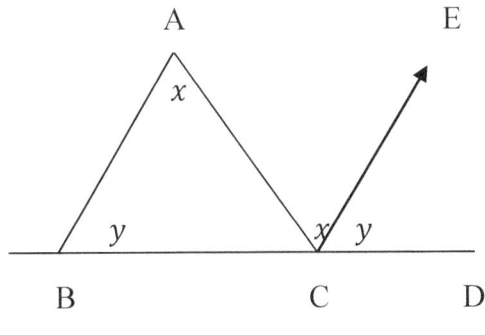

Figure 18.12

In Figure 18.12, CE is drawn parallel to AB. $\angle BAC$ and $\angle ACE$ are alternate angle, so $\angle BAC = \angle ACE = x°$, and, $\angle ABC$ and $\angle ECD$ are corresponding angle, so $\angle ABC = \angle ECD = y°$. Hence, $\angle ACD = x° + y°$. Therefore, the exterior angle of a triangle is equal to the sum of the two interior opposite angles.

Example

In each diagram, calculate the size of the angles indicated:

(a)   (b)

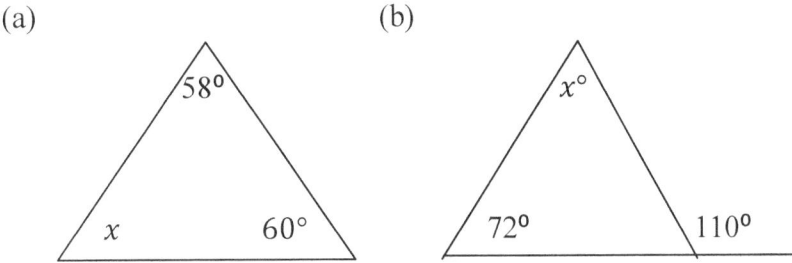

(a) The sum of the interior angles is $180°$. Therefore

$x + 58 + 60 = 180$

$x = 180 - 118$

$$x = 62$$

Therefore angle x is 62^0

(b) The exterior angle is equal to the sum of the two interior opposite angles. Therefore

$$x + 72 = 110$$

$$x = 110 - 72$$

$$x = 38$$

Therefore angle x is 38^0

Try this 6

In each diagram, calculate the size of the angles indicated

(a)   (b)

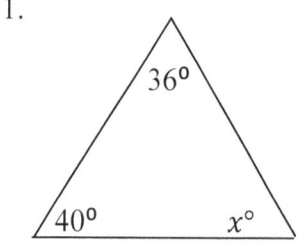

Exercise 18.2(a)

Find the value of x

1.   2.

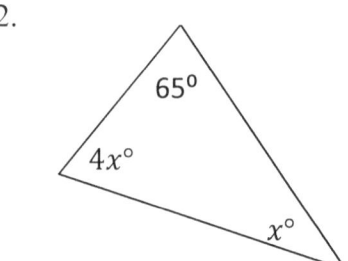

3.

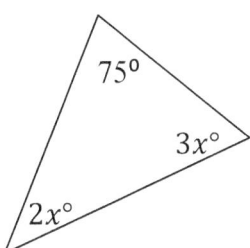

4.

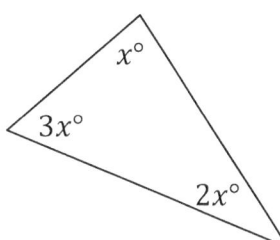

5.

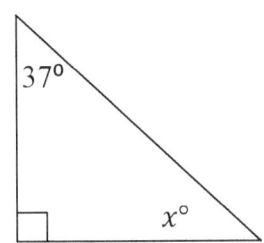

6.

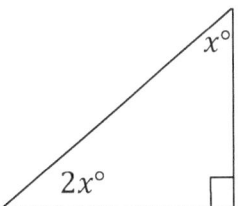

7.

8.

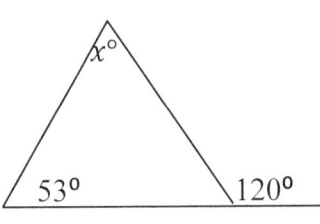

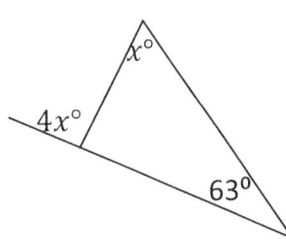

9.

10.

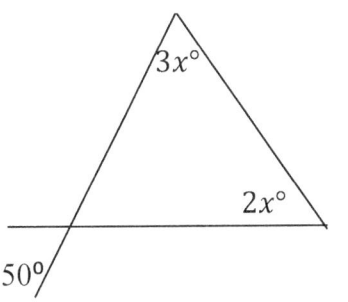

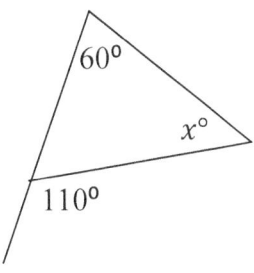

11.   12.

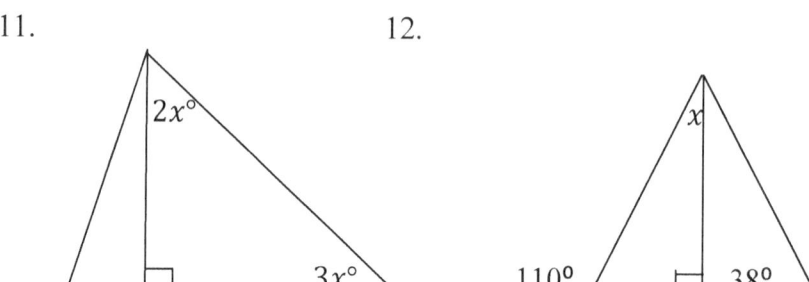

13. 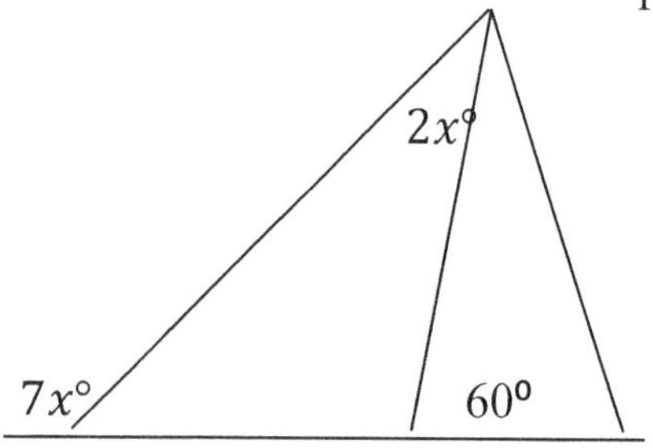   14.

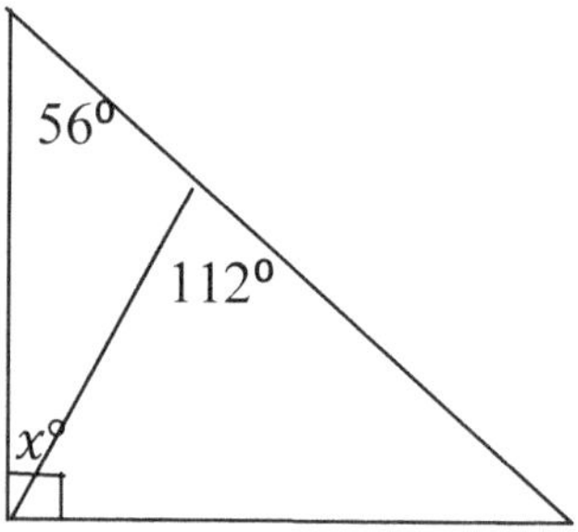

15. 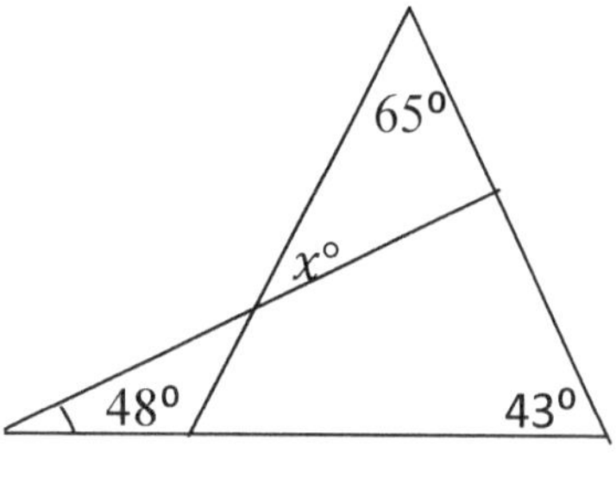   16

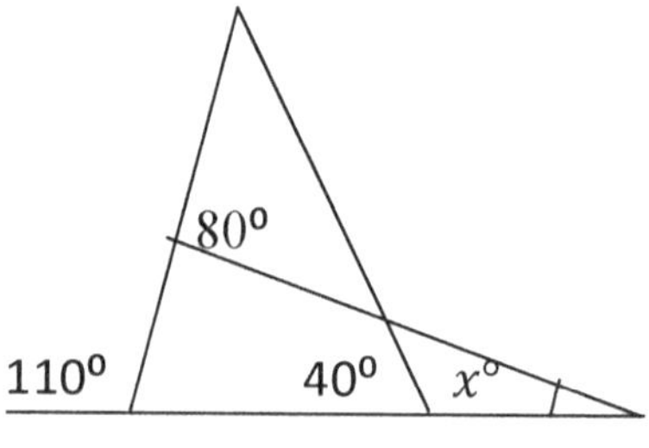

The Pythagoras' Theorem

The Pythagoras' Theorem states that in any right- angled triangle, the square of the hypotenuse is equal to the sum of the squares of the other two sides.

Recall that the hypotenuse is the side opposite the right angle.

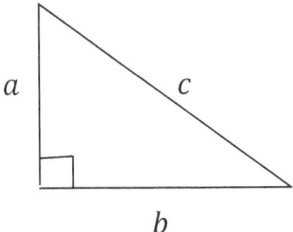

Figure 18.13

The Pythagoras' Theorem can be written in one short equation

$$a^2 + b^2 = c^2$$

If you know the lengths of any two sides, you can use the Pythagoras' theorem to find the length of the third side.

Finding Lengths of Sides in Right Angled Triangles

Example

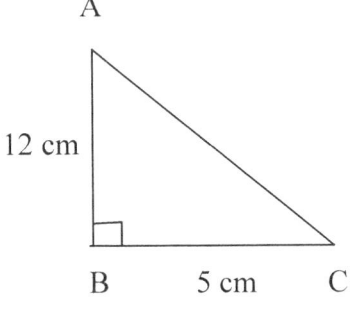

Figure 18.14

In Figure 18.14, ABC is a right- angled triangle with right angle at B, and AB = 12 cm and BC = 5 cm. Find the length of AC.

AC is the hypotenuse so

$$AC^2 = AB^2 + BC^2$$

$$= 12^2 + 5^2$$

$$= 169$$

$$AC = 13$$

The length of AC is 13 cm

Try this 7

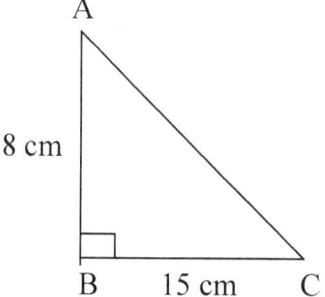

The figure shown above is a right- angled triangle with right angle at B, AB = 8 cm and BC = 15 cm, what is the length of the side AC.

Using the Pythagoras' Theorem to solve problems

Example

A ladder is placed against a building at a height that is twice the distance of its foot from the base of the building. If the length of the ladder is 15 m, how far from the base of the building should the foot of the ladder be placed?

First draw a sketch. Let x represent the distance from the foot of the ladder to the base of the building.

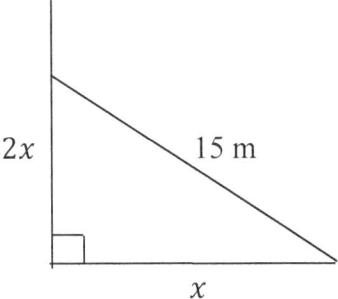

From the diagram, we have

$$(2x)^2 + x^2 = 15^2$$

$$5x^2 = 225$$

$$x = 6.7$$

The foot of the ladder should be placed 6.7 m from the base of the building

Try this 8

A ladder 25 m long is placed against a wall. If the foot of the ladder is 7 m from the base of the building, how far up the wall does the ladder reach?

Exercise 18.2(b)

1. Find the value of x

(a)   (b)

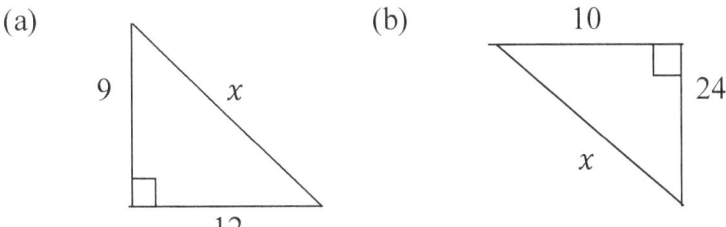

(c)    (d)

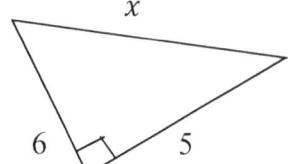

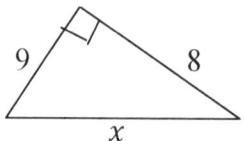

2. Find the value of x

(a)   (b)

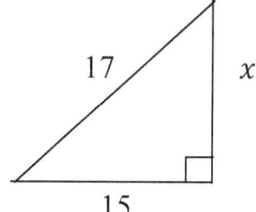

(c)   (d)

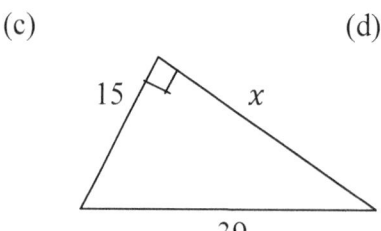

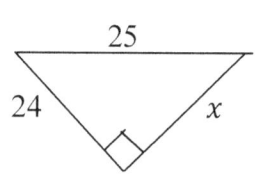

3. Find the value of x

(a)   (b)

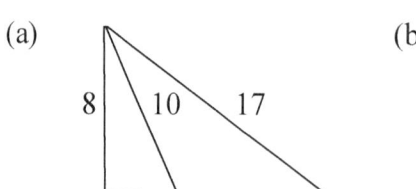

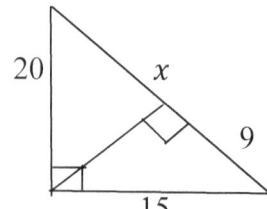

(c)

(d)

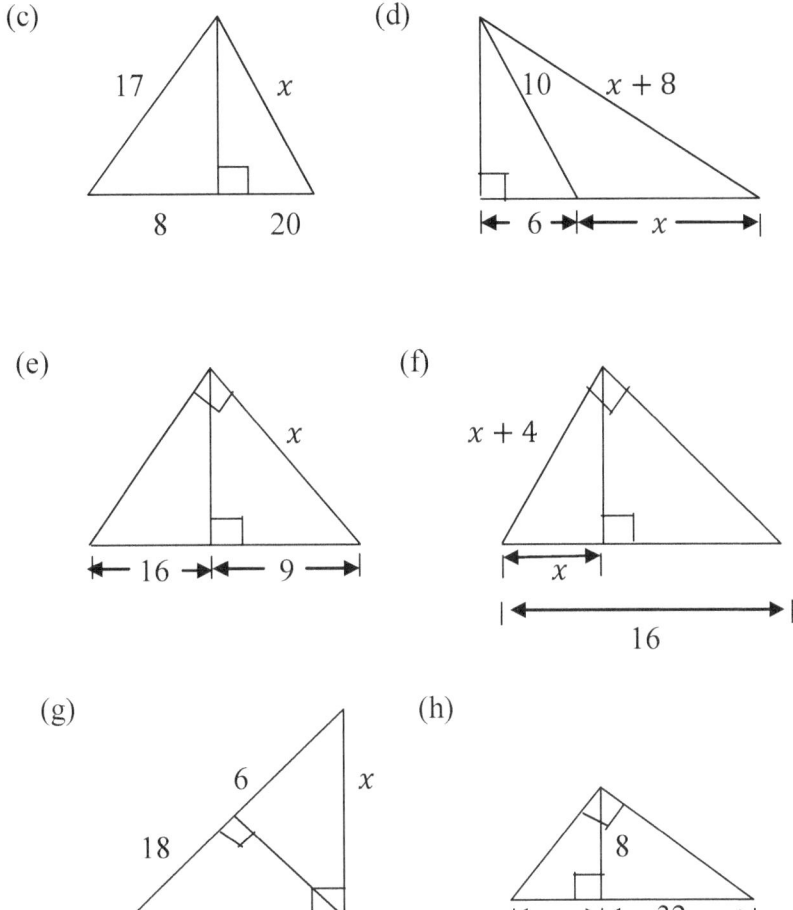

(e)

(f)

(g)

(h)

4. A fence is to have a gate 45 cm wide. The gate is 56 cm high. Find the length of a diagonal brace for the gate to the nearest centimetre.

5. A boy walked 5.8 m west and then 6.4 m north. How far was he from its starting point?

6. A man walked 1.5 km due north, and then walks due east until he

is 3 km from his starting point, how many kilometres did he walk due east?

7. A car travels 30 km due north, 40 km due east, and then 50 km due south. To the nearest tenth of a kilometre, how far is the car from its starting point?

8. A car travels 80 km due north, 60 km due west and then 70 km due north. How far is the car from its starting point?

9. A ladder 15 m long is learning against a wall. If the foot of the ladder is 6 m from the wall how far up the wall does the ladder reach?

10. A ladder 15 m long is learning against a wall and reaches 8 m up the wall. How far is the foot of the ladder from the base of the wall?

11. A flag post is 3.6 m high. A rope is attached two-thirds of the way up the flag post and is attached to the ground at a point 0.7 m from the bottom of the flag post. What is the length of the rope?

12. An aircraft took off from an airport with a speed of 30 km s^{-1}. After 5 seconds, the aircraft was vertically above a town 120 km away. Find the height of the aircraft above the town.

Properties of Special Triangles

Isosceles Triangles

Recall that isosceles triangles have two sides of equal length. Additional properties of isosceles triangles are:

1. The angles opposite the equal sides are equal

2. The line of symmetry (altitude) bisects the base at right angles

3. The line of symmetry bisects the angle at the vertex

Figure 18.15 illustrates the properties of isosceles triangles

Vertex

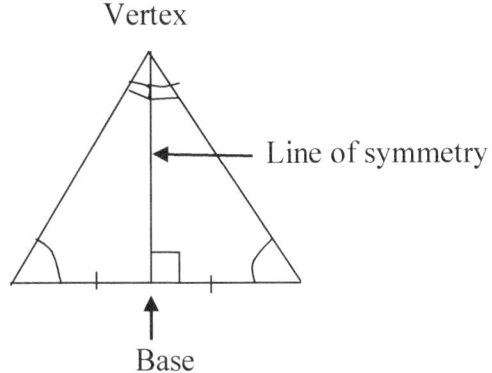

Line of symmetry

Base

Figure 18.15

Equilateral triangles

You may recall that equilateral triangles have three sides of equal length. Equilateral triangles have these additional properties:

1. The size of all the three angles are the same. Each angle is 60 degrees.

2. They have three lines of symmetry

3. The lines of symmetry bisect the sides at right angles

Example

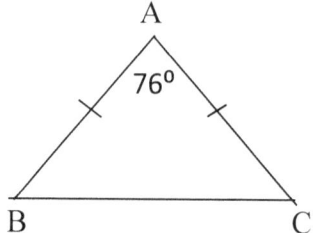

In the diagram, ABC is an isosceles triangle with $\angle BAC = 76°$ and $AB = AC$. Find $\angle ABC$

The angles opposite the equal sides are equal.

Therefore $\angle ABC = \angle BCA$.

Let $\angle ABC = \angle BCA = x$.

Then $x + x + 76° = 180°$

$$2x = 104°$$

$$x = 52°$$

Therefore $\angle ABC = 52°$

Try this 9

In the isosceles triangle ABC, $AB = AC$. If $\angle ABC = 65°$, find $\angle BAC$

Exercise 18.2(c)

Find the value of x in each of the following isosceles triangles

1.

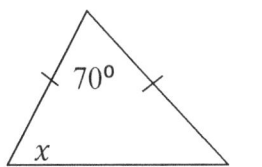

2.

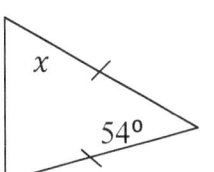

3.

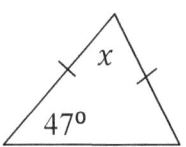

4.

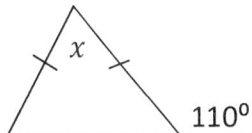

5.

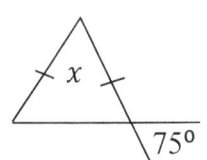

6.

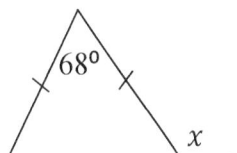

7.

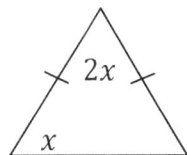

8.

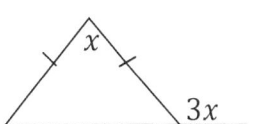

9.

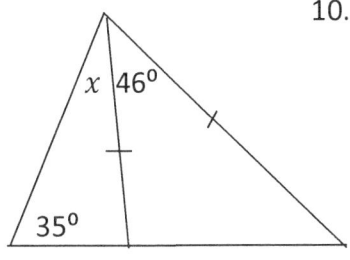

10.

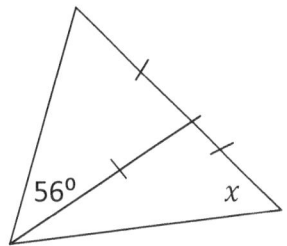

11.

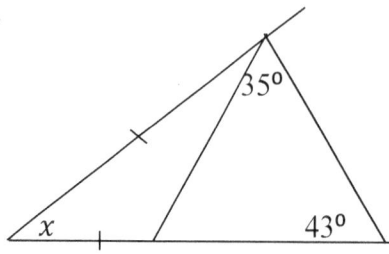

12.

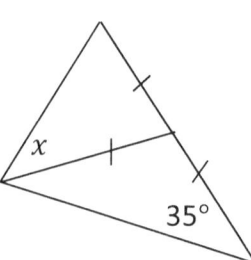

13.

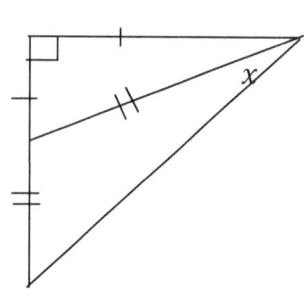

14.

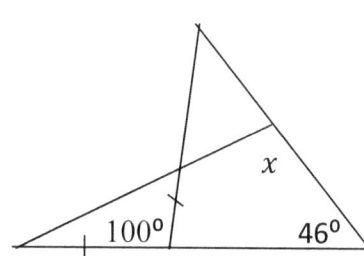

15.

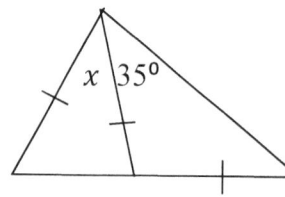

16.

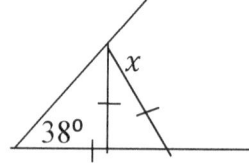

Find the value of x in each of the following equilateral triangles.

17.

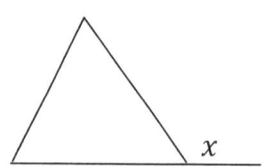

18.

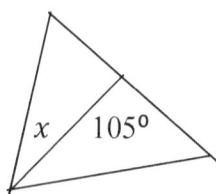

19.   20.

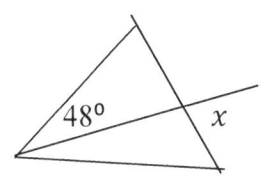   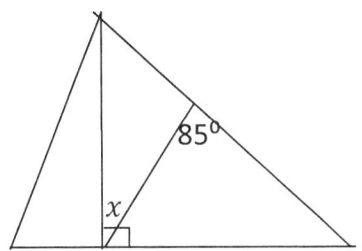

18:3 Congruent and Similar Triangles

Congruent Triangles

Congruent triangles are triangles that have the same size and shape. They have corresponding sides of the same length and corresponding angles which are equal.

Two triangles are congruent if:

1. Three sides of one triangle are equal to the corresponding sides of the other triangle (S. S. S)

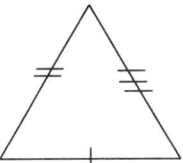   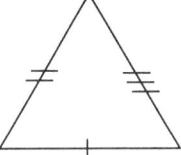

2. Two sides and the included angle of one triangle are equal to the corresponding sides and angle of the other triangle (S. A. S)

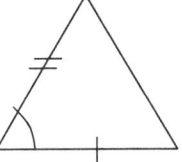   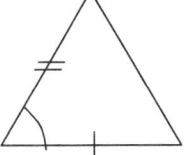

3. Two angles and the included side of one triangle are equal to the corresponding angles and side of the other triangle (A. S. A)

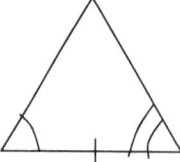 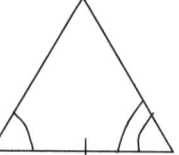

4. Two angles and the non-included side of one triangle are equal to the corresponding angles and side of the other triangle(AAS).

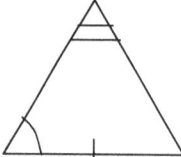 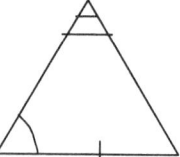

5. The hypotenuse and one side of one right-angled triangle are equal to the corresponding hypotenuse and side of the other right-angled triangle

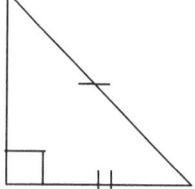 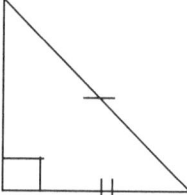

Note that when shapes are congruent, all corresponding sides and angles are also congruent.

Congruent and similar triangles are named by listing corresponding vertices in the same order.

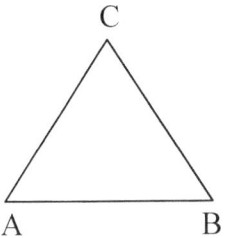 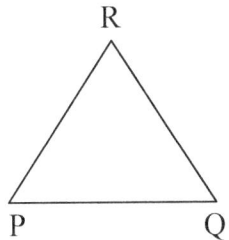

If triangle ABC is congruent to triangle PQR, the vertex labelled A corresponds to the vertex labelled P, vertex B correspond to Q, and vertex C corresponds to R. Therefore, AB corresponds to PQ, BC corresponds to QR and AC corresponds to PR.

Try this 10

State whether the following pairs of triangles are congruent or not.

(a)

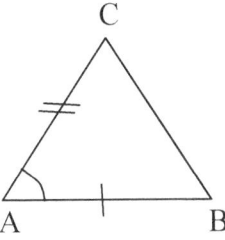 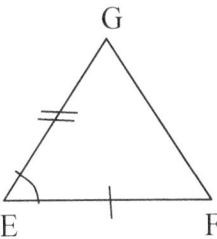

(b)

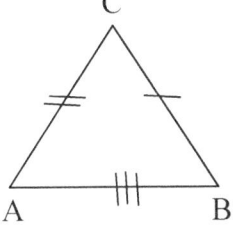 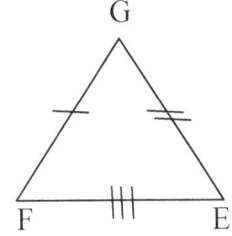

(c)

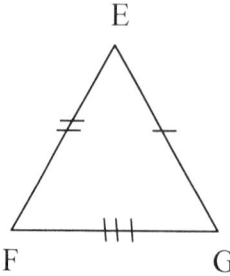   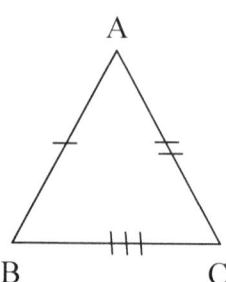

(d)

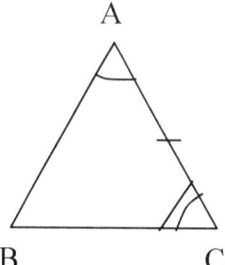   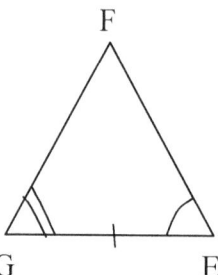

(e)

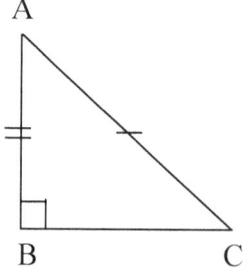   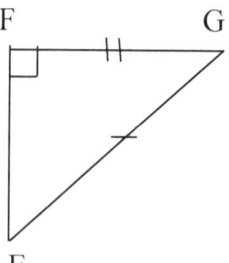

(f)

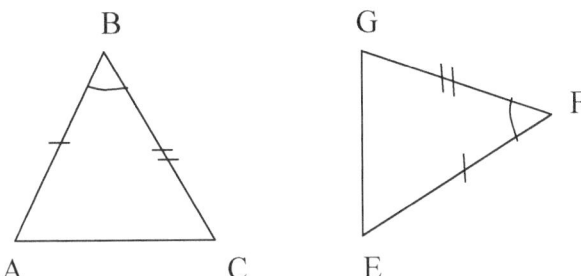

Similar Triangles

Two triangles can have the same shape and the same angles but be of different sizes. We say that the two triangles are similar. Two triangles are similar triangles if:

  1. Two angles of one triangle are the same size as the corresponding angles of another triangle

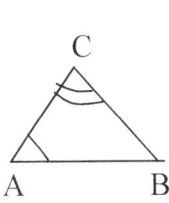 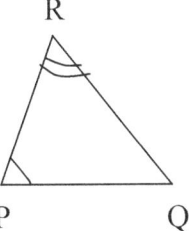

Notice that the third angles of the triangles are also equal, so two triangles are similar if all three pairs of corresponding angles are equal.

  2. The lengths of corresponding sides of two triangles are proportional.

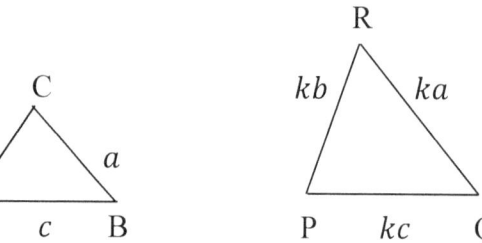

In the similar triangles ABC and PQR

$$\frac{PQ}{AB} = \frac{QR}{BC} = \frac{PR}{AC} = k$$

3. The lengths of two sides of a triangle are proportional to the lengths of two corresponding sides of another triangle and the size of the included angles are the same.

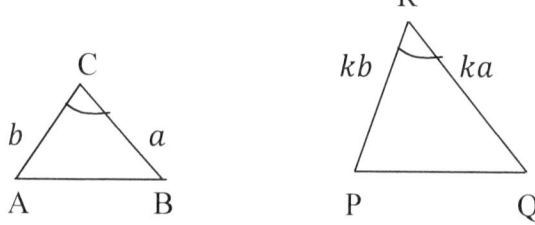

Note that if two triangles are similar then the ratios of the lengths of one triangle are equal to the ratios of the corresponding lengths of the other triangle.

Example

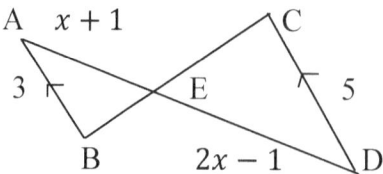

In the diagram $AB = 3$, $CD = 5$, $AE = x + 1$ and $ED = 2x - 1$, find the value of x, and the lengths of AE and DE.

In the figure $\overline{BA}//\overline{DC}$

Therefore, $\angle BAE = \angle CDE$ and $\angle ABE = \angle DCE$ (alternate angles)

$\angle AEB = \angle DEC$ (vertically opposite angles)

$\triangle ABE$ and $\triangle DCE$ are similar triangles

The ratios of the lengths of any pair of corresponding side are equal. We have

$$\frac{AB}{DC} = \frac{AE}{DE}$$

$$\frac{3}{5} = \frac{x + 1}{2x - 1}$$

$$3(2x - 1) = 5(x + 1)$$

$$6x - 3 = 5x + 5$$

$$x = 8$$

Now find AE and ED

$$AE = x + 1$$

$$= 8 + 1$$

$$= 9$$

and $ED = 2x - 1$

$$= 2(8) - 1$$

$$= 15$$

Try this 11

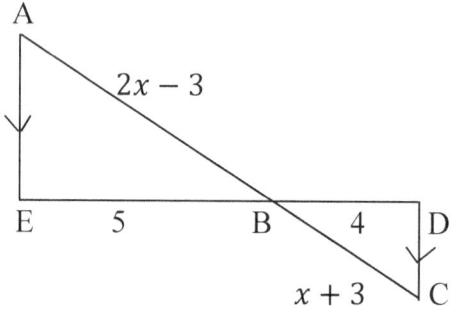

In the diagram above find length of AB and BC.

You can use similar triangles to find lengths, widths or heights of objects that are too difficult to measure directly.

Example

A pole 3 metre long cast a shadow 1.2 metre long. What is the height of a tree whose shadow is 1.8 metres long?

The sun's ray form two similar triangles as shown below

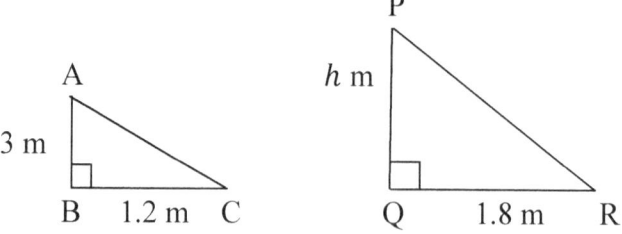

You can write the following proportion

$$\frac{height\ of\ the\ tree}{height\ of\ the\ pole} = \frac{tree\ shadow\ length}{pole\ shadow\ length}$$

Now substitute the known values and let h be the height of the tree

$$\frac{h}{3} = \frac{1.8}{1.2}$$

$$h = \frac{3(1.8)}{1.2}$$

$$= 4.5$$

The tree is 4.5 metres long.

Try this 12

A model of a tower 1.2 metre cast a shadow 0.9 metre long. If the tower's shadow is 240 metres long, what is the height of the tower?

Exercise 18.3

1. In the following diagrams, identify pairs of congruent triangles. Give reasons for your answers.

(a)   (b)

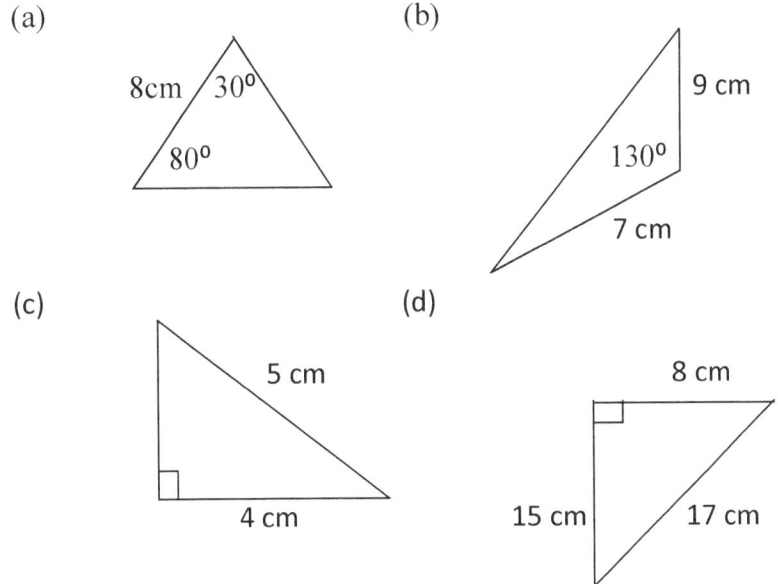

(c)   (d)

(e)

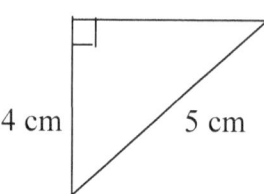

4 cm   5 cm

(f)

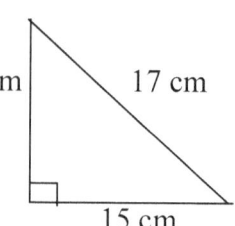

8 cm   17 cm

15 cm

(g)

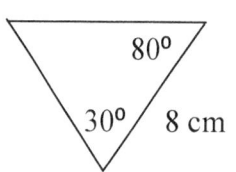

80⁰

30⁰ 8 cm

(h)

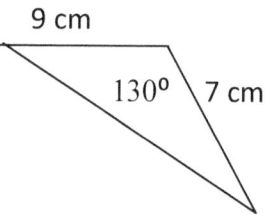

9 cm

130⁰ 7 cm

2. Identify the similar triangles in each figure

(a)

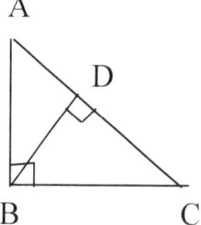

A
D
B   C

(b)

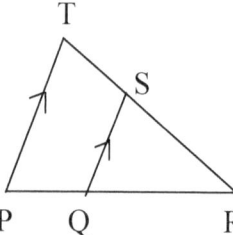

T
S
P   Q   R

(c)

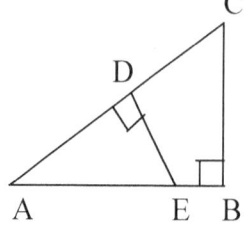

C
D
A   E B

3. Identify the similar triangles, and find the value of x

(a)

(b)

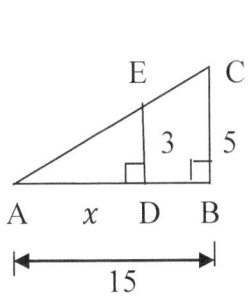

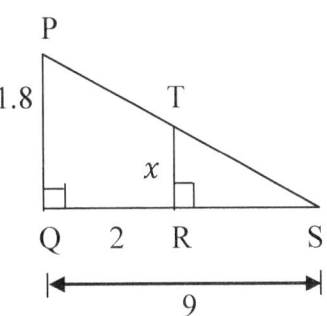

(c)

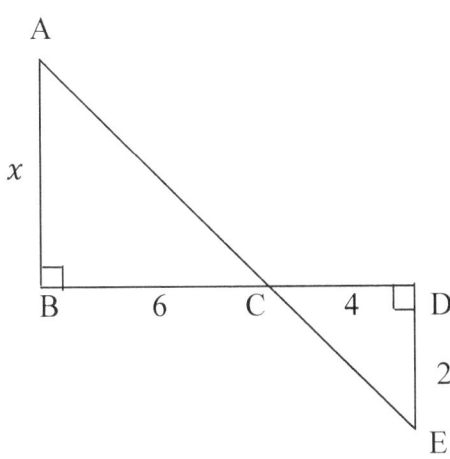

4. Identify the similar triangles, and find the value of x

(a)

(b)

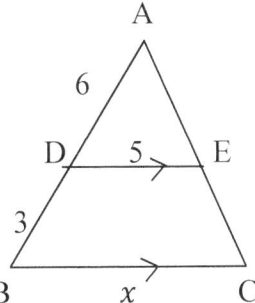

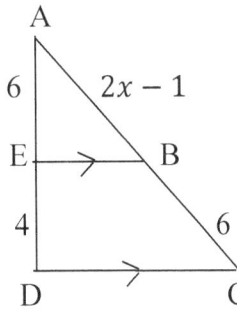

(c)   A

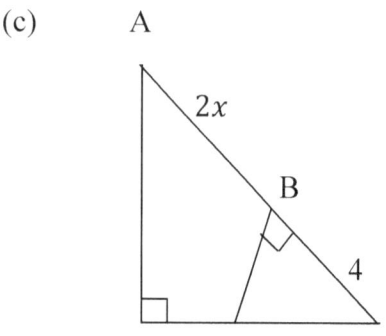

2x

B

4

E   3 D x + 2 C

5. Identify the similar triangles, and find the length of the indicated sides

(a)   AC and CE   (b)   QR and US

$$QR = x - 2 \quad US = x + 1$$

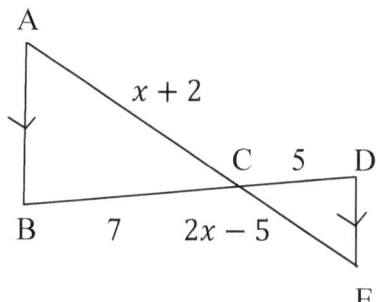

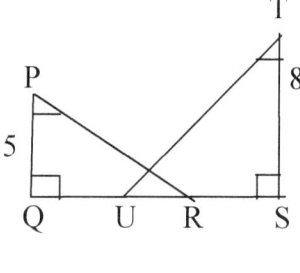

(c)   AC and AE

$$AC = 2x - 1$$

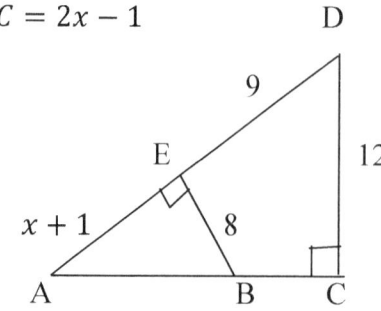

6. At 12 noon the shadow of a 2.7 metre pole is 30 centimetres. What is the height of a tree whose shadow was 1.5 metres?

7. Kofi used a camera he made from a box to photograph his father. The distance from the lens of the camera to the film is 15 centimetres. If the image on the film is 12 centimetres, and the distance of his father from the camera lens is 2 metres, how tall is his father?

8. Dzifa is having her portrait taken. The film is 1.5 centimetres from the camera lens and can have a 4.5 centimetre image. If she is 1.74 metres tall, how far from the lens can she be for a full length picture?

9. Kwame wants to measure the height of a building. He sees the top of the building in a mirror that is placed face upward 95 metres from the building. If he is 40 centimetres from the mirror, and the distance from his eyes to the ground is 1.8 metres, how tall is the building?

18.4 Polygons

A polygon is a closed plane figure bounded by three or more straight lines. Examples of polygons include triangles and quadrilaterals.

Quadrilaterals

Quadrilaterals are polygons bounded by four sides

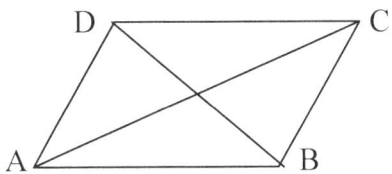

Figure 18.16

In Figure 18.16, points A, B, C and D are vertices, and AB, BC, CD and DA are the sides of the quadrilateral ABCD. A polygon is usually named by listing the consecutive vertices in order. A diagonal of a polygon is a line segment joining any two nonadjacent vertices. AC and BD are diagonals.

Types of Quadrilaterals

There are several types of quadrilaterals. The diagrams below show six main quadrilaterals.

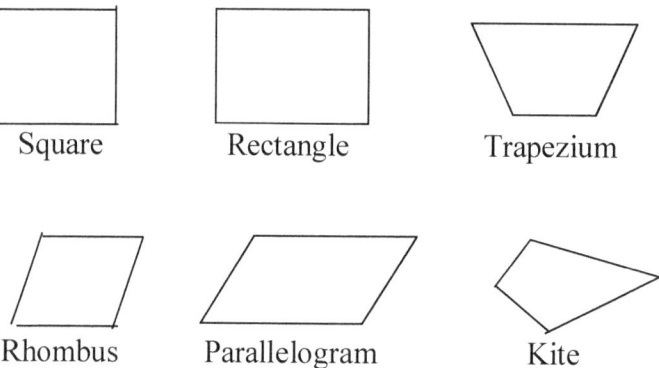

Square   Rectangle   Trapezium

Rhombus   Parallelogram   Kite

Angle properties of quadrilaterals

The properties of six main types of quadrilateral are described below:

Parallelogram

A parallelogram is a quadrilateral with both pairs of opposite sides parallel

Properties

1. The opposite sides are equal

2. The opposite angles are equal

3. The diagonals bisect each other

Trapezium

A trapezium is a quadrilateral with one pair of opposite sides parallel

A trapezium is called an isosceles trapezium if it has one pair of parallel sides and the other two sides are equal in length.

Rhombus

A rhombus has four sides of equal length

Properties

1. The opposite sides are parallel

2. The opposite angles are of equal size

3. The diagonals intersect at right angles

4. The diagonal bisect the angles

Rectangle

A rectangle has two pairs of sides of equal length

Properties

1. The size of each interior angle is 90°

2. The opposite sides are parallel

3. The diagonal bisects each other

4. The diagonals are equal

Square

A square has four sides of equal length, and the size of each interior angle is 90°.

Properties

1. The opposite sides are parallel

2. The diagonals bisect each other

3. The diagonals intersect at right angles

4. The diagonals are equal

5. The diagonals bisect the angles

Kite

A kite has two pairs of adjacent sides equal in length

Properties

1. One pair of equal angles.

2. The diagonals intersect at right angles

Other Polygons

Polygons are named by the number of sides they have. The table below list the name of some common polygons.

Number of sides	Name
5	pentagon
6	hexagon
7	heptagon
8	octagon
9	nonagon
10	decagon
11	hendecagon
12	dodecagon

Regular Polygon

A polygon is called a regular polygon if all its sides are of equal length and all its angles are of equal size. Examples of regular polygons are equilateral triangles and squares.

Angles of Polygon

Sum of Interior Angles

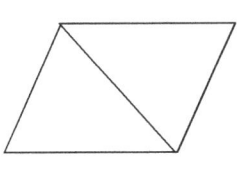

Quadrilateral

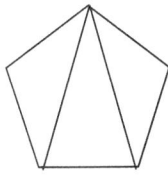

Pentagon

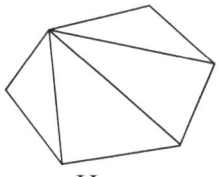

Hexagon

Notice that in each case, the polygon is separated into triangles. The number of triangles is two less than the number of sides of the polygon. In general, if a polygon has n sides, then there will be $(n - 2)$ triangles.

Recall that the sum of the interior angles of a triangle is 180^0. The sum of the interior angles of a polygon is 180^0 times the number of triangles.

Polygon	Number of sides	Number of triangles	Sum of interior angles
triangle	3	1	180
quadrilateral	4	2	360
pentagon	5	3	540
hexagon	6	4	720
heptagon	7	5	900
octagon	8	6	1080
nonagon	9	7	1260
decagon	10	8	1440

In general, the sum of the interior angles of a polygon is $180(n-2)$ degrees, where n is the number of sides of the polygon.

Examples

Find the size of each interior angle of a regular 12-sided polygon

The sum of interior angles $= 180(12 - 2)$

$$= 180 \times 10$$

$$= 1800°$$

The interior angles in a regular polygon are of equal size.

Therefore each angle is $\frac{1800}{12} = 150°$

Try this 13

Find the size of each interior angle of a regular 11-sided polygon

A regular polygon is such that each interior angle is 108^0. Find the number of sides of the polygon.

The size of each interior angle of a regular polygon is $\frac{180(n-2)}{n}$.

Therefore, $\frac{180(n-2)}{n} = 108$

$$180n - 360 = 108n$$

$$72n = 360$$

$$n = 5$$

The polygon has 5 sides

Try this 14

A regular polygon is such that each interior angle is 135^0. Find the number of sides of the polygon?

Exercise 18.4(a)

1. Find the sum of the interior angles of the polygon having the following number of sides

   (a) 14   (b) 16   (c) 18   (d) 20   (e) 36

2. Find the size of each interior angle of the regular polygon having the following number of sides

   (a) 5   (b) 8   (c) 12   (d) 15   (e) 25

3. Find the number of sides of the regular polygon with the following interior angles

(a) 140⁰   (b) 120⁰   (c) 160⁰   (d) 144⁰   (e) 156⁰

4. Find the value of x in each of the following figures

(a)

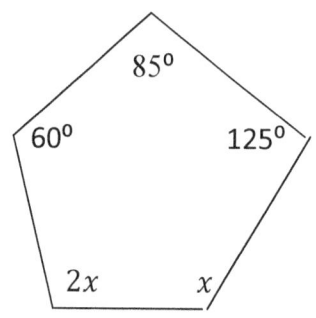

(b)

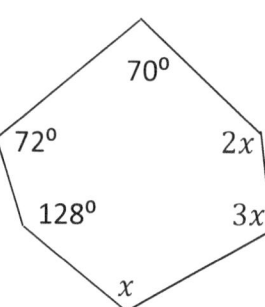

(c)

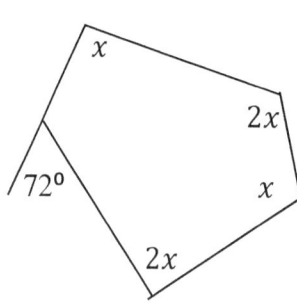

(d)

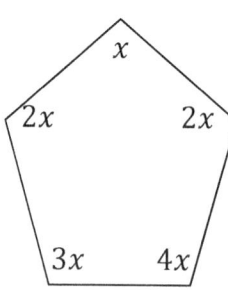

(e)

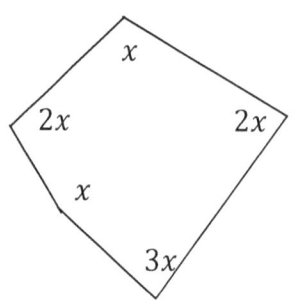

(f)

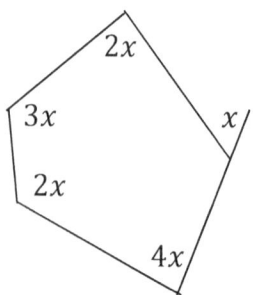

5. Three angles of an irregular octagon are 90^0, 120^0 and 150^0. The remaining angles are equal. Find the size of each of the remaining angles

6. Two angles of a heptagon are equal. Each of the other five angles has a size twice that of the other two angles. Find the size of each angle.

Sum of the Exterior Angles

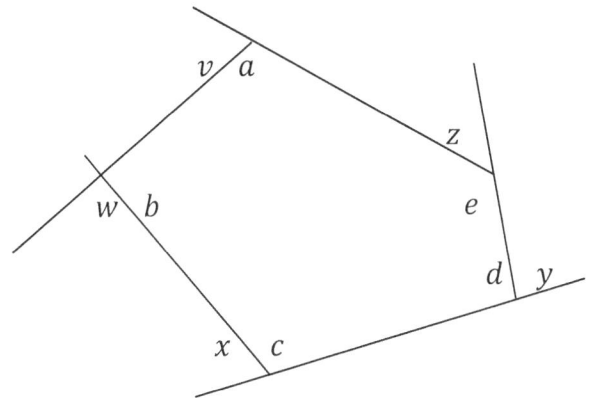

Figure 18.17

In Figure 18.17, the interior angles are labelled a, b, c, d and e, the exterior angles are labelled x, y, z, v and w.

The sum of the angles at each vertex is 180^0. Hence

$$(a + v) + (b + w) + (c + x) + (d + y) + (e + z) = 5 \times 180^0$$

$$(v + w + x + y + z) + (a + b + c + d + e) = 900°$$

That is, the sum of the exterior angles + sum of interior angles
= 900º

The sum of the interior angles of a pentagon is 540º

Therefore the sum of the exterior angles = 900º – 540º

$$= 360º$$

In general, the sum of the exterior angles of any polygon is 360º.

If the polygon is regular and has n sides, then each interior angle is $\dfrac{360°}{n}$.

Examples

Calculate the number of sides a regular polygon has if each interior angle is 144º

The sum of the interior angle and exterior angles is 180º.

Therefore, the size of each exterior angle is 180º - 144º = 36º

$$n = \frac{360}{36} = 10$$

Therefore, the polygon has 10 sides

Try this 15

Calculate the number of sides a regular polygon has if each interior angle is 165º

Find the size of each interior angle of a regular nonagon

The sum of the exterior angles is 360º.

Hence, the size of each exterior angle of a regular nonagon is

$\frac{360}{9} = 40°$

The size of each interior angle is $180° - 40° = 140°$

Try this 16

Find the size of each interior angle of a regular 12 – sided polygon

Exercise 18.4(b)

1. Find the number of sides of the regular polygon having the

   following exterior angle

   (a) $30°$ (b) $45°$ (c) $72°$ (d) $60°$ (e) $36°$

2. Find the size of each exterior angle of the following regular

   polygons

   (a) octagon   (b) heptagon   (c) pentagon   (d) dodecagon

3. Find the number of sides of the regular polygon with the

   following interior angles

   (a) $156°$ (b) $120°$ (c) $150°$ (d) $135°$

4. Find the value of x in each of the following figures

   (a)   (b)

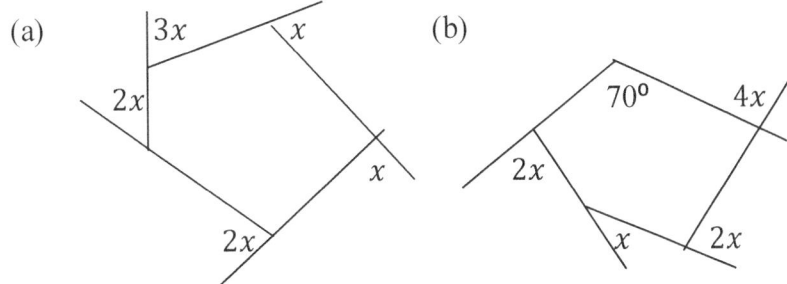

(c)   (d)

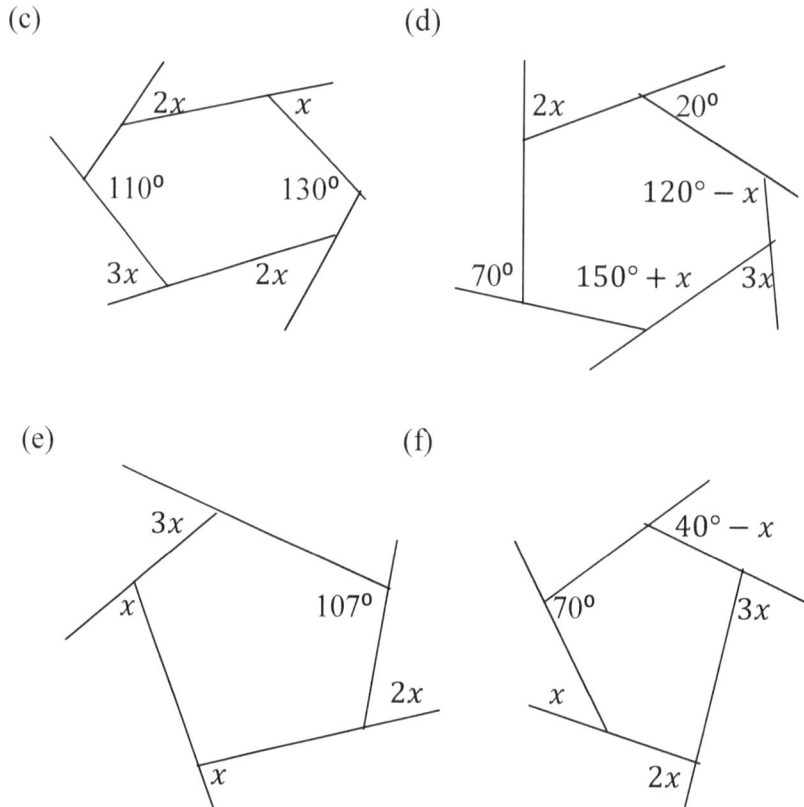

(e)   (f)

18.5 Angles in Circles

Circles

A circle is the set of all points that are a fixed distance from a fixed point in the plane called its centre. The perimeter of a circle (or the distance around the circle) is called the circumference.

Parts of a Circle

Radius

Any line segment joining the centre to any point on the circle is called a radius (plural radii).

Chord

Any line segment that joins a point on the circle to another point on the circle is called a chord. A chord that passes through the centre is called the diameter. The diameter is twice the radius

Any chord that is not a diameter divides the circle into two segments, called the major and minor segments (see Figure 18.18)

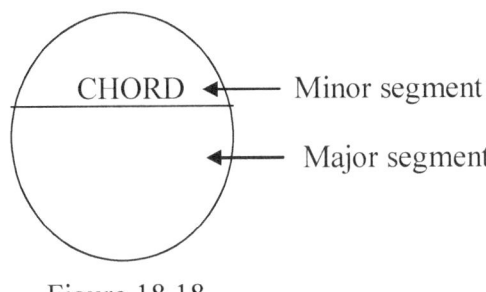

Figure 18.18

Arc of a circle

An arc is a part of a circle. For example, in Figure 18.19 the points A and B form a minor arc called arc AB. The diameter of a circle separates a circle into two arcs called semi-circles.

Sector of a circle

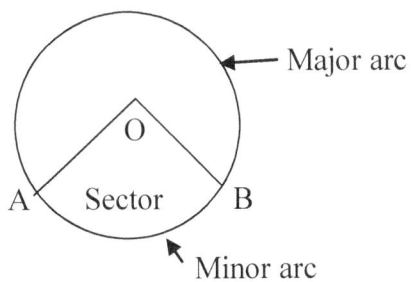

Figure 18.19

The area bounded by two radii and an arc is called a sector (see Figure 18.23)

Arc and Angles

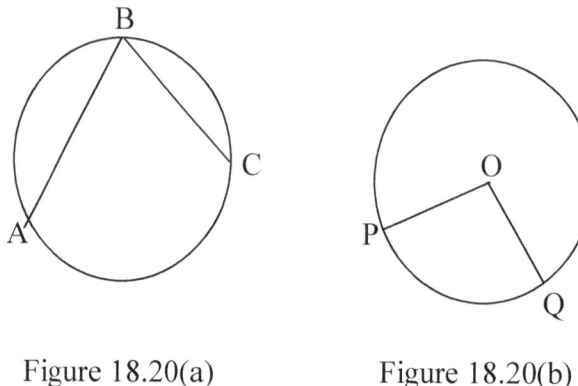

Figure 18.20(a)   Figure 18.20(b)

In Figure 18.20(a), the arc AC subtends an angle ABC at the circumference of the circle. Notice that the angle ABC is formed from the endpoints of arc AC, and the vertex B is on the circumference of the circle.

In Figure 18.20(b), the arc PQ subtends an angle POQ at the centre O. Notice that the angle POQ is formed from the end points of arc PQ, and the vertex is at the centre of the circle.

Try this 17

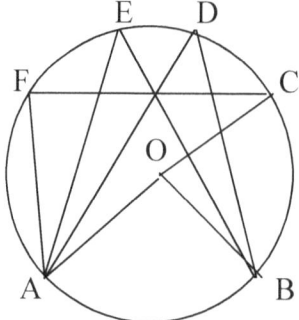

Name the angles subtended

(a) at the circumference by minor arc AB

(b) at the circumference by minor arc AC

(c) at the circumference by minor arc DE

(d) at the centre by the minor arc AB

(e) at the centre by the minor arc BC

(f) at the centre by the minor arc AC

Angles in a segment

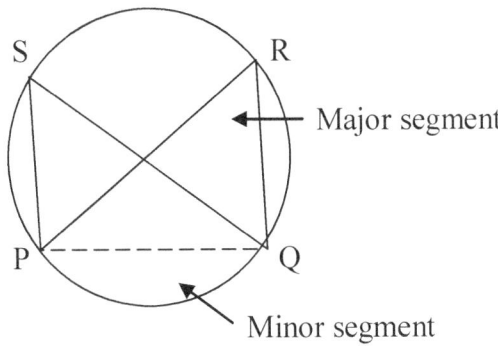

Figure 18.21

In Figure 18.21, the minor arc subtends angles PSQ and PRQ in the major segment.

Try this 18

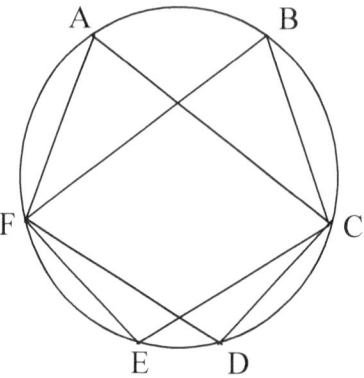

(a) Name the angles in the major segment

(b) Name the angles in the minor segment

Angles subtended by the same arc

The angle at the centre of a circle is twice the angle at the circumference

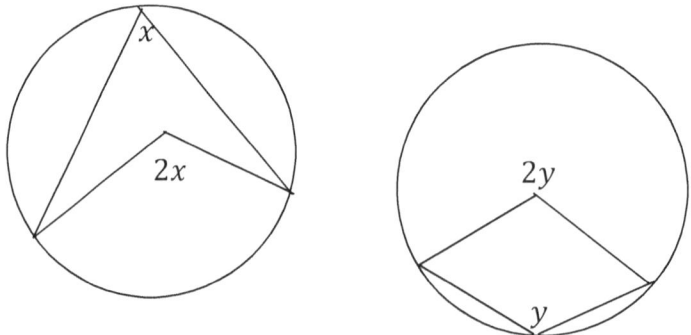

Examples

Find the size of the angle marked in the following diagrams. O is the centre of the circle

(a)   (b)

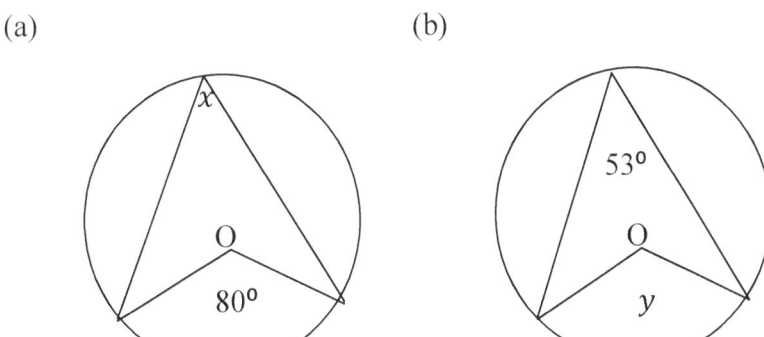

(a) The angle at the centre is twice the angle at the circumference

Therefore, $2x = 80°$

$$x = 40°$$

(b) y is the angle subtended at the centre

Therefore, $y = 2(53) = 106°$

Try this 19

Find the size of the angle marked in each diagram. The point O is the centre of the circle

(a)   (b)

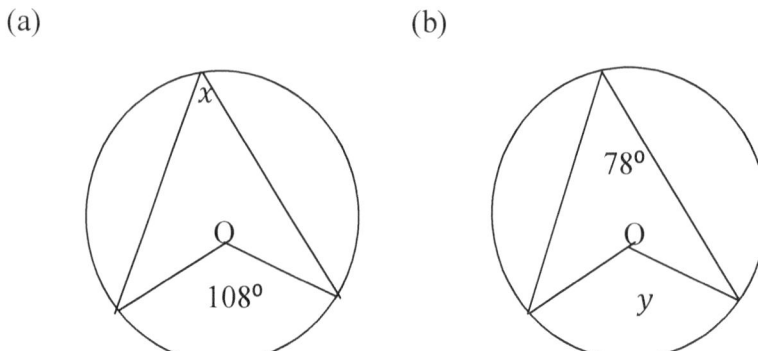

Find the value of y. The point O is the centre of the circle

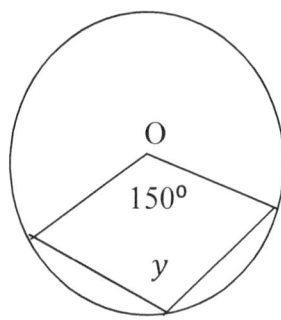

The angle subtended by the major arc at the centre is

$360° - 150° = 210°$

The angle y is one- half the angle at the centre subtended by the major arc

Therefore, $y = \frac{1}{2}(210) = 105°$

Try this 20

Find the value of x. The point O is the centre of the circle.

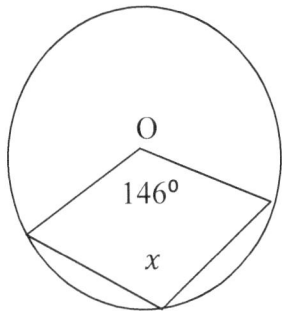

Angles in a semicircle

The angle in a semi-circle is a right angle

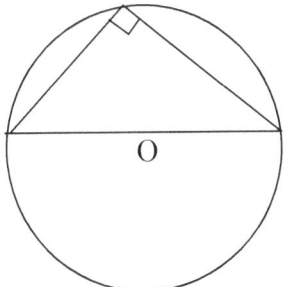

Example

Find the value of x. The point O is the centre of the circle.

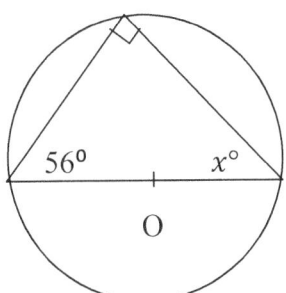

Since the angle in a semi-circle is 90^0, we have

$$x + 56 = 90$$

$$x = 34$$

Try this 21

Find the value of x. The point O is the centre of the circle

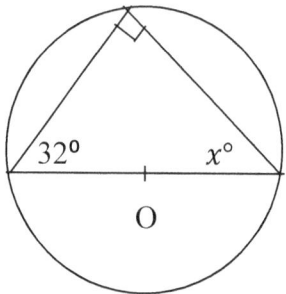

Angles in the same segment

The angles in the same segment are equal

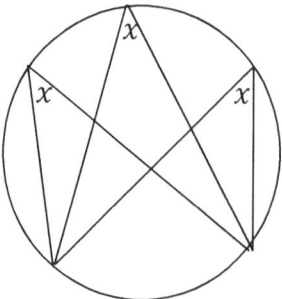

Example

Find the values of x and y. The point O is the centre of the circle.

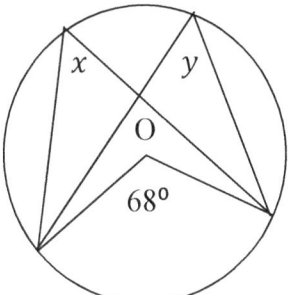

The angle at the circumference is one-half the angle at the centre.

Therefore, $x = 34°$.

Since angle x and angle y are in the same segment $y = 34°$

Try this 22

Find the values of x and y. The point O is the centre of the circle.

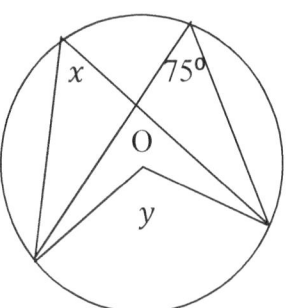

Circle Quadrilateral

A cyclic quadrilateral is a four-sided figure in a circle, with each vertex of the quadrilateral touching the circumference of the circle as shown in the diagram below.

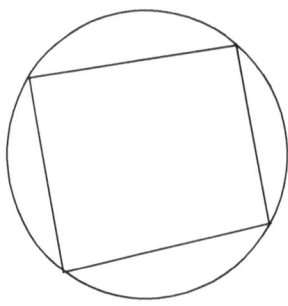

Angles in a cyclic quadrilateral

Opposite angles of a cyclic quadrilateral

The opposite angles in a cyclic quadrilateral add up to 180^0 (the angles are supplementary). Therefore, $x + y = 180°$

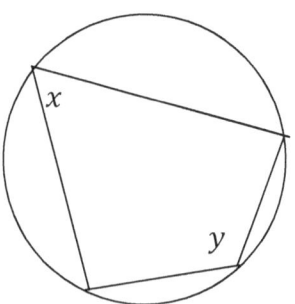

Example

Find the values of x and y. The point O is the centre of the circle

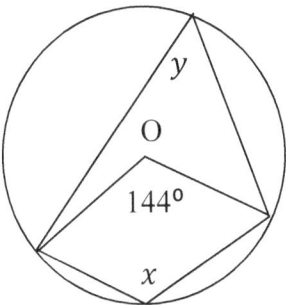

(a) The angle y is one-half the angle at the centre

Therefore, $y = \frac{1}{2}(144) = 72°$

Because the opposite angles of a cyclic quadrilateral are supplementary

$x + 72° = 180°$

$x = 108°$

Try this 23

Find the values of x and y. The point O is the centre of the circle

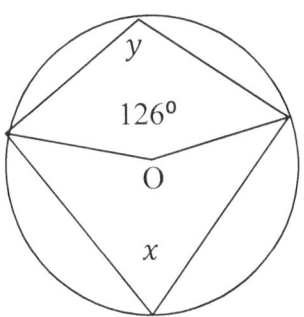

Exterior angles of a cyclic quadrilateral

The exterior angle of a cyclic quadrilateral is equal to the interior opposite angle. Therefore, $x = y$.

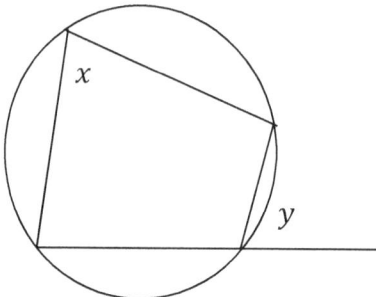

Example

Find the values of x and y. The point O is the centre of the circle

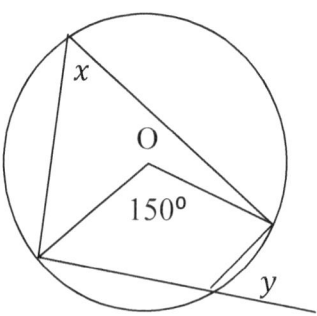

The angle x is one-half the angle at the centre, therefore

$$x = \frac{1}{2}(150) = 75°$$

Because the exterior angle of a cyclic quadrilateral is equal to the interior opposite angle

$$y = 75°$$

Try this 24

Find the values of x and y. The point O is the centre of the circle

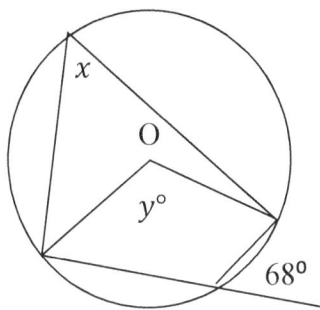

Exercise 18.5(a)

1. Find the values of x and y. The point O is the centre of the circle

(a)   (b)

(c)   (d)

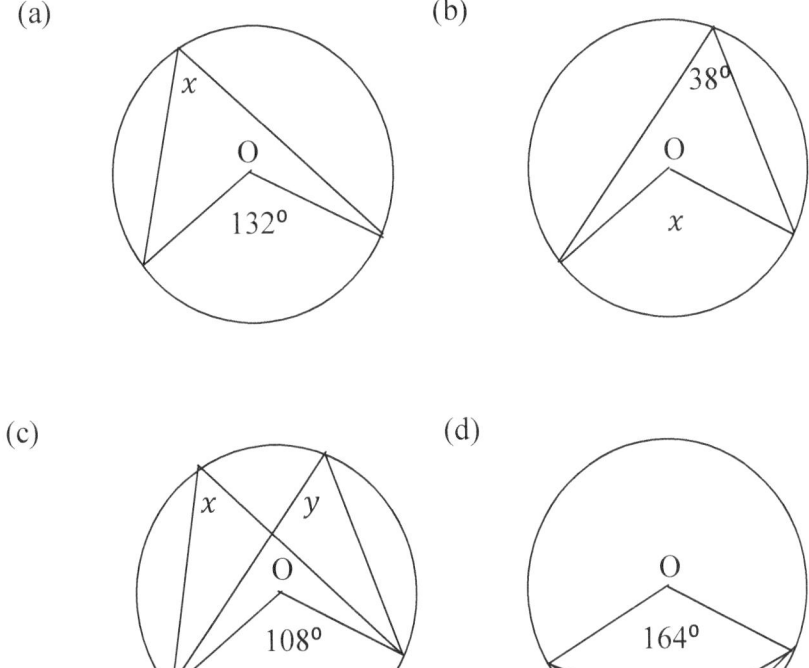

(e)   (f)

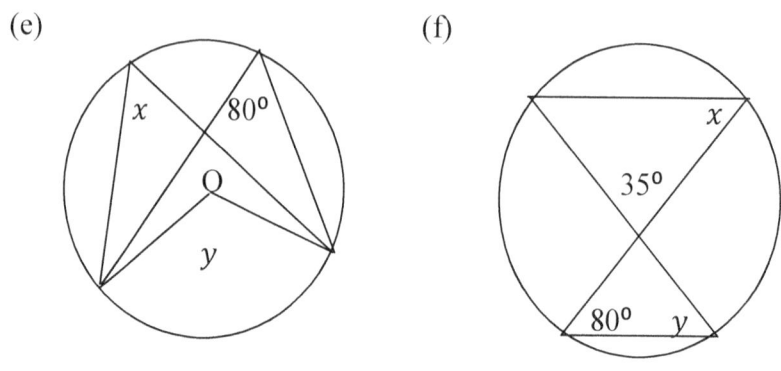

(g)   (h)

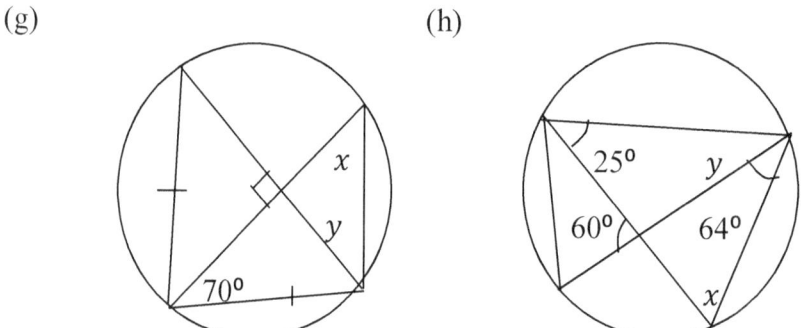

2. Find the values of x and y. The point O is the centre of the circle

(a)   (b)

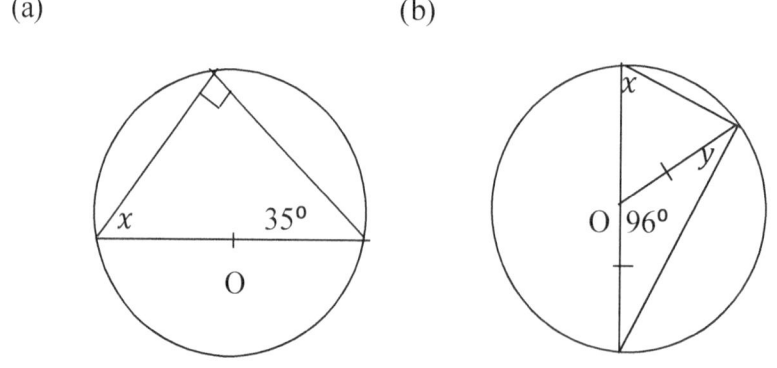

(c)

(d)

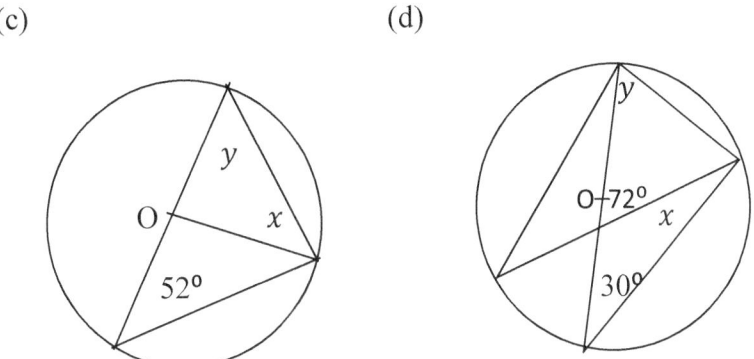

3. Find the values of x and y. The point O is the centre of the circle

(a)

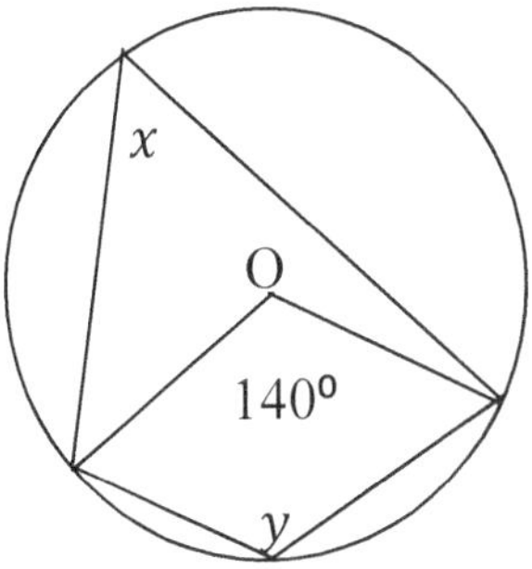

(b)

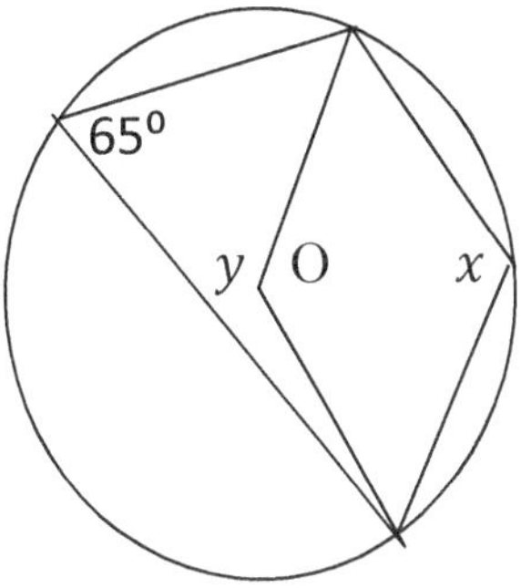

(c)

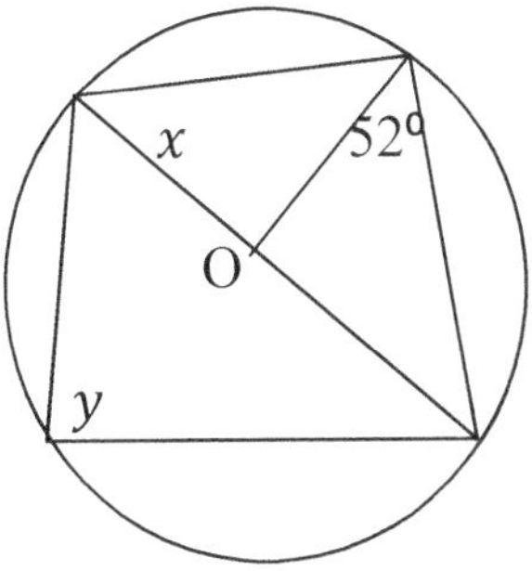

(d)

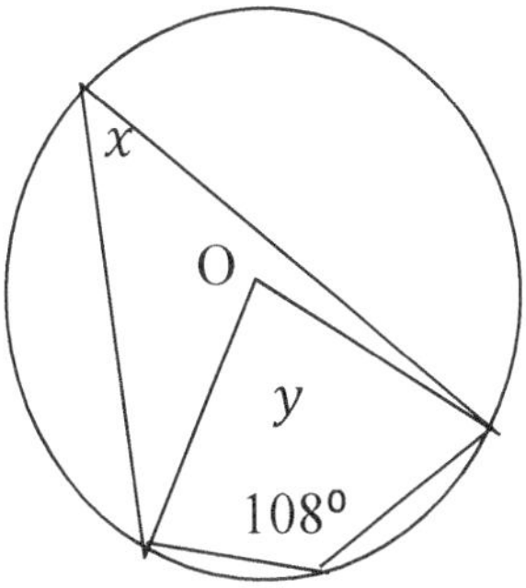

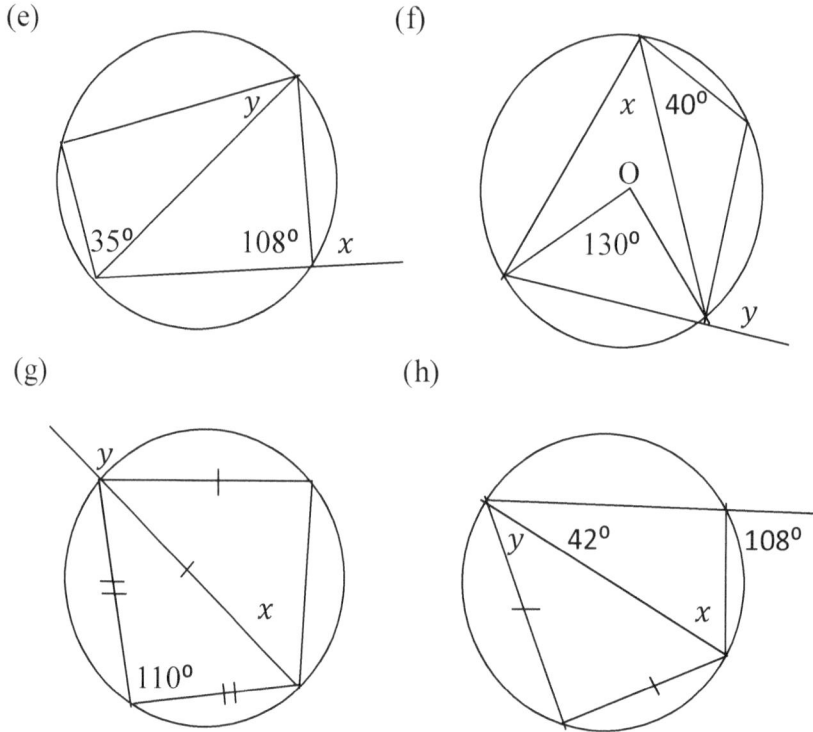

Tangents

A tanget to a circle is a straight line which touches the circle at only one point. Note that a tanget does not cross the circle, it just tonches it as shown in Figure 18.22

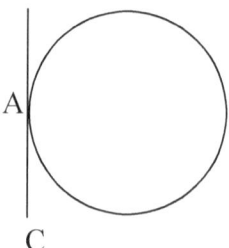

Figure 18.22

The line AC is the tangent. The point A is called the point of contact.

Angle in alternate segment

The angle which a chord drawn through the point of contact of a tangent makes with the tangent is equal to the angle subtended at the circumference in the alternate segmennt of the circle.

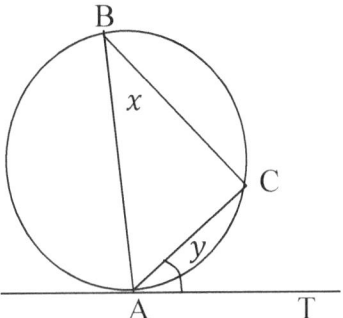

In the diagram above AT is a tangent to the circle. The angle x is equal to the angle y formed by AC and AT i.e. $x = y$.

Example

Find the values of x and y. O is the centre of the circle

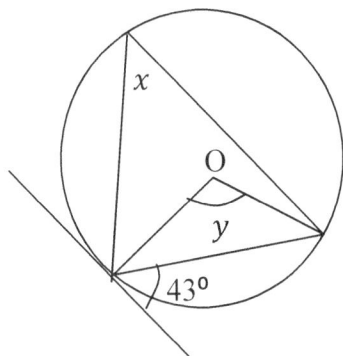

Because the angle made by the chord and the tangent is equal to the angle in the alternate segment,

$x = 43°$.

The angle at the centre is twice the angle at the circumference, therefore

$y = 2(43°) = 86°$

Try this 25

Find the values of x and y. O is the centre of the circle

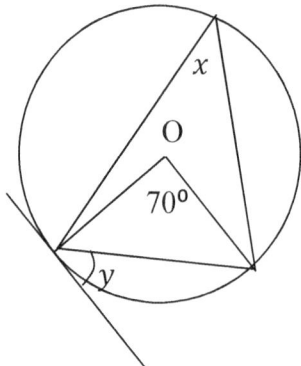

The angle between a tangent and a radius of a circle

The radius drawn to the point of contact of a tangent to a circle is perpendicular to the tangent.

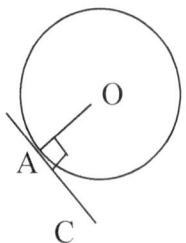

Figure 18.23

In Figure 18.23, O is the centre of the circle and A is the point of contact. OA is a radius of the circle, and angle OAC is a right angle.

Example

Calculate the values of x and y in the diagram below. O is the centre of the circle.

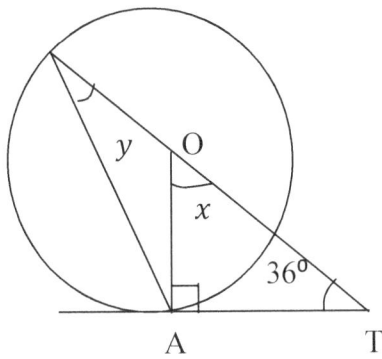

In the figure, $x + 36 = 90$

$$x = 54$$

Because y is one-half of x

$$y = \frac{1}{2}(54) = 27$$

Try this 26

Find the value of x. O is the centre of the circle.

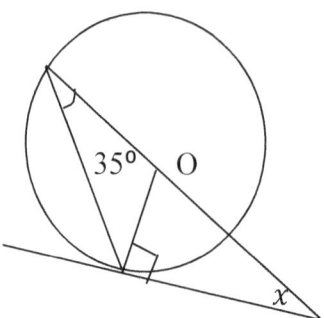

Exercise 18.5(b)

1. O marks the centre of each circle. Find the size of the angles marked x and y in each circle.

(a)   (b)

(c)   (d)

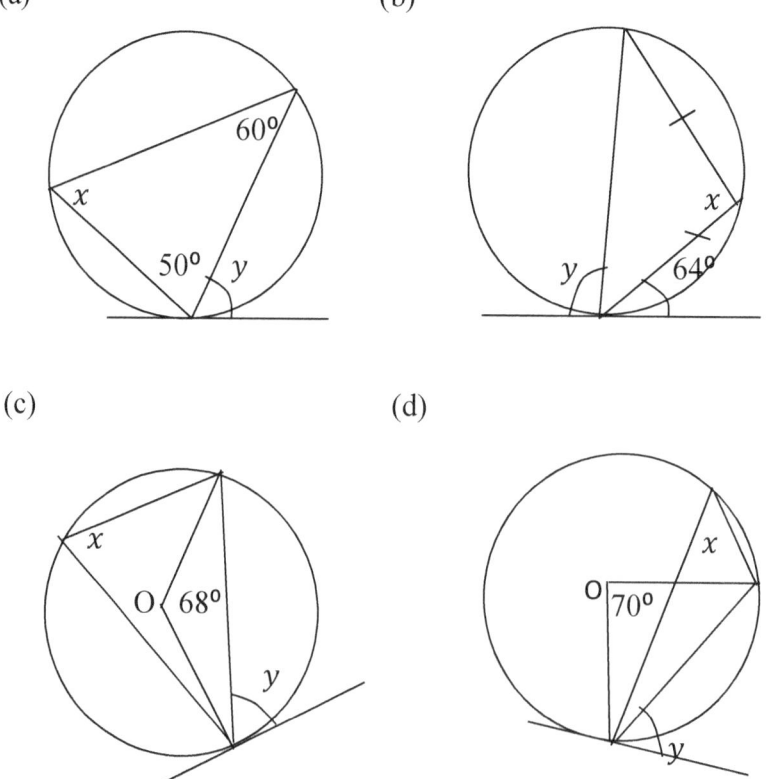

2. O marks the centre of each circle. Find the size of the angle marked x in each case.

(a)   (b)

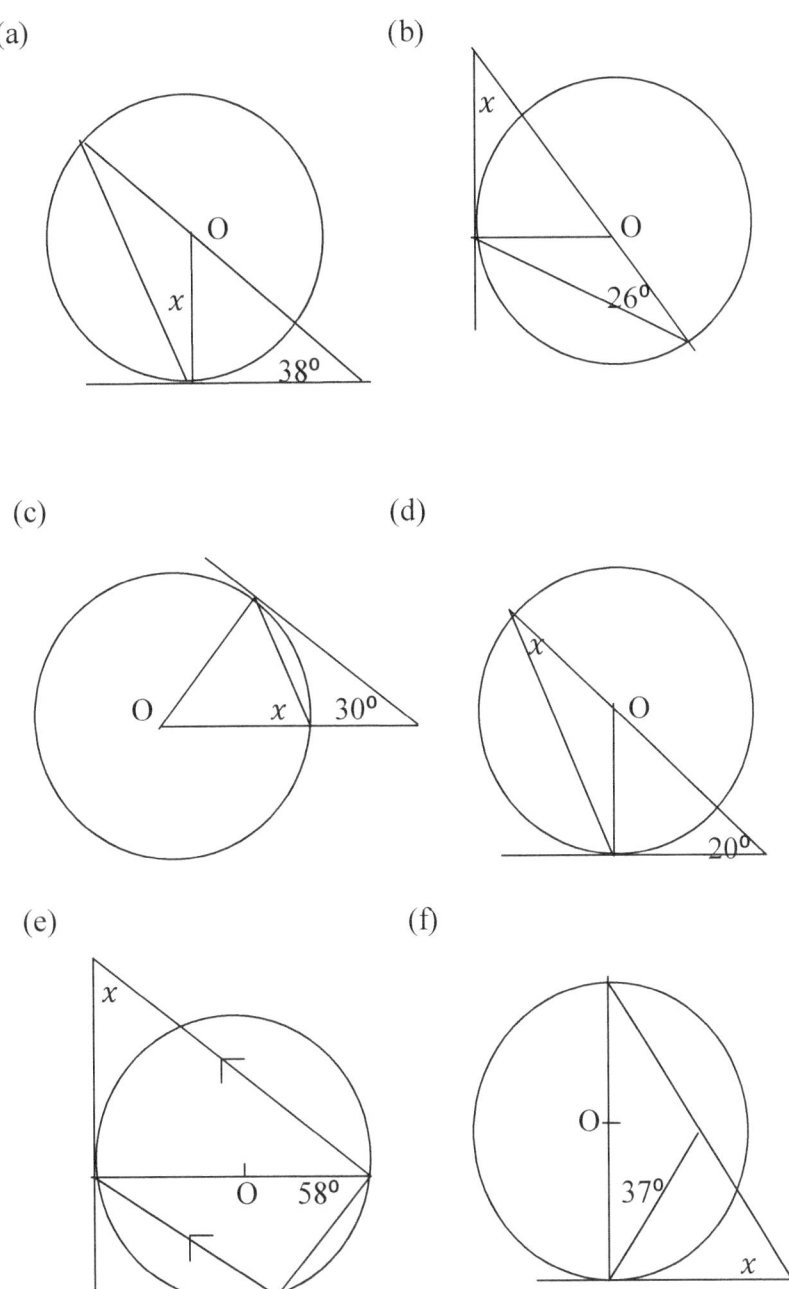

(c)   (d)

(e)   (f)

3. O marks the centre of each circle. Find the length of the side

marked x in each diagram.

(a)   (b)

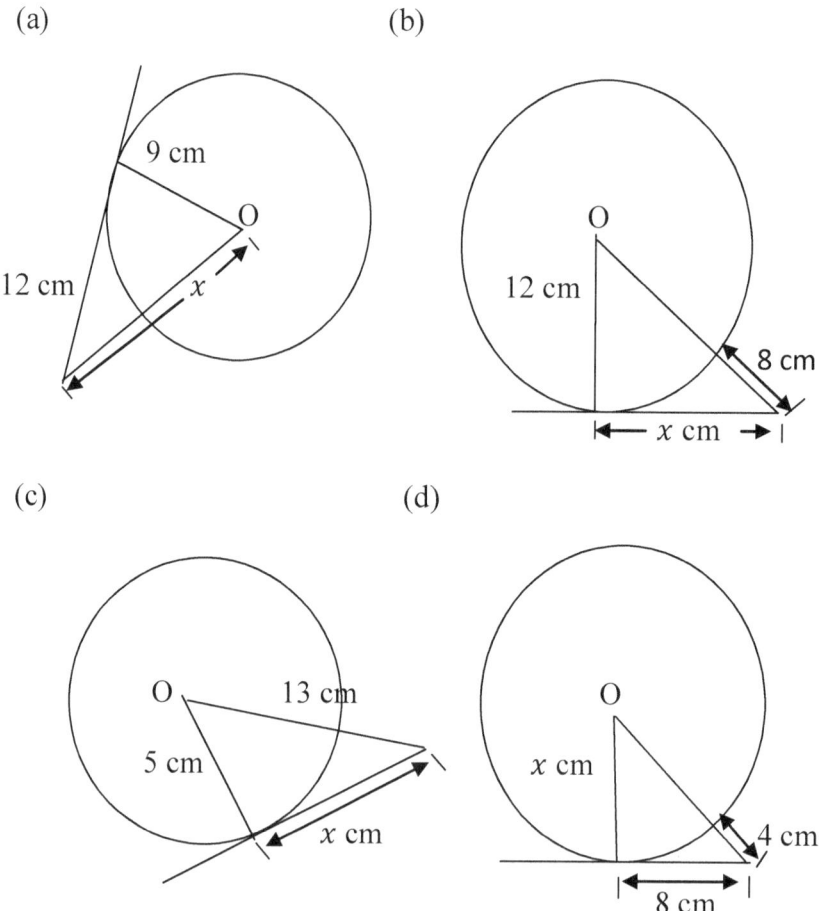

(c)   (d)

Review exercise 18

1. Use a protractor to draw the following angles

(a) 87^0 (b) 98^0 (c) 132^0

2. Find the size of the angles marked x and y in the following diagrams:

(a)

(b)

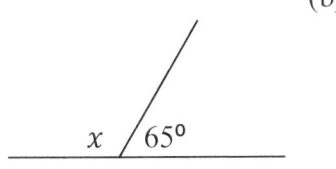

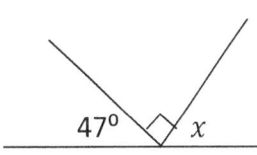

(c)

(d)

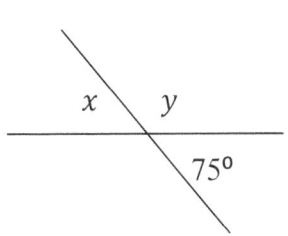

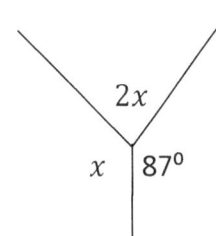

3. Find the values of x, y and z

(a)

(b)

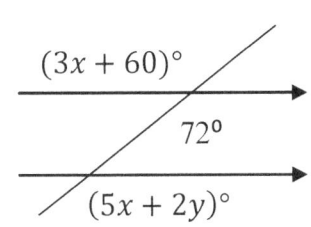

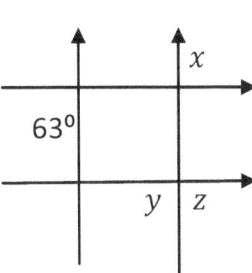

(c)

(d)

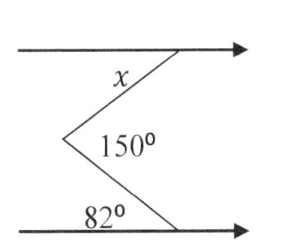

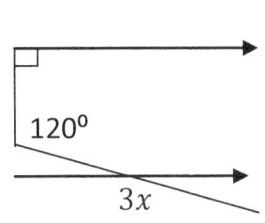

4. Find the values of x and y

(a)   (b)

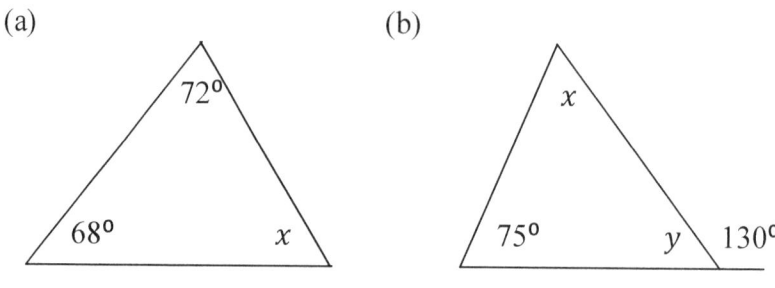

(c)   (d)

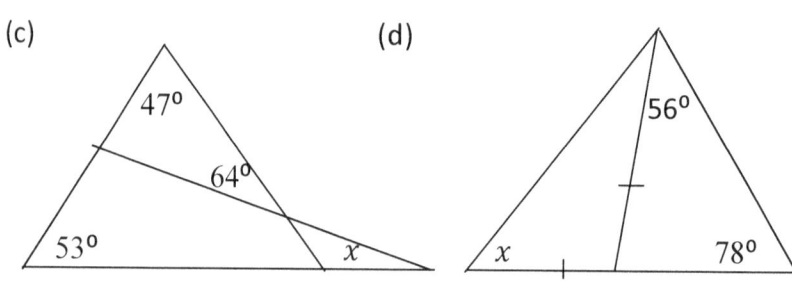

5. Find the value of x

(a)   (b)

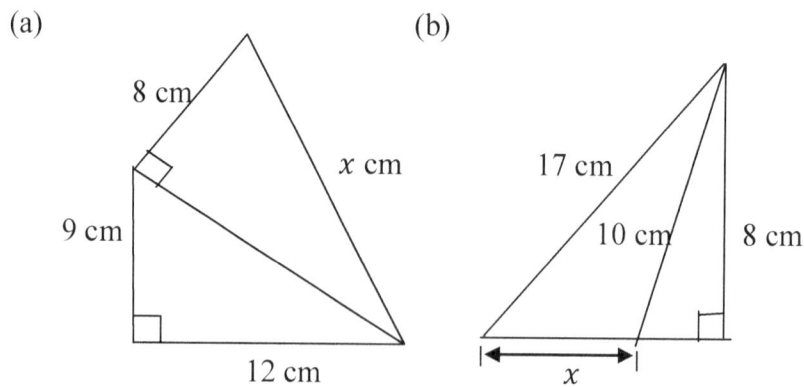

(c)   (d)

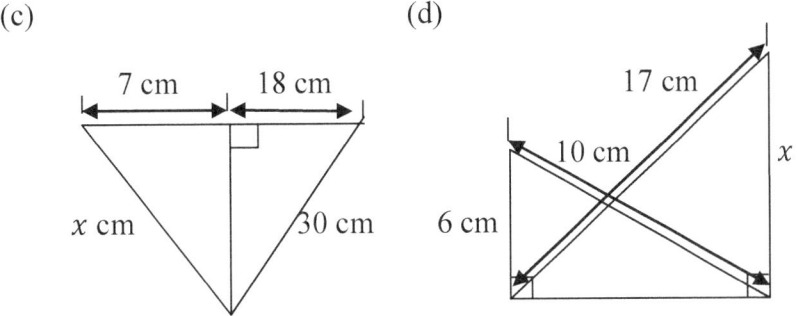

6. A ladder 18 m long is learning against a wall. If the foot of the ladder is 9 m from the wall, how far up the wall does the ladder reach?

7. A rope is attached to the top of a pole and is attached to the ground at a point 8 m from the bottom of the pole. If the length of the rope is 12 m, what is the length of the pole?

8. Find the sum of the interior angles of the polygon having the following number of sides

(a) 11   (b) 16   (c) 25

9. Find the size of each interior angle of the regular polygon having the following number of sides

(a) 6   (b) 9   (c) 16

10. Find the number of sides of the regular polygon with the following interior angles

(a) 150^0 (b) 162^0

11. Find the exterior angle of each regular

   (a) hexagon   (b) hendecagon

12. At 2 pm the shadow of a 2.5 metre pole is 30 centimetres. What is the height of a tree whose shadow was 15 metres?

13. Kweku is having his portrait taken. He is 3.6 metres from the camera lens and the film is 1.8 centimetres from the lens. If he is 1.8 metres tall, what is the height of the image on the film?

14. In the following diagrams, identify pairs of congruent triangles.

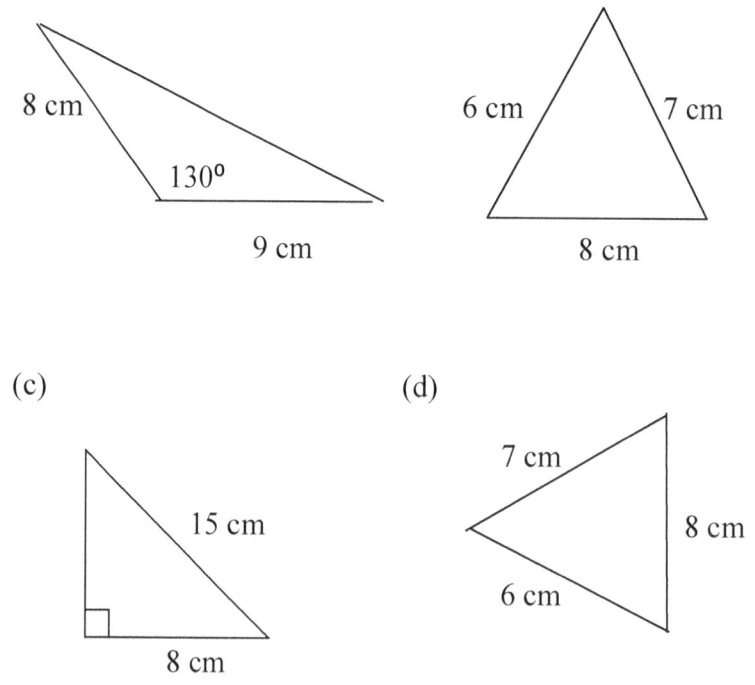

(a)

8 cm 130° 9 cm

(b)

6 cm 7 cm 8 cm

(c)

15 cm 8 cm

(d)

7 cm 8 cm 6 cm

(e)   (f)

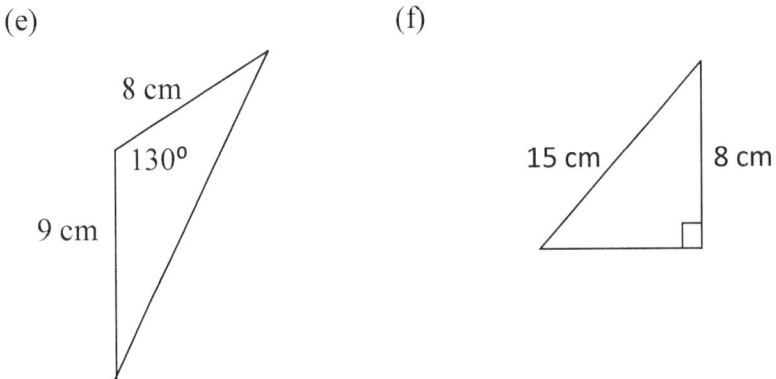

15. Identify the similar triangles. Find x and the length of the side

Indicated

(a)   (b)

AB and BC   DE

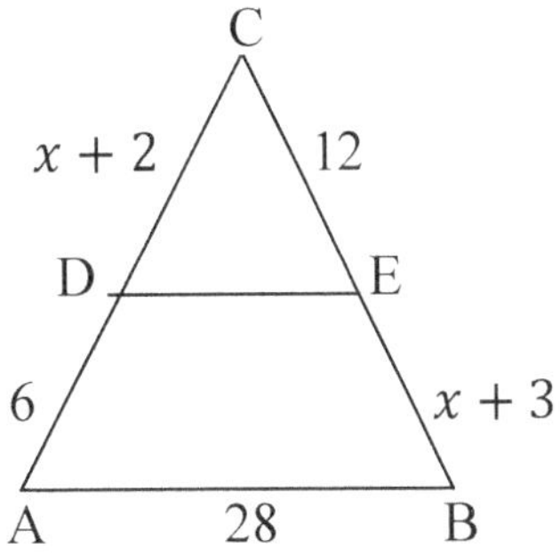

(c)   A   AB and ED

A 2x

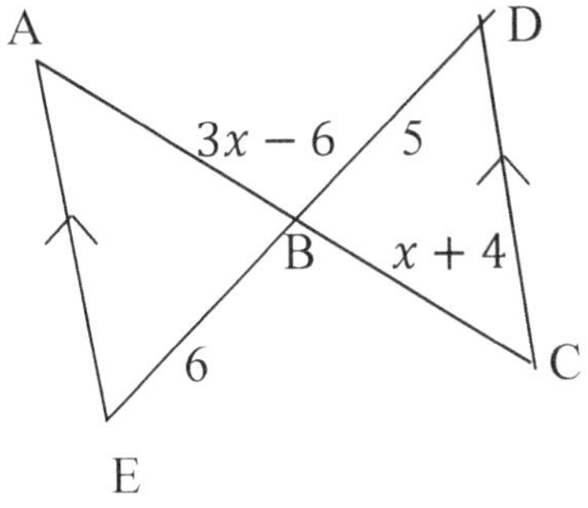

2x − 3 E

10

B   D   C

12

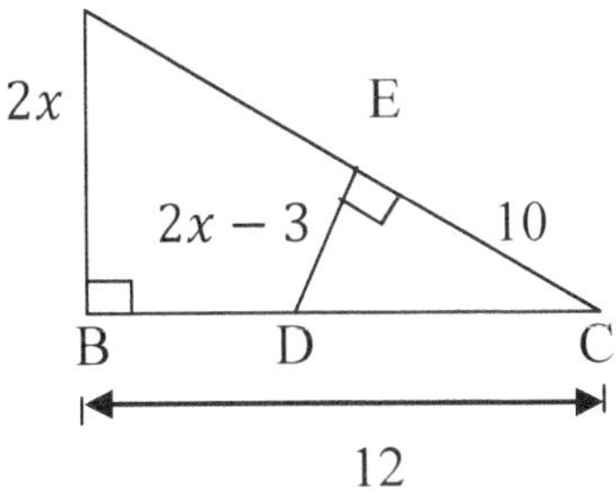

16. Find the value of x in each of the following diagrams

(a)   (b)

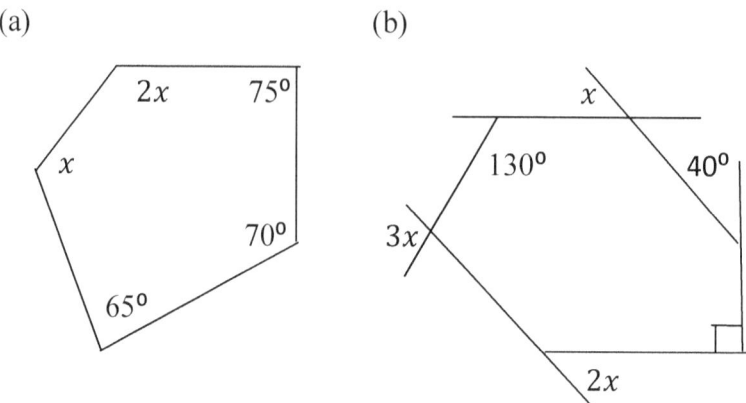

17. Find the values of x and y. O is the centre of the circle.

(a)   (b)

(c)   (d)

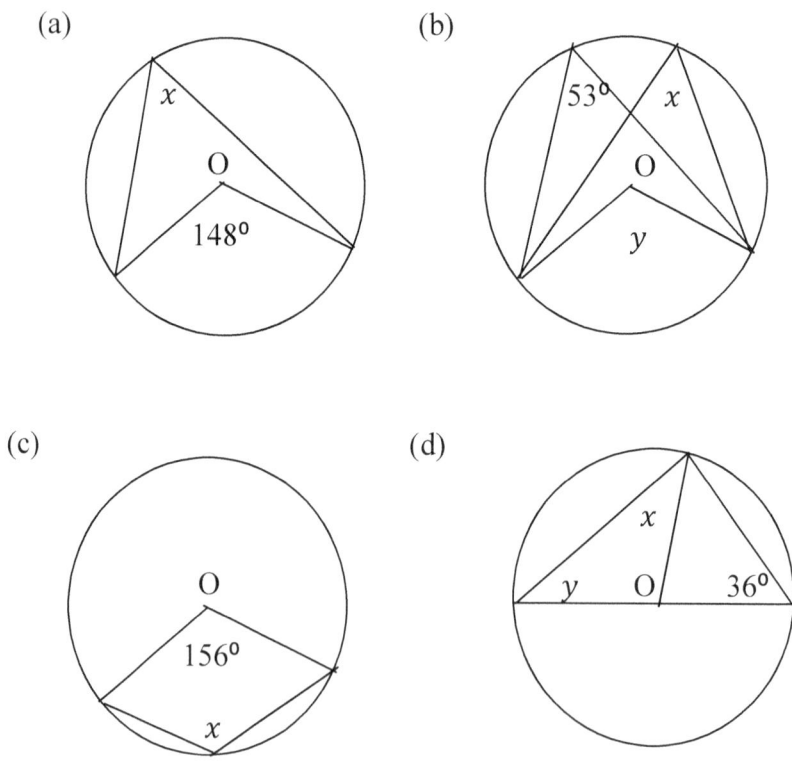

18. Find the values of x and y. O is the centre of the circle.

(a)   (b)

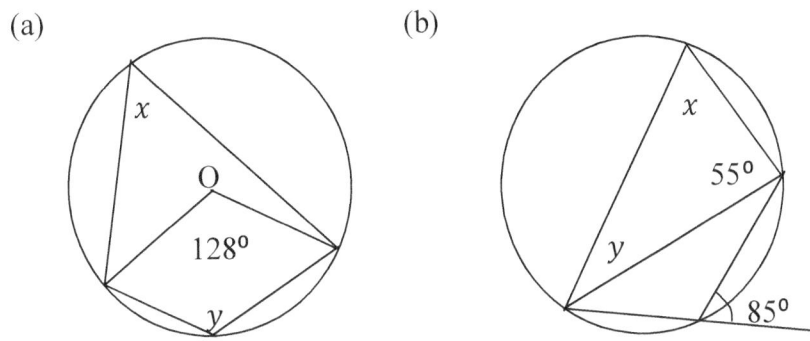

19. Find the values of x and y. O is the centre of the circle.

(a)   (b)

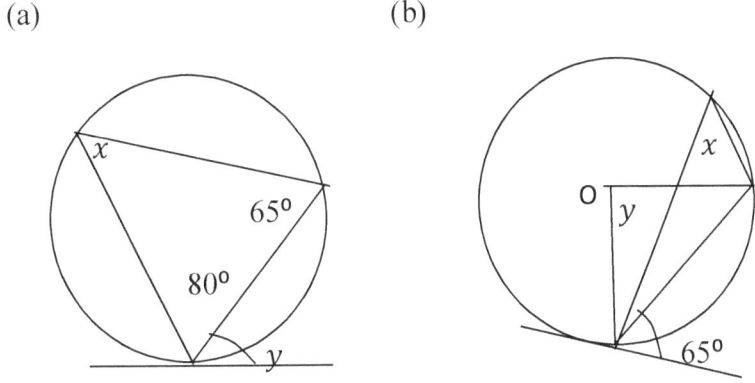

20. Find the value of x. O is the centre of the circle.

(a)   (b)

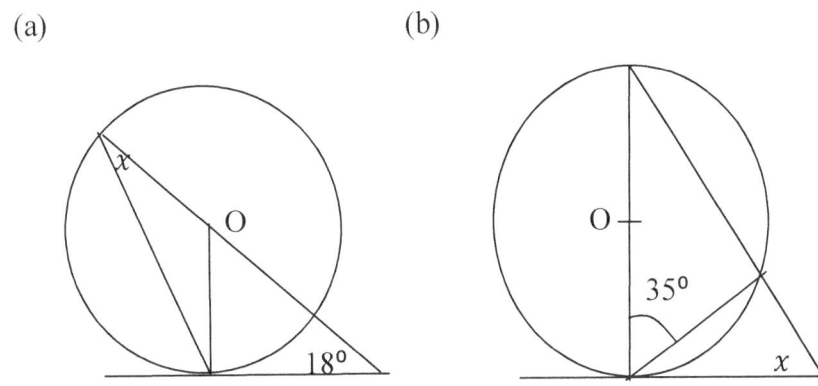

(a)   (b)

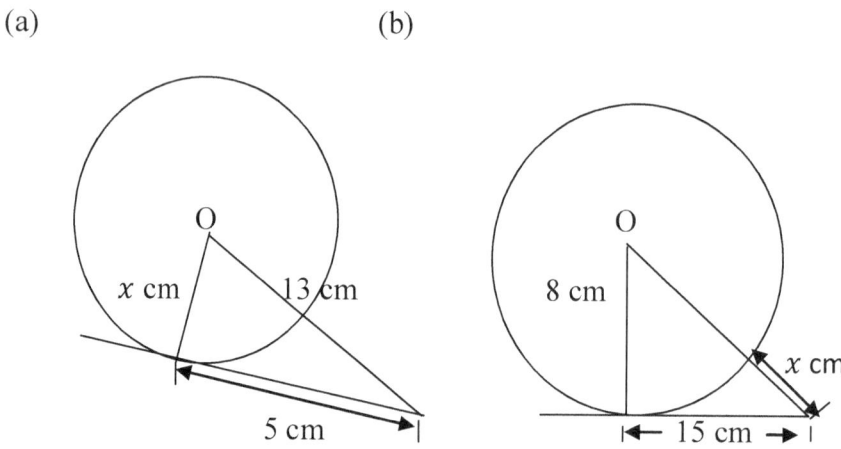

Chapter Test 18

Take this test as you would take a test in class. After you are done,
check your work against the answers in the back of the book

1. In the diagrams below AB and CD are straight lines. Find the

   values of x and y.

(a)   (b)

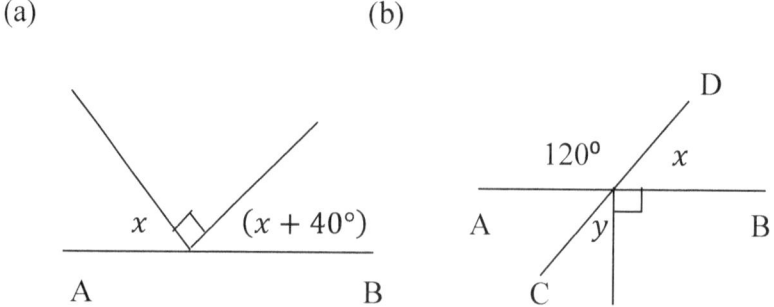

2. Find the values of x and y

(a)

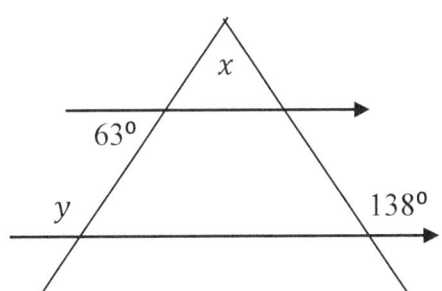

(b)

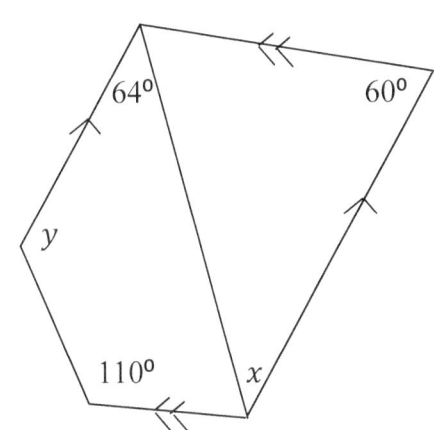

3. Find the values of x and y.

(a)   (b)

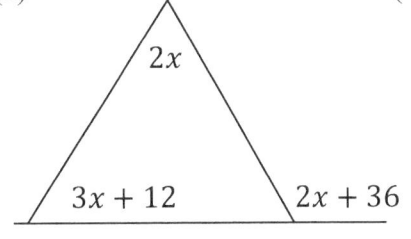
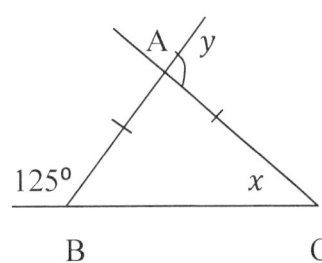

In the diagram AB = AC

4. Find the values of x and y.

(a)   (b)

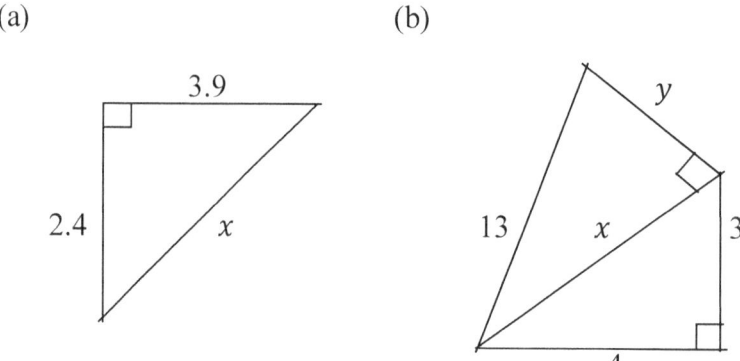

(c) Two buses leave a terminal at the same time. One travels
    80 km h^{-1} due west and the other bus travels due south at
    60 km h^{-1}. Calculate how far apart the two buses are after
    4 hours.

5. Identify the similar triangles, and find the value of x

(a)   (b) $PQ = 12$

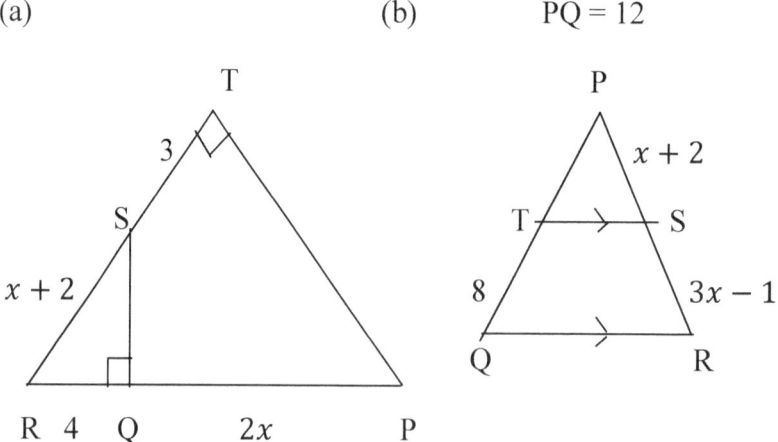

6. The shadow of a model of a tower 1.5 metre at 1 pm was 90 centimetres. If the shadow of the tower was 240 metres, how tall is the tower?

7. Find the values of x and y.

(a)   (b)

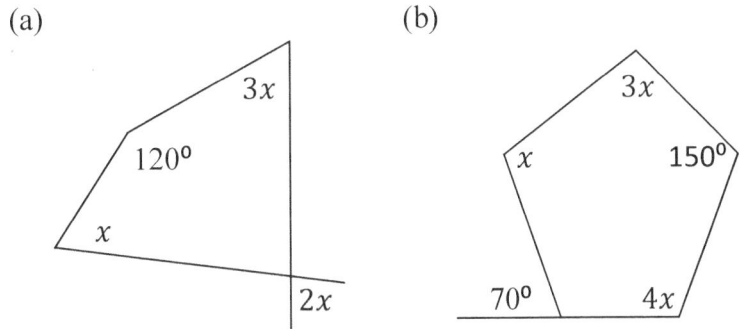

8 In the diagram below, O is the centre of the circle, and RS is perpendicular to PQ. Find the values of x and y.

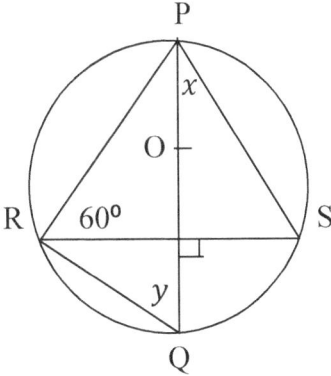

9. In the diagram below O is the centre of the circle, and AT is

a tangent. Find the values of x and y.

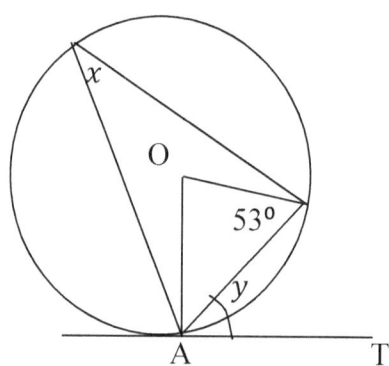

10.

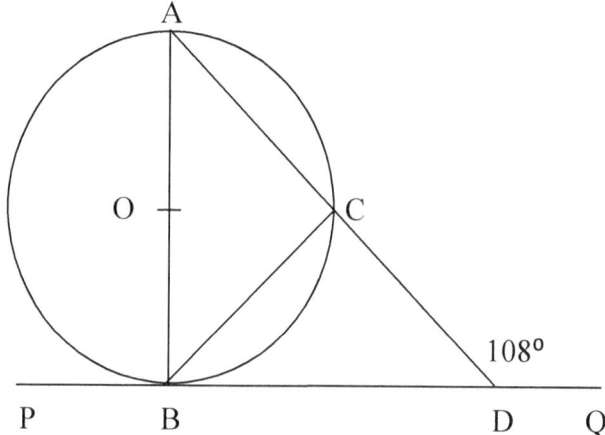

O is the centre of the circle and PQ is a tangent to the circle at B.

(a) Find the size of angle BAD

(b) Find the size of angle ABC

Answers to exercises

Answers to exercises

Chapter 1

Try this

1. $A = \{12, 13, 14\}$

2. $A = \{x : x \text{ is a positive even number less than } 10\}$

3. $\{14\}$ 4. $-2 \notin N$ 5. empty set

6. $A = \{4, 5, 6, 7, 8\}$ $n(B) = 5$ 7. Equivalent   8. $A = \{3, 6\}$

9(a) infinite   (b) finite   10(a) subset   (b) not a subset

11(a) ϕ, $\{2\}$, $\{3\}$, $\{5\}$, $\{2, 3\}$, $\{2, 5\}$, $\{3, 5\}$, $\{2, 3, 5\}$ (b) 512

12. $\{a, b, c, d, e, f\}$ $\{a, b, c, d, e, f\}$

13. $\{2, 4, 6, 7, 8, 10\}$ $\{2, 4, 6, 7, 8, 10\}$

14. $\{3, 6\}$ 15. $\{2\}$ 16. Not disjoint   17. $\{2, 5, 6, 8, 9\}$

18. (a)   (b)

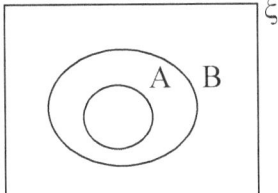

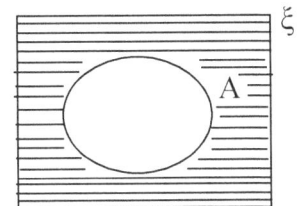

(c)

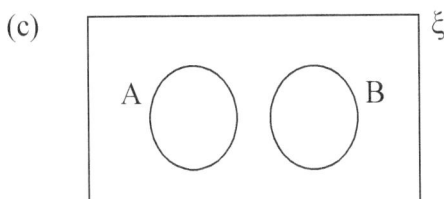

19

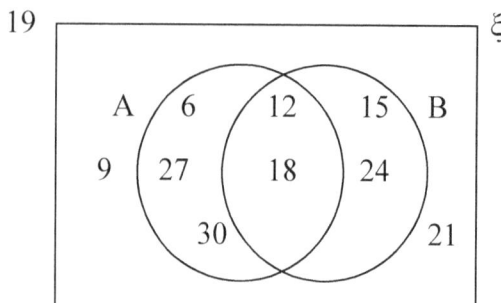

20.

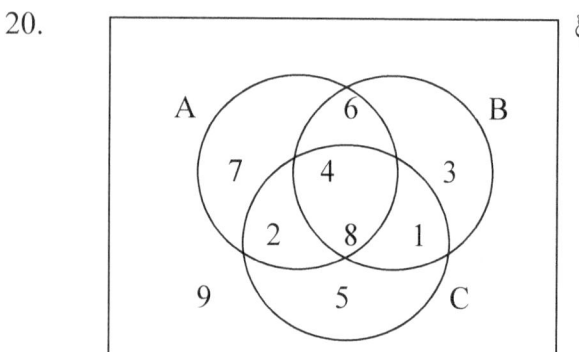

21. 8   22. 31

Exercise 1.1(a)

1(a) {1, 2, 3, 4, 5, 6}   (b) {23, 29}(c) {1, 2, 3, 4, 6, 12}

 (d) {2,5}   (e) {2, 4, 6, 8, 10, 12, 14}

2(a) $\{x: x \in Z \text{ and } x > 12\}$ (b) $\{x: x \in N \text{ and } x \text{ is odd}\}$

 (c) $\{x: x \text{ is a multiple of } 5\}$ (c) $\{x: x \text{ is a factor of } 20\}$

3(a) {8, 10, 12, 14, 16}   (b) {1, 2, 3, 4, 6, 9, 18, 36}

 (c) {11, 12, 13, 14, 15, 16}   (d) {8, 10, 12, 14}

4(a) $3 \notin E$ (b) $8 \in M$ (c) $7 \notin A$ (d) $6 \in F$ (e) $4 \notin P$

5(a) $\in$ (b) $\notin$ (c) $\in$ (d) $\notin$ (e) $\in$

Exercise 1.1(b)

1. empty set   2. Empty set   3. not empty

4. not empty   5. Empty set

Exercise 1.1(c)

1. equal   2. Equal   3. not equal   4. not equal   5. not equal

6. equal   7. equivalent   8. not equivalent   9. equivalent

10. not equivalent   11. equivalent

Exercise 1.1(d)

1. {multiples of 2}   2. {integers}   3. {triangles}

4. {quadrilaterals}   5. {integers}   6. {factors of 36}

Exercise 1.1(e)

1. infinite   2. finite   3. finite

4. infinite   5. infinite   6. finite

Exercise 1.1(f)

1(a) $\supset$ (b) $\subset$ (c) $\supset$

2. 8 ϕ, {5}, {6}, {7}, {5,6}, {5,7}, {6,7}, {5,6,7}

3. 10   4. 7   5. 4

Exercise 1.2(a)

1. {1, 2, 3, 4, 5, 6}   2. {1, 2, 3, 4, 5, 6, 8}, {1, 2, 3, 4, 5, 6, 8}

3. $\{1, 2, 3, 4, 5, 6, 7, 8, 9, 10, 11, 12\}$

4. $\{1, 2, 3, 4, 6, 8, 9\}, \{1, 2, 3, 4, 6, 8, 9\}$ 5. $\{1, 2, 3, 4, 5, 6, 7, 8\}$

6. $\{1, 2, 3, 5, 7, 9\}$ 7. $\{1, 2, 4, 5, 6, 8, 10, 20\}$

Exercise 1.2(b)

1. $\{f\}$ 2. $\{1, 2\}$ 3. ϕ 4. $\{6\}$ 5. $\{3, 6\}$ 6. $\{6\}$ 7. $\{6, 9\}$

8. $\{1, 2, 4\}, 4$ 9. $\{1, 2, 3, 4, 5, 6\}$ $\{1, 2, 3, 4, 5, 6\}$

10. $\{1, 2, 3, 5\}$ $\{1, 2, 3, 5\}$

11(a) disjoint   (b) not disjoint   (c) not disjoint

   (d) disjoint   (e) disjoint

Exercise 1.2(c)

1. $\{2, 4, 6, 8\}$ 2. $\{1, 3, 6, 8\}$ 3. $\{5, 7, 8, 9, 10, 11\}$

4. $\{1, 2, 3, 4, 5\}$ 5. $\{7\}$ $\{5\}$ 6. $\{d, f\}$ $\{d, f\}$

7. $\{a, b, f\}$ $\{a, b, f\}$ 8. ϕ, U   9. $\{1, 3, 10\}$ $\{1, 3, 10\}$

10. $\{3, 4, 6, 7, 8\}, \{3, 4, 6, 7, 8\}$ 11. $\{1, 2, 3, 4, 5, 6, 7, 8, 9, 10\}$

12. ϕ

Exercise 1.3

1(a) $A \cup B'$ (b) $(A \cap B') \cup B'$

 (c) $(A \cup B) \cap C'$ (d) $A \cup (B \cup C)'$

2(a) $A' \cap B$ (b) $(A \cup B)'$ $A' \cap B'$

3(a)

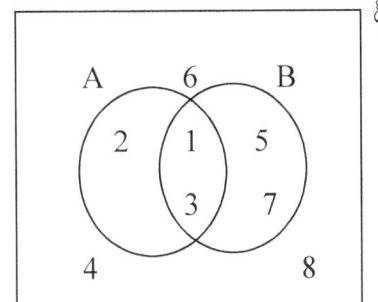

(b) (i) $\{b, d, e, f\}$ (ii) $\{g\}$

4(a) $\{1, 2, 3\}$ (b) $\{4, 5, 6, 7, 8\}$ (c) $\{1, 3, 5, 7\}$ (d) $\{2, 4, 6, 8\}$

(e) $\{1, 3\}$ (f) $\{2, 4, 5, 6, 7, 8\}$ (g) $\{1, 2, 3, 5, 7\}$ (h) $\{4. 6, 8\}$

(i) $\{2, 4, 5, 6, 7, 8\}$ (j) $\{4, 6, 8\}$ $(A \cup B)'$, $A' \cap B'$

5(a)

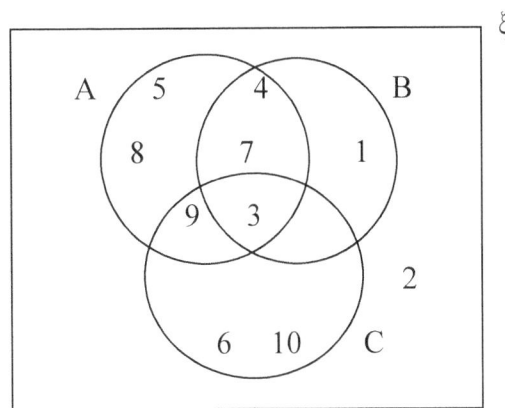

(b) (i) $\{1,, 4, 5, 7, 8\}$ (ii) $\{2\}$

6(a) $\{b, c, d\}$ (b) $\{a, b, c, d, e\}$ 7(a) $P' \cap (Q \cup R)$ (b) $\{1, 6\}$

8.

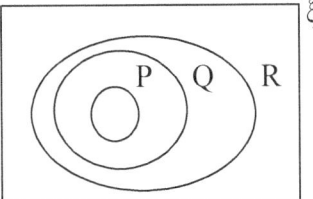

Exercise 1.4

1. 20   2. 8   3. 6   4. 4   5. 15, 120   6. 32   7. 40   8. 3

9. 6   10. 18, 9   11. 30, 12   12. 15

Review exercise 1

1(a) {8, 10, 12, 14, 16, 18}   (b) {2, 3, 4, 5, 6, 7, 8}

 (c) {- 1, 0, 1, 2, 3, 4, 5, 6}   (d) {2, 3, 5, 7}

2(a) equal   (b) equivalent   (c) equivalent   (d) equal

3(a) finite   (b) infinite   (c) infinite   (d) finite

4(a) false   (b) true   (c) false   (d) true   (e) false   (f) true

 (g) true   (h) false   (i) false   (j) false

6(a) 32   (b) 256

7(a) {3, 8}   (b) {3, 4, 5, 6, 7, 8}

 (c) {1, 2, 5, 7, 9}   (d) {1, 2, 4, 6, 9}

8(a) {5, 6}   (b) {2, 3, 4, 5, 6, 7, 8, 10}

 (c) {1, 7, 8, 9, 10}   (d) {1, 2, 3, 4, 9}

9(a) {5}   (b) {1, 2, 3, 4, 5, 6, 7}   (c) {1, 2, 3, 4}   (d) {6, 7}

 (e) {5, 6, 7}   (f) {1, 2, 3, 4}

10(a) A   (b) A   (c) A   (d) ϕ 11(a) A   (b) B   (c) B

14. 105   15. 21   16. 40, 49, 20   17. 49   18. 45   19. 46

20. 60   21. 161   22. 36

Chapter Test 1

1(a) {- 2, - 1, 0, 1, 2, 3, 4, 5, 6, 7, 8, 9, 10}

(b) {1, 2, 5, 10, 25, 50}   (c) {2, 3}

2(a) equal   (b) equivalent   (c) equal   (d) equivalent

3(a) infinite   (b) finite   (c) infinite

4(a) $B \subset A$ (b) $A \subset B$ (c) $A = B$

5(a) ϕ, {0}, {1}, {3}, {0,1}, {0,3}, {1,3}, {0,1,3}   (b) 128

6(a) {1, 2, 3, 4, 5, 6, 7, 8}   (b) {2, 4, 8}

7(a) {1, 2, 3, 4, 5, 7, 9, 10} {3, 4, 7, 9, 10}

(b) {1}   {1, 3, 4, 6, 7, 9, 10}

8(a) {1, 3, 6, 7, 8, 10}   (b) {2, 5, 7, 9}   (c) {2, 5, 9}

(d) {2, 4, 5, 7, 9}

9(a) {1, 3, 5, 7, 8, 9, 10, 11}   (b) {5, 7, 9, 11}   (c) {5, 7, 9, 11}

10.

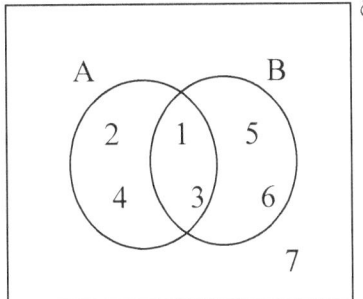

11.

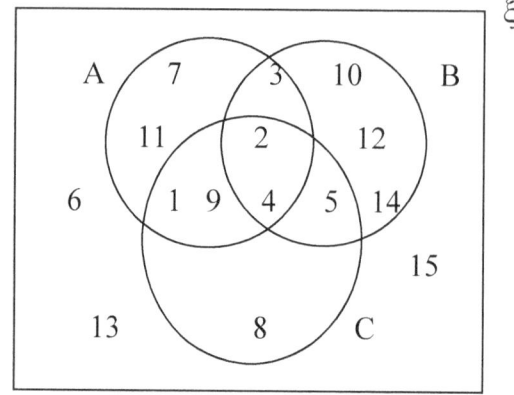

12. 12   13. 70   14. 158   15. 9, 12

Chapter 2

Try this

1.

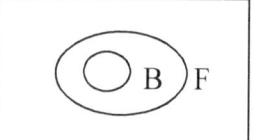

B = {boys}

F = {people who play football}

2

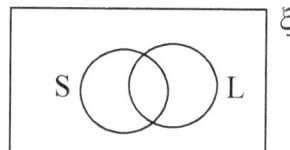

S = {students}

L = {students who are lazy}

3.

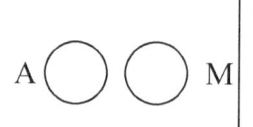

A = {animals}

M = {men}

4.

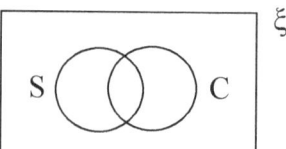

S = {students}

C = {students who are clever}

5. 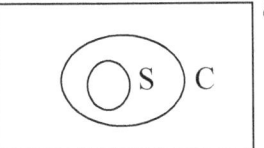   S = {science students}

C = {people who are clever}

Valid

6. 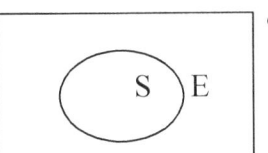   S = {students who study hard}

E = {students who pass their exams}

Not valid

Exercise2.1

1. 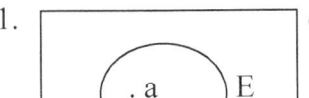   a = Ama

E = {students who pass their exams

2. 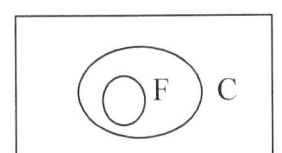   F = {friends}

C = {people who go to church}

3. 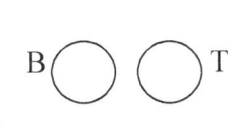   B = {boys}

T = {boys who fail the test}

4. 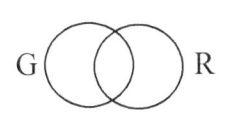   G = {girls}

R = {people who like reading}

5. 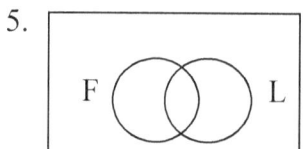   ξ   F = {friends}

L = {people who are loyal}

6. 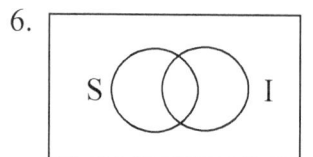   ξ   S = {students]

I = {people who are intelligent}

7. 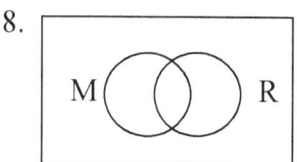   ξ   B = {boys}

T = {people who are tall}

8. 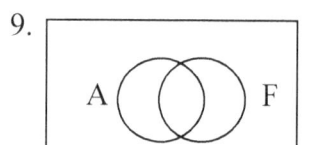   ξ   M = {men}

R = {people who are rich}

9.    ξ   A = {adults}

F = {people who play football}

10.    ξ   F = {football players}

P = {people who are poor}

Exercise 2.2

1. Valid   2. Valid   3. Not valid   4. Valid   5. Not valid

6. Not valid   7. Valid   8. Valid

9. Mr Sam wears glasses   10. Kwesi wears helmet

11. Policemen are not public officers   12. Some adults can read

Review exercise 2

1(a) 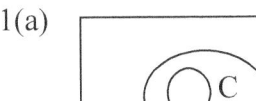   F = {animals that fly}

C = {cats}

(b) 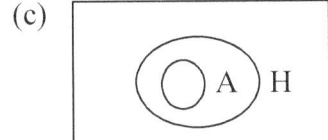   B = {brown coloured animals}

D = {dogs}

(c) 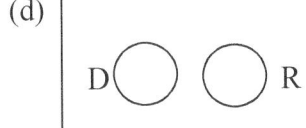   H = {hard workers}

A = {ants}

(d) 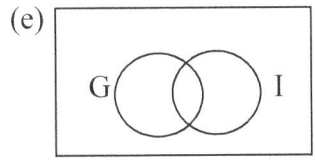   R = {object that read}

D = {dogs}

(e) 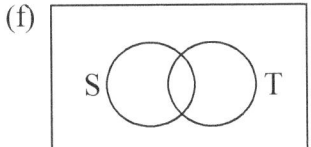   I = {people who eat ice cream}

G = {girls}

(f) 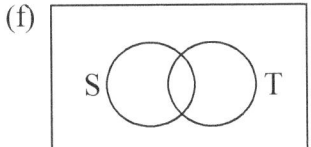   T = { student who passed the test}

S = {student who took the test}

2. Not valid   3. Valid   4. Valid   5. Not valid

6. Valid   7. Not valid   8. Not valid

9. Esi is intelligent   10. Mr Adu do not steal

11. All students are rich   12. No policeman is a students

Chapter Test 2

1. 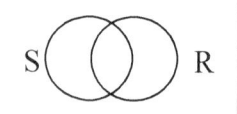   ξ   D = {objects having four doors}

C = {cars}

2. 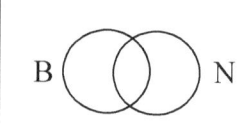   ξ   J = {people who like jazz music}

B = {boys}

3. 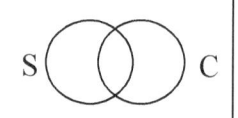   ξ   R = {people who read everyday}

S = {students}

4.   ξ   N = {new books}

B = {books in the library}

5.   ξ   C = {people who are clever}

S = {students in the class}

6.  F = {animals that fly}

D = {dogs}

7. Valid   8. Not valid   9. Valid   10. Not valid

11. Mr Mensah is not homeless   12. Some doctors are healthy

Chapter 3

Try this

1. (a) {5, 12}   (b) { - 3, - 2, 0, 5, 12}

(c) $\{-\frac{9}{2}, -3, -2, 0, 5, \frac{17}{2}, 12\}$ (d) $\{-\sqrt{3}, -\frac{\pi}{2}, \sqrt{7}\}$

3(a) >   (b) <   (c) >   (d) <

4(a) 31   (b) 6   5(a) 15.7   (b) 83.98   (c) 125.790

6(a) 32.5   (b) 642.0   (c) 0.00072   (d) 0.40

7. $\frac{2}{3}$ 8. $\frac{6}{11}$ 9. $\frac{7}{15}$ 10(a) – 16   (b) – 37

11(a) – 11   (b) 3   12(a) – 6   (b) 7   13(a) 56   (b) 60

14(a) – 54   (b) – 45   15(a) 3   (b) 4   16(a) – 6   (b) – 3

17. 1470   18. 55   19. – 3   20. 9   21. 650   22. - 20° C

23. $\frac{1}{5}$ 24. $\frac{13}{20}$ 25. $\frac{13}{24}$ 26. $1\frac{7}{45}$ 27. $3\frac{2}{15}$ 28. $\frac{4}{7}$

29. $4\frac{1}{5}$ 30. $\frac{5}{6}$ 31. $1\frac{4}{7}$ 32. $3\frac{1}{2}$ 33. GH¢ 420   34. – 3

35(a) No   (b) c   (c) a, inverse b,   b, inverse a   c, inverse c

(d) $d * (c * b) = (d * c) * b$

Exercise 3.1(a)

1(a) $\{1, 8\}$ (b) $\{-12, 0, 1, 8\}$ (c) $\{-12, -\frac{2}{3}, -\frac{1}{4}, 0, \frac{1}{8}, 1, 8\}$

(d) $\{-\sqrt{7}, \sqrt{3}, 4\pi\}$

2(a) $\{8, 12\}$ (b) $\{0, 8, 12\}$ (c) $-\frac{9}{2}, -\frac{3}{8}, 0, \frac{10}{3}, 8, 12\}$

(d) $\{-\sqrt{6}, \sqrt{13}\}$

3(a) $\{\ \}$ (b) $\{-\sqrt{4}\}$ (c) $\{-3.6, -\sqrt{4}, -\frac{1}{2}, -0.\dot{3}, -0.3, 26.4\}$

(d) $\{\sqrt{5}, 3\pi\}$

4(a) $\{3, 20\}$ (b) $\{0, 3, 20\}$ (c) $\{-\sqrt{25}, -0.11, -\frac{5}{2}, 0, 0.75, 3, 20\}$

(d) $\{-\sqrt{6}\}$

7(a) <   (b) >   (c) <   (d) >

8(a) <   (b) >   (c) >   (d) <

9(a) >   (b) >   (c) <   (d) >

Exercise 3.1(b)

1. 12   2. 85   3. 12   4. 15   5. − 17   6. 18   7. − 16   8. − 25

9. − 20   10. − 31   11. 7   12. <   13. =   14. =   15. <   16. >

Exercise 3.1(c)

1(a) 3.6   (b) 7.5   (c) 1.0

2(a) 15.0   (b) 0.78   (c) 7.10

3(a) 24.70   (b) 0.807   (c) 2.487

4(a) 12.310   (b) 5.175   (c) 6.413

5(a) 1.9   (b) 27   (c) 15.7

6(a) 25.0   (b) 456.7   (c) 700

7(a) 3000   (b) 0.070   (c) 0.20

8(a) 0.507   (b) 40.7   (c) 0.0090

9(a) 7800   (b) 80.0   (c) 6500

Exercise 3.1(d)

1. $\frac{1}{3}$ 2. $\frac{4}{9}$ 3. $\frac{7}{9}$ 4. $\frac{4}{33}$ 5. $\frac{5}{11}$

6. $\frac{5}{9}$ 7. $\frac{8}{11}$ 8. $\frac{4}{37}$ 9. $\frac{5}{27}$ 10. $\frac{13}{27}$

11. $\frac{32}{37}$ 12. $\frac{9}{11}$ 13. $\frac{1}{6}$ 14. $\frac{7}{18}$ 15. $\frac{17}{30}$

16. $\frac{5}{6}$ 17. $\frac{131}{200}$ 18. $\frac{7}{12}$ 19. $\frac{5}{12}$ 20. $\frac{11}{60}$

Exercise 3.2(a)

1(a) 2   (b) 3   (c) 12

2(a) 7   (b) 8   (c) 4

3(a) 6   (b) 7   (c) 7

4(a) 5   (b) 13   (c) 9

5(a) − 3   (b) − 12   (c) − 3

6(a) − 8   (b) − 5   (c) − 5

7(a) − 5   (b) − 8   (c) − 15

8(a) – 3   (b) – 11   (c) – 9

9(a) – 16   (b) – 28   (c) – 14

10(a) – 20   (b) – 35   (c) – 40

11(a) – 23   (b) – 35   (c) – 40

12(a) – 51   (b) – 60   (c) – 56

Exercise 3.2(b)

1 (a) 15   (b) 25   (c) 29

2(a) – 14   (b) – 10   (c) – 25

3(a) 28   (b) – 37   (c) – 97

4(a) 37   (b) – 100   (c) 86

5(a) 3   (b) 4   (c) 19

6(a) – 2   (b) – 1   (c) – 5

7(a) 5   (b) – 7   (c) – 4

8(a) 3   (b) – 12   (c) 12

9(a) - 7   (b) 4   (c) - 16

10(a) – 5   (b) – 8   (c) – 30

11(a) – 8   (b) 7   (c) – 5

12(a) 5   (b) – 25   (c) – 15

13(a) – 3   (b) 10   (c) – 48

14(a) 12   (b) – 25   (c) – 32

15(a) 4   (b) − 26   (c) 100

Exercise 3.2(c)

1(a) 42   (b) 72   (c) 96

2(a) 60   (b) 36   (c) 36

3(a) 56   (b) 84   (c) 90

4(a) 45   (b) 140   (c) 180

5(a) − 30   (b) − 70   (c) − 96

6(a) − 44   (b) − 36   (c) − 26

7(a) − 56   (b) - 70   (c) − 48

8(a) − 45   (b) − 120   (c) − 165

Exercise 3.2(d)

1(a) 2   (b) 4   (c) 2

2(a) 3   (b) 7   (c) 4

3(a) 3   (b) 5   (c) 4

4(a) 2   (b) 9   (c) 3

5(a) - 5   (b) - 3   (c) - 5

6(a) - 5   (b) - 6   (c) - 6

7(a) - 5   (b) - 5   (c) - 8

8(a) - 4   (b) − 7   (c) − 2

Exercise 3.2(e)

1(a) 100   (b) 100   (c) 1200

  (d) 160   (e) 2500   (f) 1080

2(a) 693   (b) 336   (c) 1182

  (d) 16983   (e) 741   (f) 1192

Exercise 3.2(f)

1. -4 2. 20   3. -48 4. 21   5. -2 6. 3

7. 9   8. -1 9. 22   10. -5 11. -3 12. 1

Exercise 3.2(g)

1. -5 2. -27 3. $-$ GH¢ 25   4. $-$ GH¢ 70   5. -5 6. -500

7. -800 m   8. $-2\,^{\circ}$C   9. $-2\,^{\circ}$C   10. $-22\,^{\circ}$C   11. 75 $^{\circ}$C

12. 985 years

Exercise 3.3(a)

1. $\dfrac{4}{5}$ 2. $\dfrac{6}{7}$ 3. $\dfrac{4}{7}$ 4. $\dfrac{2}{3}$ 5. $1\dfrac{1}{5}$

6. $\dfrac{1}{4}$ 7. $\dfrac{2}{9}$ 8. $\dfrac{1}{3}$ 9. $\dfrac{2}{5}$ 10. $\dfrac{1}{5}$

Exercise 3.3(b)

1. $1\dfrac{4}{15}$ 2. $1\dfrac{7}{12}$ 3. $\dfrac{22}{45}$ 4. $\dfrac{23}{30}$ 5. $1\dfrac{2}{3}$

6. $\dfrac{1}{15}$ 7. $\dfrac{11}{56}$ 8. $\dfrac{11}{24}$ 9. $\dfrac{1}{4}$ 10. $\dfrac{1}{24}$

Exercise 3.3(c)

1. $6\frac{1}{4}$ 2. $7\frac{5}{6}$ 3. $7\frac{31}{45}$ 4. $12\frac{7}{40}$

5. $4\frac{23}{30}$ 6. $1\frac{3}{8}$ 7. $2\frac{3}{8}$ 8. $\frac{11}{18}$

9. $2\frac{21}{40}$ 10. $1\frac{1}{3}$

Exercise 3.3(d)

1. $1\frac{1}{4}$ 2. $3\frac{3}{4}$ 3. $\frac{3}{4}$ 4. $\frac{5}{9}$

5. $\frac{3}{10}$ 6. 22   7. 14   8. $7\frac{1}{2}$

9. $4\frac{1}{7}$ 10. $31\frac{1}{2}$

Exercise 3.3(e)

1. 18   2. $\frac{1}{20}$ 3. $1\frac{1}{4}$ 4. $1\frac{1}{6}$

5. $\frac{5}{6}$ 6. $\frac{5}{8}$ 7. 8   8. $3\frac{3}{4}$

9. 2   10. $3\frac{1}{2}$

Exercise 3.3(f)

1. $1\frac{13}{50}$ 2. $\frac{1}{2}$ 3. $2\frac{3}{5}$ 4. $13\frac{1}{3}$

5. 2   6. $4\frac{29}{60}$ 7. $4\frac{1}{8}$ 8. $\frac{3}{5}$

9. 2   10. $\frac{2}{5}$ 11. 7   12. $2\frac{2}{5}$

13. $9\frac{1}{3}$ 14. $\frac{5}{12}$ 15. 3

Exercise 3.3(g)

1. GH¢ 1500   2. GH¢ 960   3. GH¢ 32000

4. 3500   5. 60   6. 375   7. GH¢ 4500

8. 300 m   9. GH¢ 35   10. GH¢ 129.60

Exercise 3.4

1. − 18   2. − 1   3. − 8   4. 19   5. − 2

6. 0   7. − 1   8. − 53   9. 49   10. 2

11(a) no   (b) 3   3 is inverse of 3; 2 is inverse of 4; 5 is inverse of 5

   (d) $(2 * 3) * 4 = 2 * (3 * 4)$

12(a) yes   (b) no identity element   (c) 1 is inverse of 1

   (d) $4 * (3 * 2) \neq (4 * 3) * 2$

Review exercise 3

1(a) {12}   (b) $\{-5,, 0, 12\}$

   (c) $\{-3.5, -\sqrt{9}, -2.3, -1, -0.75, 0, \frac{19}{2}, 12\}$ (d) $\{\sqrt{3}, 2\pi\}$

2(a) {9, 23}   (b) {0, 9, 23}   (c) $\{-\frac{9}{2}, -\frac{3}{4}, 0, \sqrt{36}, \frac{15}{2}, 9, 23\}$

4(a) <   (b) >   (c) <   (d) >

5(a) 8   (b) 7.3   (c) − 32   (d) − 15

6(a) <   (b) >   (c) =

7(a) 7.33   (b) 1.0   (c) 0.900   (d) 13.74

8(a) 0.007   (b) 65.0   (c) 0.061   (d) 1400

9(a) $\frac{2}{3}$ (b) $\frac{8}{11}$ (c) $\frac{11}{30}$ (d) $\frac{7}{25}$

10(a) 11   (b) 2   (c) 0   (d) – 4

(e) 24   (f) – 4   (g) – 8   (h) – 10

11(a) – 24   (b) – 36   (c) 78   (d) – 120

(e) – 8   (f) 8   (g) – 3   (h) 4

12(a) 300   (b) 4150   (c) 1379   (d) 1488

13(a) 26   (b) 7   (c) 28   (d) 21

(e) 22   (f) – 8   (g) 7   (h) 12

14. –GH¢ 85   15. – 6 m   16. – 13 ⁰ C

17(a) $1\frac{1}{4}$ (b) $\frac{3}{5}$ (c) $\frac{14}{15}$ (d) $\frac{13}{36}$

(e) $1\frac{1}{4}$ (f) $7\frac{11}{12}$ (g) $2\frac{11}{12}$ (h) $6\frac{8}{15}$

18(a) 6   (b) $\frac{5}{9}$ (c) $25\frac{1}{2}$ (d) $7\frac{3}{5}$

(e) 9   (f) $\frac{5}{6}$ (g) $1\frac{1}{2}$ (h) $2\frac{1}{14}$

19(a) $1\frac{1}{2}$ (b) $\frac{3}{4}$ (c) $\frac{19}{35}$ (d) $\frac{1}{3}$ (e) $\frac{8}{27}$ (f) $1\frac{1}{6}$

20(a) – 11   (b) – 29   (c) 18   (d) 21

21(a) 4   (b)   (c) 5

Chapter Test 3

1(a) {2, 5}   (b) {0, 2, 5}

(c) $\{-6.7, -\sqrt{25}, -\frac{9}{2}, -1.2, 0, 2, \frac{7}{2}, 5\}$ (d) $\{-\pi, \sqrt{7}\}$

3(a) 9   (b) 9   (c) - 3

4(a) – 3   (b) – 54   (c) – 3

5. – 3 º C   6. 60 m below sea level

7(a) Associative   (b) Distributive   (c) Commutative

8(a) 20.0   (b) 0.00071   (c) 0.90

9(a) $\frac{2}{11}$ (b) $3\frac{4}{5}$

10(a) 10   (b) – 8

11(a) $2\frac{4}{5}$ (b) $4\frac{1}{10}$

12. – 11

13(a) Yes   (b) 0

(c) – 1 , inverse 1, 0, inverse 0   1, inverse – 1, 2 inverse 2

(d) 0

Chapter 4

Try this

1(a) $11x$ (b) $-2y$ (c) $-15xy$

2(a) $9x + 5y$ (b) $-a + 4b$

3(a) $14ab$ (b) $-10xy$ (c) $12pq$

4(a) $15x^3y^2$ (b) $-14a^4b^3$ (c) $9x^4y^6$

5(a) $2a$ (b) $2x^2y^2$ (c) $-2pq$ 6. 9

7(a) $6x$ (b) $5 + y$ (c) $a - 7$ (d) $8 - x$

(e) $3m$ (f) $\dfrac{3}{n}$ (g) $\dfrac{1}{2}x$ (h) $9 + y$

8(a) $-6x - 3y$ (b) $-2x + 6y$

9(a) $6p^2 - 3pq$ (b) $-2a^2b - 6ab^2$

10(a) $3x - 12y$ (b) $-3x + 2y$ 11. $2x - 3y$

12(a) $3a(a + 5)$ (b) $7ab(a - 2b)$

13(a) $-2ab(3a^2 - 4b)$ (b) $-3xy(5y + 2x)$

14. $(x + 4)(3x + 2y)$ 15. $(x - 3)(5x - 2y)$

16. $x^2 + 2x - 15$ 17. $x^2 - 5x + 6$

18(a) $x^2 - 14x + 49$ (b) $9x^2 + 12x + 4$

19(a) $x^2 - 36$ (b) $x^2 - 4y^2$

20. $(x + 2)(x + 9)$ 21. $(x - 2)(x - 4)$

22. $(x - 4)(x + 6)$ 23. $(x - 4)(x + 3)$

24. $(2x - 3)(x + 5)$ 25. $(3x - 4)(x - 2)$ 26. $(x + 4)^2$

27. $(x - 9)^2$ 28. $(x - 5)(x + 5)$

29. $(4x^3 - 5y^2)(4x^3 + 5y^2)$

30. $2(2x - 5)(2x + 5)$ 31. 292   32. 2491

33(a) $\dfrac{6x}{5y^3}$ (b) $\dfrac{y}{2z^2}$

34(a) $\dfrac{y-3}{2y}$ (b) $\dfrac{y}{2x+3}$

35(a) $\dfrac{4x}{x+8}$ (b) $\dfrac{x-2}{3y}$

36(a) $\dfrac{2y^3}{5x}$ (b) $\dfrac{5x}{x+2}$

37(a) $\dfrac{y^2}{6x^2}$ (b) $\dfrac{2x^2}{x-2}$

38(a) $30xyz$ (b) $60x^3y^3$ (c) $2x(x-2)(x+3)$

39(a) $\dfrac{x+y}{x}$ (b) $\dfrac{x+11}{(3x-2)(2x+1)}$ (c) $\dfrac{5x+3}{(x-1)(x+2)(x+3)}$

Exercise 4.1(a)

1(a) $6x$ (b) $8y$ (c) $9a$ (d) $13b$ (e) $12x^2$

2(a) $5x$ (b) $4y$ (c) $-a$ (d) $-3b$ (e) $4x^2$

3(a) $5x$ (b) $-4a$ (c) $3y$ (d) $2x$ (e) $-y^2$

4(a) $-9x$ (b) $-4y$ (c) $5x$ (d) $-5ab$ (e) $-7xy^2$

Exercise 4.1(b)

1(a) $8x+7y$ (b) $5a+3b$ (c) $7x+4y$

(d) $4a+5b$ (e) $8x^2+5y$

2(a) $-6a$ (b) $2x-8y$ (c) $5x-9y$

(d) $2a-8b$ (e) $3y^2-16y$

3(a) $10x^2+13$ (b) $8ab^2+7a^2b$ (c) $9xy+7yz$

(d) $2x^3+3x^2$ (e) $-3x^2y+5xy^2$

4(a) $2x^2+3x$ (b) $3ab^2+2a^2b$ (c) $2x^3-5x^2$

(d) $2x^2y-4xy^2$

Exercise 4.1(c)

1(a) $12ab$ (b) $10xy$ (c) $12pq$

(d) $56uv$ (e) $5st$

2(a) $-6rs$ (b) $-8pq$ (c) $-15ab$

(d) $-24xy$ (e) $-21pq$

3(a) $4ab$ (b) $10xy$ (c) $42pq$

(d) $24rs$ (e) $18xy$

Exercise 4.1(d)

1(a) $2y^2$ (b) $4y^2$ (c) $9x^4y^2$ (d) $-8x^3y^6$ (e) $-4x^3y^2$

2(a) $3a^5b^4$ (b) $-24a^3b^2$ (c) $-36c^3d^2$ (d) $-5b^5$

(e) $6a^3b^5$

3(a) $6a^2$ (b) $6a^2b$ (c) $30a^4b^2$ (d) $8x^5y^3$ (e) $2a^3b^2c$

4(a) $-6a^2b$ (a) $8x^2y$ (c) $-15x^2y^3$

(d) $6x^3y^5$ (e) $-6a^3b^4c^2$

5(a) $6x^2y^7$ (b)$6p^4q^3$ (c) $20a^5b^7$

(d) $15x^3y^3$ (e) $42a^3b^4$

6(a) $-6x^5y^6$ (b) $10a^7b^3$ (c) $-7b^5c^7$

(d) $32x^6y^6$ (e) $21a^5b^4c^5$

Exercise 4.1(e)

1(a) $4b$ (b) $5z$ (c) $3b$ (d) $3y$ (e) $5x$

2(a) $2y$ (b) $4xy$ (c) $3z^2y$ (d) $\dfrac{a}{3b}$ (e) $\dfrac{5xz}{6y}$

3(a) $-3x^2y$ (b) $\frac{2y}{x}$ (c) $-\frac{3}{2y}$ (d) $-\frac{5y}{z}$ (e) $\frac{4a^2z}{b}$

4(a) $3y$ (b) ab (c) $\frac{2ab}{3}$ (d) $\frac{6}{b}$ (e) $\frac{4y}{x}$

Exercise 4.1(f)

1. -42 2. 120   3. 162   4. 100   5. 15

6. 9   7. 13   8. 11   9. 3   10. 42

11. 24   12. -4 13. 20   14. $3\frac{1}{3}$ 15. 4

Exercise 4.1(g)

1. $x+y$ 2. $x+3$ 3. $a+2$

4. $p+5$ 5. $a-c$ 6. $s-7$

7. $7-z$ 8. pq 9. $6b$

10. $2x-8$ 11. $3m+1$ 12. $2(a+b)$ 13. $2(x+y)$

14. $3(x-y)$ 15. $\frac{a-b}{3}$ 16. $\frac{2x+y}{5}$ 17. $2+3x$

18. $3(x-4)$ 19. $2(x-3)$ 20. $\frac{2a-5}{3}$

Exercise 4.2(a)

1(a) $3x+6y$ (b) $12a+20b$ (c) $8c-8d$ (d) $8r-6s$

(e) $14p-21q$ (e) $10x+5y$

2(a) $-6a+6b$ (b) $-6x+9y$ (c) $-5c+15d$

(d) $-8a-4b$ (e) $-9q-18p$ (f) $-21x-14y$

3(a) $8a+16$ (b) $30-10b$ (c) $52+4x$

(d) $-24 + 14a$ (e) $-12p - 36$ (f) $-21a + 28$

4(a) $12a + 8b + 4c$ (b) $-5x + 10y + 15z$

(c) $3a - 3b + 6$ (d) $12 + 6a - 4b$

(e) $14x + 7y - 21$ (f) $-8 + 12b - 4c$

5(a) $6x^2 + 12xy$ (b) $-12y + 8y^2$

(c) $5a^2 - 10ab$ (d) $6r^2 - 8rs$

(e) $-2x^2y^2 + 3x^2y$ (f) $-3p^3q^2 - 3p^2q^3$

6(a) $15x^2 + 10x^2y - 5xy$ (b) $12a^2b - 14ab^2 + 2abc$

(c) $-2x + 4x^2y + 8xy^2$ (d) $-3a^2 - 6a^3 + 9a^4$

(e) $8r^3 - 12r^2s + 16r^4t$ (f) $-8p^2 + 16p^2q - 24pq$

Exercise 4.2(b)

1(a) $10a + 12$ (b) $17x + 20$ (c) $9x + 13$ (d) $12a + 7$

(e) $9p + 2q$ (f) $26x - 3$ (g) $-3m + 5$ (h) $5y - 8$

2(a) $3r + 6$ (b) $-10x + 9$ (c) $-a + 6$ (d) $-2a + 8b$

(e) $10p - 7q$ (f) $-4r + 12$ (g) $2p - 9q$ (h) $4x + 10$

3(a) $5x^2 - 8xy$ (b) $-4y^2 + 14y$ (c) $rs + 2s^2$

(d) $-xy^2 + x^2y$ (e) $3p^2 + 2p^2q$ (f) $2r^2 - 2r^3 + 2r^4$

(g) $p^2 + 6p$ (h) $e^2 - 11ef$

Exercise 4.3(a)

1(a) 5   (b) 3   (c) 4   (d) $4a$ (e) $2p$ (f) $7y$

(g) $5q$ (h) xy (i) $2a^2$ (j) $3rs$ (k) $4y$ (l) $5r^2$

2(a) $a + 2$ (b) $2 - y$ (c) $c + 2d$ (d) $2r - s$

(e) $1 + 2x$ (f) $p - 2$ (g) $a + b$ (h) $ax - 1$

(i) $x - 2y$ (j) $3b + 2c$ (k) $3p - 4q$ (l) $2p + q$

3(a) $3(a + 4)$ (b) $4(y - 2)$ (c) $5(p + 2q)$

(d) $7(x - 3y)$ (e) $4y(2x - 3z)$ (f) $3p(2p - 5r)$

(g) $6pq(p - 3q)$ (h) $xy(7y + 4x)$

4(a) $-2xy(2x + 3y)$ (b) $-5a(2a - 3b)$ (c) $-4ab(3a - 2b)$

(d) $-5xy(y + 2x)$ (e) $-4p^2q(3pq - 2)$ (f) $-p(2p + 3q)$

(g) $-7xy(5y + 2x)$ (h) $-7a^2b^2(a - 3b)$

Exercise 4.3(b)

1. $(b + c)(a + d)$ 2. $(r + s)(p - q)$

3. $(x + y)(5a + 2b)$ 4. $(a + 2b)(3c + 4d)$

5. $(y + 2z)(2x - 3y)$ 6. $(2y - 3z)(3x - 4y)$

7. $(3x - y)(a + b)$ 8. $(5a - b)(2y + 6)$

9. $(r - 2s)(4p - 5q)$ 10. $2(a + b)(2u + 3v)$

11. $(x - 2y)(7x - 3z)$ 12. $2(x + y)(5a + 4b)$

Exercise 4.4(a)

1. $x^2 + 9x + 20$ 2. $x^2 + 13x + 42$

3. $x^2 + 10x + 9$ 4. $x^2 + 17x + 60$

5. $x^2 + 12x + 20$ 6. $x^2 + 18x + 45$

7. $2x^2 + 11x + 12$ 8. $6x^2 + 7x + 2$

9. $10x^2 + 43x + 28$ 10. $18x^2 + 39x + 20$

11. $x^2 + 2x - 15$ 12. $x^2 + 3x = 70$

13. $x^2 + 4x - 32$ 14. $2x^2 + x - 15$

15. $x^2 + x - 72$ 16. $3x^2 - x - 30$

17. $7x^2 - x - 30$ 18. $x^2 - 4x - 45$

19. $x^2 - 3x - 70$ 20. $x^2 - 6x - 72$

21. $x^2 + 5x - 36$ 22. $x^2 + 5x - 24$

23. $x^2 + 5x - 14$ 24. $x^2 + 8x - 65$

25. $x^2 - 3x - 40$ 26. $x^2 - 7x - 44$

27. $x^2 - 3x - 54$ 28. $6x^2 + 5x - 6$

29. $8x^2 - 32x - 21$ 30. $5x^2 + 38x - 16$

31. $x^2 - 7x + 10$ 32. $x^2 - 12x + 32$

33. $x^2 - 10x + 21$ 34. $x^2 - 13x + 30$

35. $x^2 - 9x + 18$ 36. $x^2 - 22x + 120$

37. $15x^2 - 29x + 12$ 38. $5x^2 - 31x + 6$

39. $6x^2 - 31x + 35$ 40. $8x^2 - 26x + 15$

Exercise 4.4(b)

1. $x^2 + 6x + 9$ 2. $x^2 + 14x + 49$

3. $x^2 + 12x + 36$ 4. $x^2 + 24x + 144$

5. $4x^2 + 12x + 9$ 6. $25x^2 + 40x + 16$

7. $x^2 - 10x + 25$ 8. $x^2 - 4x + 4$

9. $x^2 - 20x + 100$ 10. $x^2 - 6x + 9$

11. $36x^2 - 12x + 1$ 12. $49x^2 - 28x + 4$

Exercise 4.4(c)

1. $x^2 - 16$ 2. $x^2 - 25$ 3. $x^2 - 9$ 4. $x^2 - 64$

5. $x^2 - 49$ 6. $4x^2 - 9$ 7. $16x^2 - 25$ 8. $9x^2 - 81$

9. $49x^2 - 16$ 10. $36x^2 - 169$

Exercise 4.5(a)

1. $(x + 4)(x + 5)$ 2. $(x + 3)(x + 7)$

3. $(x + 6)(x + 8)$ 4. $(x + 2)(x + 10)$

5. $(x + 5)(x + 9)$ 6. $(x + 3)(x + 12)$

7. $(x + 4)(x + 8)$ 8. $(x + 3y)(x + 6y)$

9. $(x + 2y)(x + 3y)$ 10. $(x + 3y)(x + 4y)$

11. $(x + 2y)(x + 5y)$ 12. $(x + 7y)(x + 8y)$

13. $(x - 2)(x - 7)$ 14. $(x - 3)(x - 5)$

15. $(x - 4)(x - 9)$ 16. $(x - 5)(x - 6)$

17. $(x - 7)(x - 10)$ 18. $(x - 5)(x - 8)$

19. $(x - 3)(x - 9)$ 20. $(x - 2y)(x - 8y)$

21. $(x - y)(x - 6y)$ 22. $(x - 4y)(x - 7y)$

23. $(x - 3y)(x - 5y)$ 24. $(x - 6y)(x - 9y)$

Exercise 4.5(b)

1. $(x - 4)(x + 5)$ 2. $(x - 3)(x + 7)$

3. $(x - 6)(x + 8)$ 4. $(x - 5)(x + 10)$

5. $(x - 7)(x + 10)$ 6. $(x - 5)(x + 8)$

7. $(x - 3)(x + 9)$ 8. $(x - 3y)(x + 5y)$

9. $(x - 2y)(x + 3y)$ 10. $(x - 4y)(x + 8y)$

11. $(x - 3y)(x + 10y)$ 12. $(x - 4y)(x + 9y)$

13.$(x - 7)(x + 2)$ 14. $(x - 10)(x + 3)$

15. $(x - 12)(x + 8)$ 16. $(x - 8)(x + 7)$

17. $(x - 6)(x + 4)$ 18. $(x - 9)(x + 6)$

19. $(x - 9)(x + 3)$ 20. $(x - 8y)(x + 3y)$

21. $(x - 7y)(x + 5y)$ 22. $(x - 5y)(x + 4y)$

23. $(x - 8y)(x + 2y)$ 24. $(x - 7y)(x + 6y)$

Exercise 4.5(c)

1.$(2x + 1)(x + 4)$ 2. $(3x + 2)(x + 3)$

3. $(5x - 4)(x + 6)$ 4. $(2x - 3)(3x + 2)$

5. $(2x - 3)(5x + 3)$ 6. $(3x - 2)(7x + 4)$

7. $(6x - 7)(4x + 3)$ 8. $(2x + 9)(3x - 5)$

9. $(2x + 5)(x + 3)$ 10. $(3x - 5)(5x + 2)$

11. $(3x - 5)(4x - 3)$ 12. $(2x + 7)(4x - 5)$

13. $(2x + 3)(3x - 5)$ 14. $(5x - 3)(x + 1)$

15. $(3x + 4)(x + 2)$

Exercise 4.5(d)

1. $(x + 3)^2$ 2. $(x + 5)^2$ 3. $(x + 7)^2$

4. $(x + 6)^2$ 5. $(x + 9)^2$ 6. $(x + 12)^2$

7. $(x + 15)^2$ 8. $(2x + 4y)^2$ 9. $(3x + 5y)^2$

10. $(4x + 3y)^2$ 11. $(5x + 4y)^2$ 12. $(2x + 5y)^2$

13. $(x - 2)^2$ 14. $(x - 6)^2$ 15. $(x - 8)^2$

16. $(x - 10)^2$ 17. $(x - 11)^2$ 18. $(x - 13)^2$

19. $(x - 6y)^2$ 20. $(x - 5y)^2$ 21. $(3x - 2y)^2$

22. $(2x - 5y)^2$ 23. $(4x - 3y)^2$ 24. $(7x - 9y)^2$

Exercise 4.5(e)

1. $(x - 4)(x + 4)$ 2. $(x - 8)(x + 8)$

3. $(x - 12)(x + 12)$ 4. $(3y - 8)(3y + 8)$

5. $(2x - 5)(2x + 5)$ 6. $(13 - x)(13 + x)$

7. $(5x - 4y)(5x + 4y)$ 8. $3(3x - 2y)(3x + 2y)$

9. $8(x - 3y)(x + 3y)$ 10. $5x(x - 3)(x + 3)$

11. $3(x - 2)(x + 2)(x + 4)$ 12. $(1 - 7x)(1 + 7x)$

13. $5x(22x - 3y)(2x + 3y)$ 14. $2y(4x - 3y)(4x - 3y)$

15. $2x(5x - 2y)(5x + 2y)$

Exercise 4.5(f)

1. 112   2. 68   3. 7300   4. 4800   5. 672000

6. 8096   7. 39984   8. 6375   9. 3591   10. 999900

Exercise 4.6(a)

1. $\dfrac{2a}{3b}$ 2. $\dfrac{5y}{6x}$ 3. $\dfrac{4}{3x}$ 4. $\dfrac{3}{a}$ 5. $\dfrac{2x-1}{x}$

6. $\dfrac{1}{y-4}$ 7. $\dfrac{x(x+4)}{x+2}$ 8. $\dfrac{x+5}{3x}$ 9. $\dfrac{x+2}{x+3}$ 10. $\dfrac{x-4}{x+6}$

11. $\dfrac{1-2x}{x+3}$ 12. $\dfrac{x-3}{x+8}$ 13. $\dfrac{x-2y}{x+2y}$ 14. $\dfrac{2-x}{x+2}$ 15. $\dfrac{1}{3y-x}$

Exercise 4.6(b)

1. $\dfrac{5y}{6x}$ 2. $\dfrac{3x}{4}$ 3. $\dfrac{3y^2}{2x^3}$ 4. $\dfrac{y}{2x}$ 5. $\dfrac{5z}{3y}$ 6. $\dfrac{x-y}{x}$

7. $\dfrac{2x^2}{y(x+2y)}$ 8. $\dfrac{3}{2}$ 9. $\dfrac{x-2y}{a}$ 10. $\dfrac{a+b}{a(a-b)}$ 11. 1   12. $\dfrac{x+3}{x+1}$

Exercise 4.6(c)

1. $\dfrac{3xz}{2y}$ 2. $\dfrac{3y}{2}$ 3. $\dfrac{5}{6xy^2}$ 4. $\dfrac{x}{4a}$ 5. $\dfrac{1}{y}$ 6. $\dfrac{4x}{x+y}$

7. $\dfrac{x+2}{x}$ 8. $\dfrac{1}{x-4}$ 9. $\dfrac{x+3}{10x}$ 10. $\dfrac{b}{a}$ 11. $-\dfrac{2}{3y}$ 12. $\dfrac{1}{a+b}$

Exercise 4.6(d)

1. $12x^2$ 2. $6x^2y^2$ 3. $(a-b)(a+b)$

4. $3(x-1)(x+1)$ 5. $(2x-y)(2x+y)$

6. $(x-2)(x+3)$ 7. $x(x+2)(x+3)$

8. $(x-2)(x-3)(x-4)$ 9. $(x+1)(x+2)$

10. $x(x-y)(x+y)$

Exercise 4.6(e)

1. x 2. $\dfrac{y}{4}$ 3. $\dfrac{13}{2x}$ 4. $\dfrac{1}{6y}$ 5. $\dfrac{7x-y}{12}$

6. $-\dfrac{5}{12}$ 7. $-\dfrac{29y}{12}$ 8. $\dfrac{x+5}{2}$ 9. $\dfrac{20x+12}{(2x-3)(2x+3)}$ 10. $-\dfrac{2}{x+3}$

11. $\dfrac{3}{x-y}$ 12. $\dfrac{x+5y}{(x-y)(x+y)}$ 13. $\dfrac{6}{(x+5)9x-6)}$ 14. $\dfrac{2x-10}{(x-3)(x+1)(x+3)}$

15. $\dfrac{3y^2-4y-10}{(y-2)(y-5)(y+3)}$ 16. $\dfrac{x+27}{(x-5)(x+2)(x+7)}$ 17. $\dfrac{9x+6}{(x-4)(x+3)^2}$

18. $\dfrac{-5x+5}{(x-3)(x+2)^2}$ 19. $\dfrac{-7x+19}{(x-2)(x+2)}$ 20. $\dfrac{x+3y}{(x-y)(x+y)}$

Review exercise 4

1(a) $15x^4y^2$ (b) $6x^3y^4$ (c) $15y^5$ (d) $28x^8$

(e) $-14x^5$ (f) $-12x^6$ (g) $32x^5y^3$ (h) $21x^6y^5$

(i) $-6x^9y^3$ (j) $4x^4y^6$ (k) $-125a^3b^6$ (m) $81x^8y^{12}$

2(a) $3x^4$ (b) $-4x^2$ (c) $3x^2$ (d) $-5mn$ (e) $-8xy$

(f) $5x^3y$ (g) $7m^3n^2$ (h) $6p^2q$ (i) $-3m^2$ (j) $3x^3y$

3(a) 72   (b) - 4   (c) -36 (d) -40 (e) 36   (f) 14

4(a) $a+b$ (b) $x+2$ (c) $x+3$ (d) $x-y$

(e) $a-6$ (f) $8x$ (g) $5(a+b)$ (h) $2(x-y)$

5(a) $7x$ (b) $-2a$ (c) $8y$

(d) $14y$ (e) $8x+y$ (f) $-9a+3b$

6(a) $5x^2+2x$ (b) $15x-6x^2$ (c) $10y^2-5y$

(d) $-15t+10t^2$ (e) $-18xy+12x$ (f) $-3y^2-2xy$

(g) $6xy+2x^2$ (h) $-5x+3y$ (i) $4x^2+8xy-12x$

(j) $3x^26xy+3x$ (k) $-2p^2+6pq-8p$

7(a) $-3x + 3$ (b) $a + 3$ (c) $5y - 5$ (d) $7y + 7$

(e) $6 + y^3$ (f) $x^3 - 12$ (g) $-5x + 10$ (h) $1 - 3x$

(i) $2y^2 - y$ (j) $-2x - 6y$

8(a) $x^2 + 9x + 18$ (b) $x^2 + 7x + 10$

(c) $x^2 + 3x - 10$ (d) $x^2 + 4x - 12$

(e) $x^2 - 13a + 42$ (f) $y^2 - 12y + 32$

(g) $x^2 - 9$ (h) $x^2 - 36$ (i) $9x^2 - 4$

(j) $x^2 + 6x + 9$ (k) $4x^2 - 12x + 9$ (m) $9x^2 + 12x + 4$

9(a) $2x(x + 3)$ (b) $5t(2t - 1)$

(c) $5x^2(x + 2)$ (d) $6y(2y - 1)$

(e) $9pq(3 + 2p)$ (f) $4x(3x + 2)$

(g) $8xy(1 - 3y)$ (h) $3xy(x + 2y)$

(i) $7xy(x - x^2)$ (j) $4x^2y^2(9x - 2y)$

10(a) $(x + 2)(x + y)$ (b) $(x - 2)(2x + z)$

(c) $(x + 1)(x + y)$ (d) $(a - 3)(a + y)$

(e) $(x - 2)(x + 2)(x + 3)$ (f) $(x + 6)(2x^2 - 5)$

11(a) $(x + 1)(x + 5)$ (b) $(x + 2)(x + 5)$

(c) $(y + 4)(y + 7)$ (d) $(y + 9)(y - 5)$

(e) $(x - 5)(x + 12)$ (f) $(x - 5)(x + 3)$

(g) $(a - 6)(a + 2)$ (h) $(x - 3)(x - 5)$

(i) $(a - 2)(a - 5)$ (j) $(y - 1)(y - 10)$

(k) $(x-6)(x+7)$ (l) $(a-7)(a+5)$

12(a)$(2x-1)(x+4)$ (b) $(2x-7)(3x-1)$

(c) $(7x+1)(x+2)$ (d) $(3x-4)(3x+2)$

(e) $(2a+3)(2a-5)$ (f) $(3x+1)(x-2)$

13(a) $(x-9)^2$ (b) $(x+7)^2$ (c) $(4x-3)^2$

(d) $(2x+3)^2$ (e) $(8x+1)^2$ (f) $(3x-5)^2$

14(a) $(x-6)(x+6)$ (b) $(x-3)(x+3)$

(c) $(3y-2)(3y+2)$ (d) $(4a-3)(4a+3)$

(e) $(4a-3b)(4a+3b)$ (f) $2(3x-2y)(3x+2y)$

(g) $(7a-4b)(7a+4b)$ (h) $3x(5y^3-x^2)(5y^3+x^2)$

15(a) $\dfrac{2x^3}{3}$ (b) $\dfrac{2y}{3x}$ (c) $\dfrac{3ab^3}{4}$ (d) $\dfrac{3x^3}{4y^2}$ (e) $\dfrac{3x}{4y}$

(f) $-\dfrac{2x^2}{3y}$ (g) $-\dfrac{2x}{3y^3}$ (h) $\dfrac{b^2}{2a^2}$ (i) $\dfrac{x+1}{3}$ (j) $\dfrac{5y}{y-5}$

(k) $\dfrac{x-4}{x-5}$ (l) $x-2$ (m) $\dfrac{x+3}{x-1}$ (n) $-\dfrac{(x+3y)}{2y+x}$ (o) $-\dfrac{(b+4)}{b+3}$

16(a) $\dfrac{x^2}{6y^2}$ (b) $\dfrac{3y^2}{2x^4}$ (c) $\dfrac{5b}{12}$ (d) $\dfrac{2}{3y}$ (e) $\dfrac{2}{3}$

(f) $x(x-2)$ (g) $\dfrac{3x}{x-8}$ (h) $\dfrac{5x}{x+2}$ (i) $\dfrac{2(x+2)}{x}$ (j) $-\dfrac{3(x-2)}{x}$

17(a) $\dfrac{3xy^2}{z}$ (b) $\dfrac{5}{6x^2y^2}$ (c) $\dfrac{9}{20}$ (d) $\dfrac{1}{3}$ (e) $\dfrac{a+3}{10a}$

(f) $\dfrac{2(p-2)}{9}$ (g) $\dfrac{x}{x+2}$ (h) $\dfrac{2x+1}{2x}$ (i) $\dfrac{x+2}{3x^2}$ (j) $\dfrac{3y}{x+2y}$

18(a) $\dfrac{1}{xy}$ (b) $\dfrac{2}{ab}$ (c) 7 (d) $\dfrac{x-y}{y}$ (e) $\dfrac{-x+3}{(x+1)(x+3)}$

(f) $\dfrac{1}{2(x-3)}$ (g) $\dfrac{3y^2+8y+10}{(y-3)(y+2)(y+5)}$ (h) $\dfrac{2a^2-7a+9}{(a-3)(a-2)(x+4)}$

(i) $\dfrac{x}{(x+2)(x+3)}$ (j) $\dfrac{7y^2-10y}{(y-5)(y-1)(y+1)}$

Chapter Test 4

1(a) $6x^3y^5$ (b) $4x^4y^6$ (c) $3xy^2$

2(a) $6xy - 3x^2$ (b) $-6p^2 + 8pq$ (c) $xy^2 - x^2y$

3(a) $4p - 3q$ (b) $x - 6$ (c) $2y + 12$

4(a) $5y$ (b) $-a + 3b$ (c) $6k + t$

5(a) $x + 7$ (b) $a - 8$ (c) $12b$ (d) $\dfrac{3a}{b}$

6(a) 0 (b) 5 (c) -4

7(a) $x^2 - x - 56$ (b) $4x^2 - 9$ (c) $9x^2 - 12x + 4$

8(a) $2x(x - 3)$ (b) $7x^2(2y + x)$ (c) $-6ab(3b - 2a)$

9(a) $(x - 2y)(x - 3)$ (b) $(x - 5)(x^2 + 3)$ (c) $(2r + s)(p - 3q)$

10(a) $(x - 2)(x + 10)$ (b) $(x - 7y)(x + 6y)$

(c) $(x + 2y)(x + 5y)$

11(a) $(x - 1)(3x + 7)$ (b) $(x + 1)(7x + 3)$

12(a) $(x + 9)^2$ (b) $(2x - 3y)^2$ (c) $(9x - 10y^3)(9x + 10y^3)$

13(a) 16 (b) 4720 (c) 946000 14(a) $\dfrac{3a}{4b}$ (b) $\dfrac{2a}{3}$ (c) $\dfrac{x+2}{x-4}$

15(a) $\dfrac{3y}{2x}$ (b) $\dfrac{x}{x+1}$ (c) $\dfrac{-2}{x+3}$

Chapter 5

Try this

1(a) 1   (b) 5   2(a) 3   (b) 6   3(a) -3 (b) 2

4. 1   5. -5 6(a) 30   (b) 10   7. 5

8. 6 50 Gp coins   (b) 8 20 Gp coins   9. 800 adults

10. 40 km h^{-1},   50 km h^{-1} 11. 8 hours   12. $x = \frac{a-b}{c}$

13. $x = \frac{a+c}{b+d}$ 14. $E = \frac{1}{2}mv^2$ 15. $R_1 = \frac{RR_2}{R_2 - R}$ 16. 30

17(a) $a \geq 15$ (b) $b \leq 8$ (c) $c > 12$ (d) $d < 10$

18

-1 0 1 2 3 4 5

19.

-4 -3 -2 -1 0 1 2

20. $x > 1$ 21. $x \leq 4$ 22. $x > 2$ 23. $x < -1$

24. $x > -4$ 25. $x > -2$ 26. $-1 < x \leq 1$

27. $x \leq 4$ 28. 14. 5 metres

Exercise 5.1(a)

1. 3   2. 2   3. 3   4. 3   5. 2   6. 2

7. -6 8. 3   9. -2 10. 2   11. $\frac{1}{4}$ 12. $1\frac{1}{2}$

Exercise 5.1(b)

1. 4   2. 1   3. 1   4. -1 5. -2 6. 2

7. 2   8. – 2   9. – 3   10. 3   11. 1   12. – 3

Exercise 5.1(c)

1. 3   2. 4   3. – 3   4. – 3   5. 3   6. 4

7. 2   8. 2   9. $1\frac{1}{3}$ 10. 2   11. – 3   12. 4

Exercise 5.1(d)

1. 7   2. 1   3. 7   4. $\frac{3}{4}$ 5. 6

6. 2   7. – 6   8. $-1\frac{1}{2}$ 9. $2\frac{1}{2}$ 10. 18

Exercise 5.1(e)

1. 18   2. 12   3. 9   4. – 10   5. $7\frac{1}{3}$ 6. $-\frac{4}{5}$

7. – 14   8. – 3   9. – 11   10. $\frac{7}{12}$ 11. 2   12. – 1

Exercise 5.2(a)

1. $2x + 3$ 2. $3y - 6$ 3. $2x - y$ 4. $\frac{a+b}{5}$ 5. $x + 5$

6. $12 - x$ 7. $x - 3$ 8. $32 - x$ 9. $\frac{x}{4}$ 10. $x + 5$

11. $x + 2$ 12. $3x + 5$ 13. $15 - 9x$ 14. $2x + 50$

Exercise 5.2(b)

1. $x + 3x = 12$ 2. $2x - x = 5$

3. $x + (x + 2) + (x + 4) + 4 = 19$ 4. $5x + 3 = 2x - 9$

5. $(30 - x) - 3 = 2(x - 3)$ 6. $2(x + 5) + 3x = 8.50$

7. $2(x + 3x) = 72$ 8. $4x + x = 25$

9. $2x + 3 = 5x - 9$ 10. $3(x + 1) = x$

Exercise 5.2(c)

1. 3, 9   2. 3   3. 13, 18   4. 2   5. 9

6. 26   7. 8, 10, 12   8. 32, 33, 34   9. 27, 29, 31   10. 6

11. width 8 cm, length 22 cm   12. width 17 cm, length 25 cm

13. 7, 35 Gp stamps 8, 15 Gp stamps   14. 35   15. 80

16. 4 at GH¢ 1.50, 6 at GH¢ 2   17. 20 years, 40 years

18. 16 years, 48 years   19. 10 years   20. 9 years

Exercise 5.2(d)

1. 40 km h^{-1},   30 km h^{-1} 2. 20 km h$^{-1,}$ 25 km h^{-1}

3. 7 p.m.   4. 5 pm   5. 4 pm   6. 4 pm   7. 3 hours

8. $2\frac{1}{2}$ hours   9. 25 km h^{-1}, 50 km h^{-1}

10. 15 km h^{-1}, 20 km h^{-1} 11. 160 km, 110 km   12. 12 noon

Exercise 5.2(e)

1. $x = \frac{b+d}{c}$ 2. $x = \frac{c-a}{b}$ 3. $x = a(c - b)$

4. $x = \frac{bc}{c-b}$ 5. $x = \frac{d}{f-e}$ 6. $x = \frac{2b-3a}{a+b}$

7. $x = \frac{bc}{a(c-b)}$ 8. $x = ac + b$ 9. $x = \frac{5a}{a+b}$

10. $x = \frac{a-c}{b-c}$ 11. $x = \sqrt{\frac{b}{1-a}}$ 12. $x = \sqrt[3]{\frac{b-1}{a}}$

13. $x = \sqrt{\frac{3f-2g}{g+f}}$ 14. $x = \frac{b^3c}{a}$ 15. $x = \sqrt{\frac{c}{b-a}}$

Exercise 5.2(f)

1. $t = \dfrac{v-u}{a}$ 2. $n = \dfrac{2S}{a+l}$ 3. $R = \dfrac{E-rl}{l}$

4. $r = \dfrac{s-a}{s}$ 5. $T = \dfrac{100I}{PR}$ 6. $v = \sqrt{\dfrac{2gk}{m}}$

7. $R = \sqrt{\dfrac{A+hr^2}{h}}$ 8. $r = \sqrt[3]{\dfrac{3V}{4\pi}}$ 9. $n = \dfrac{IR}{E-Ir}$

10. $t = \sqrt{\dfrac{2s}{g}}$ 11. $g = \dfrac{4\pi^2 l}{T^2}$ 12. $C_1 = \dfrac{CC_2}{C_2-C}$

13. $v = \sqrt{\dfrac{2(E-V)}{m}}$ 14. $v = \dfrac{mg\cos\alpha}{k^3}$ 15. $P = \dfrac{m+R^2 Q}{R^2}$

Exercise 5.2(g)

1. 10.5   2. 4   3. 200   4. 4

5(a) $l = g\dfrac{T^2}{4\pi^2}$ (b) 1.12   6. 60 ohms   7. 4.47 m s^{-1}

8. 45^0 9(a) $P = \dfrac{rk}{Q} - ms$ (b) 27   10(a) $v = \dfrac{fu}{u-f}$ (b) 4

Exercise 5.3(a)

1.

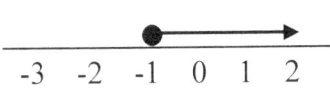

-3 -2 -1 0 1 2

2.

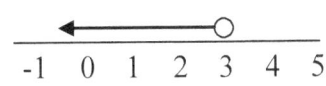

-1 0 1 2 3 4 5

3.

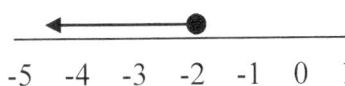

-5 -4 -3 -2 -1 0 1

4.

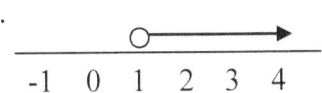

-1 0 1 2 3 4

5.

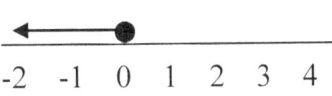

-2 -1 0 1 2 3 4

6.

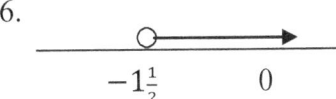

$-1\tfrac{1}{2}$ 0

7.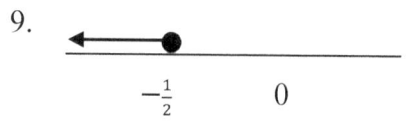

8.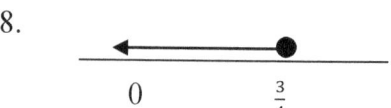

9.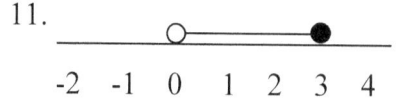

10.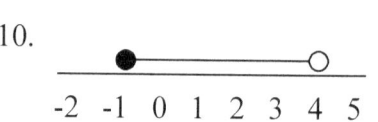

11.

-2 -1 0 1 2 3 4

12.

-3 -2 -1 0 1 2 3 4 5 6

Exercise 5.3(b)

1. $x > 3$ 2. $x < 2$ 3. $x \leq 2$ 4. $x \leq 3$

5. $x \geq -4$ 6. $x > 5$ 7. $x \geq 1\frac{1}{2}$ 8. $x \geq 1$

9. $x > -1$ 10. $x \leq 2$ 11. $x > 1\frac{1}{2}$ 12. $x < -20$

Exercise 5.3(c)

1. $x > -1$ 2. $x < 3$ 3. $x \geq 1\frac{1}{3}$ 4. $x < -2$

5. $x \geq -8$ 6. $x > 1\frac{1}{2}$ 7. $x \leq -5$ 8. $x < -3$

9. $x > 3$ 10. $x \leq -2\frac{1}{2}$ 11. $x > -5$ 12. $x \geq -2$

Exercise 5.3(d)

1. $8 < x < 10$ 2. $-2 \leq x < -1$

3. $-3 \leq x < 2$ 4. $-5 < x < 1$

5. $-5 < x < 11$ 6. $-4 < x < 1$

7. $1 < x < 3$ 8. $2 \leq x \leq 4$

9. $-23 < x < -11$ 10. $2 \leq x < 5$

Exercise 5.3(e)

1. $x < 2$ 2. $x \leq 7$ 3. $x \geq -4$

4. 5   5. At least 92 metres   6. at most 12 cm

7. $10 < x < 17$ 8. 4   9. 12 years

10. At least 97   11. At least 82   12. 20

Review Exercise 5

1(a) 3   (b) 5   (c) 2   (d) 3   (e) – 5   (f) – 4

2(a) 5   (b) $2\frac{1}{2}$ (c) 4   (d) – 4   (e) 10   (f) 10

3(a) 2   (b) 2   (c) 2   (d) 8   (e) 17   (f) 8

4(a) $a = \frac{2A-bh}{h}$ (b) $P = \frac{A}{1+rt}$ (c) $c = \sqrt{\frac{E}{m}}$

(d) $h = \frac{A-2\pi r^2}{2\pi r}$ (e) $r = \sqrt[3]{\frac{3V}{4\pi}}$ (f) $d = \frac{ac}{2a+c}$

5. 3.75   6. 5 cm   7. 68^0 F

8(a) $x + 5 = 9$ (b) $x - 7 = 15$ (c) $3x - 4 = 2x$

(d) $2(x + 3) = x + 12$

9(a) 13   (b) 16   (c) 20 21 22   (d) 33 35 37

(e) 40 42 44   (f) 1550, 1710   (g) 9 years old   (h) £820

10(a)   (b)

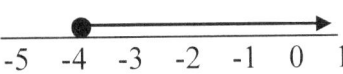   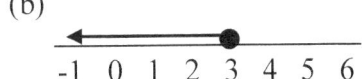

(c)   (d)

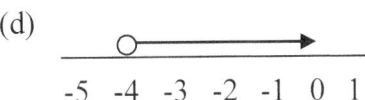

(e)

-5   -4   -3   -2   -1   0   1

(f)

$-\frac{3}{4}$ 0

11(a) $x \geq -4$ (b) $x < 7$ (c) $x < 8$ (d) $x > 8$ (e) $x > 9$

(f) $x \geq 8$ (g) $x < -2\frac{1}{2}$ (h) $x < 1\frac{1}{2}$ (i) $x > -\frac{2}{3}$

12(a) $3x + 5 < 7$ (b) $x - 3 > 5$ (c) $2x + 7 \geq 12$

(d) $60 < x < 80$ (e) $x \geq 45$ (e) $x \leq 1200$

(g) $x \leq 100$ (h) $x \geq 2.50$

13(a) $2\frac{1}{2}$ hours   (b) 6   (c) At least 84   (d) At most 12 m

(e) at least 6 hours   (f) at least 5 minutes

Chapter Test 5

1(a) -12 (b) 1   (c) $2\frac{4}{5}$

2(a) $x > -1$

-2   -1   0   1   2   3

(b) $x \geq 4$

-1   0   1   2   3   4   5   6

(c) $-1\frac{1}{2} < x \leq 2\frac{1}{2}$

$-1\frac{1}{2}$ 0 $2\frac{1}{2}$

3(a) $r = \sqrt{\dfrac{3V}{\pi h}}$ (b) $A = \dfrac{\pi r^2 S}{360}$ (c) $s = \dfrac{gt^2}{2}$

4. 0.4   5. 85 cm^2 6. 33.75 m^2 7. 100   8. $\frac{1}{3}$ hour

9. At least GH¢ 25, 000   10. Not more than 25

Chapter 6

Try this

1. $x = -1, y = 2$ 2. $x = 1, x = -1$ 3. $x = 2, y = -1$

4. $x = 2, y = 1$ 5. $x = 2, y = -1$

6. Price of pen drive is GH¢ 5   Price of compact disk is GH¢ 2

7. 45 5 Gp coins, 20 20 Gp coins

8. 300 ml of 30 % alcohol   200 ml of 40 % alcohol

Exercise 6.1(a)

1. $x = 2, y = 3$ 2. $x = 1, y = 2$ 3. $a = 1, b = -1$

4. $x = 3, y = -3$ 5. $a = 1, b = 4$ 6. $x = 3, y = 2$

7. $m = 3, n = -2$ 8. $x = 1, y = 4$ 9. $x = 1, y = -3$

10. $a = 1, b = 1$ 11. $x = 2, y = 3$ 12. $m = -2, n = 1$

13. $a = -1, b = -5$ 14. $x = 3, y = 2$ 15. $x = -1, y = 2$

16. $m = 3, n = 4$ 17. $a = 2, b = -\frac{1}{2}$ 18. $x = 1, y = 1$

19. $x = 3, y = 2$ 20. $x = 2, y = -1$ 21. $x = 2, y = -2$

22. $x = 1, y = 1$ 23. $x = 1, y = 2$ 24. $a = 5, b = 3$

25. $a = 0, b = -6$ 26. $a = -2, b = -3$ 27. $x = 3, y = 1$

28. $x = 3, y = 4$ 29. $a = 2, b = 3$ 30. $x = 1, y = 2\frac{1}{2}$

Exercise 6.1(b)

1. $x = -1, y = 2$ 2. $a = 3, b = 4$

3. $s = 1, t = -2$ 4. $m = 0, n = -3$

5. $a = 1, b = 1$ 6. $a = 3, b = 2$

7. $x = 4, y = -3$ 8. $p = 3, q = -1$

9. $x = 1, y = 2$ 10. $x = 1, y = 1$

Exercise 6.2

1. 10, 15   2. 8, 9   3. 3, 4   4. 7, 3

5. Pencil cost GH¢ 1,   Erasers cost GH¢ 0.50

6. Disk cost 25 GP, pen drive cost 50 GP   7. 12 m, 18m

8. 4 m, 9 m   9. 5 cm, 13 cm   10. 12 kg, 18 kg

11. 12 years, 36 years   12. 11 years, 35 years

13. 13 5 Gp coins   9 20 Gp coins

14. 250 students tickets, 150 adult tickets

15. 300 adults, 400 children

16. Fixed amount is GH¢ 100 and fixed amount GH¢ 20

17. 80 ml of 50 % acid solution and 120 ml of 25 % acid solution

18. 200 ml of 15 % alcohol and 100 ml of 45 % alcohol

19. GH¢ 7000 at 8 % and GH¢ 5000 at 9 %

20. GH¢ 5000 at 12 % and GH¢ 3000 at 8 %

Review exercise 6

1. $x = 3, y = 2$ 2. $x = 1, y = 1$ 3. $x = 3, y = 2$

4. $x = 1, y = -2$ 5. $x = 1, y - 1$ 6. $x = 2, y = 1$

7. $x = 4, y = 3$ 8. $x = -2, y = -1$ 9. $x = 2, y = 1$

10. $x = -2, y = 2$ 11. $x = 7, y = -2$ 12. $x = 2, y = -5$

13. $x = 4, y = 1$ 14. $x = 6, y = 3$ 15. $x = \frac{1}{3}, y = 2$

16. $x = 10, y = 1$ 17. $x = -3, y = -5$ 18. $x = 3, y = -1$

19. 13, 37   20. 85^0, 95^0

21. Desk cost GH¢ 1800 and chair cost GH¢700

22. 90 boys and 160 girls   23. 12 10 Gp coins and 10 5 Gp coins

24. 300 GH¢ 5 tickets and 200 GH¢ 8 tickets

25. 3 cm wide and 13 cm long   26. 5 cm , 7 cm, 7cm

27. GH¢ 500 at 12 % and GH¢ 300 at 8 %

28. GH¢ 720 and GH¢ 480

29. 200 ml of 15 % acid solution and 300 ml of 25 % acid solution

30. speed of plane is 120 km h^{-1} and wind speed is 60 km h^{-1}

Chapter Test 6

1(a) $x = -2, y = 2$ (b) $x = 10, y = 3$ (c) $x = 1, y = -1$

2(a) $x = 4 , y = 5$ (b) $x = 4, y = 1$ (c) $x = 2, y = 4$

3. $x = 10, y = 28$ 4. 35^0 and 55^0

5. 200 children tickets and 300 adult tickets

6. GH¢ 200 at 9 % and GH¢ 300 at 8 %

7. 100 ml of 30 % alcohol and 300 ml of 50 % alcohol

8. speed of boat is 12 km h^{-1} and speed of current is 3 km h^{-1}

Chapter 7

Try this

1. 2 and 5   2. $-$ 2 and 3   3. 3 twice   4. ± 5

5. $-$ 4 and 0   6. $-1\frac{1}{2}$ and 2   7. 3 and 5   8. $-$ 2 and $\frac{1}{3}$

9. $-$ 5 and $1\frac{1}{2}$ 10. $-2\frac{1}{2}$ and 3

12. The length is 16 m and the width is 14 m   12. 3 metres

Exercise 7.1(a)

1. $-$ 6 and 2   2. 5 and 2   3. $-$ 9 and 2

4. $-$ 5 and 6   5. $-$ 4 and 5   6. $-$ 8 and $-$ 4

7. $-$ 6 and 10   8. 5 and 9   9. $-$ 7 and 4

10. $-$ 8 and $-$ 7   11, 2 and 25   12. $-$ 3 and 18

13. $-$ 2 and 7   14. $-$ 12 and 2   15. $-$ 3 and 20

16. $-$ 6 and 7   17. $-$ 5 and 3   18. $-$ 7 and 0

19. 0 and $1\frac{1}{3}$ 20. $\pm 2\frac{1}{2}$ 21. $\pm 1\frac{1}{3}$

22. $-$ 2 and 9   23. 4 and 5   24. 3 and 4

25. – 2 and 5   26. – 9 and – 3   27. – 2 and 9

28. – 7 and 2   29. – 3 and 10   30. – 8 and 2

Exercise 7.1(b)

1. $-\frac{1}{3}$ and 7   2. $-1\frac{1}{2}$ and $-\frac{1}{4}$ 3. $-\frac{1}{2}$ and $1\frac{2}{3}$

4. $1\frac{1}{3}$ and 3   5. $-1\frac{1}{3}$ and $3\frac{1}{2}$ 6. $\frac{2}{7}$ and 5

7. – 8 and $1\frac{1}{2}$ 8. $-1\frac{1}{2}$ and $\frac{1}{2}$ 9. $\frac{2}{5}$ and 3

10. $-1\frac{2}{3}$ and 3   11. $3\frac{1}{2}$ and 5   12. – 1 and $1\frac{2}{3}$

13. – 1 and $1\frac{1}{5}$ 14. $-2\frac{1}{2}$ and $1\frac{1}{3}$ 15. $-2\frac{1}{2}$ and $1\frac{1}{2}$

Exercise 7.1(c)

1. – 3 and 1   2. – 3 and 4   3. – 1.16 and 5.16

4. – 6.73 and – 3.27   5. – 2.14 and 5.14   6. – 2.37 and 4.37

7. $\frac{1}{2}$ and 1   8. 0.24 and 2.76   9. – 8.87 and – 0.56

10. $-\frac{3}{5}$ and 2   11. $-1\frac{1}{2}$ and 5   12. – 0.72 and 1.72

Exercise 7.1(d)

1. 1 and 3   2. – 2 and 5   3. – 4.41 and – 1.59

4. 0.44 and 4.56   5. – 10 and 3   6. $-\frac{1}{3}$ and 2

7. $-\frac{2}{7}$ and 1   8. – 7.16 and – 0.84   9. 0.34 and 1.26

10. – 2 and $\frac{3}{4}$ 11. – 1.09 and 0.34   12. – 0.15 and 6.46

Exercise 7.2

1. 15 and 16   2. 13 and 15   3. 12 and 14   4. -2 and 4

5. 13 cm wide and 26 cm long   6. -8 and -5 or 5 and 8

7. 6 cm wide and 8 cm long

8. The height is 7 cm and the length base is 10 cm

9. The height is 18 m and the length base is 9 m

10. 12 m   11. 6 m   12. 2 hours

Review exercise 7

1. 1 and 6   2. -5 and -1 3. -7 and 3

4. -9 and 2   5. -7 and -2 6. -5 and -3

7. -6 and 0   8. 0 and 8   9. -3 and 0

10. 0 and 4   11. 1   12. 4

13. -5 and -1 14. 3   15. 1 and 4

16. $-\frac{5}{3}$ and 4   17. -1 and $\frac{5}{3}$ 18. $-\frac{1}{4}$ and $\frac{2}{3}$

19. -1 and $\frac{2}{3}$ 20. -1 and $\frac{6}{5}$ 21. -4 and $\frac{5}{2}$

22. -7 and $\frac{5}{2}$ 23. $\frac{1}{5}$ and 5   24. -4 and $\frac{3}{2}$

25. -4 and $\frac{4}{3}$ 26. $-\frac{7}{4}$ and 2   27. -3 and $\frac{6}{5}$

28. $-\frac{3}{2}$ and 2   29. -4 and 3   30. $\frac{1}{2}$ and $\frac{3}{4}$

31. 14 and 15   32. 14 and 16   33. -19 and -17

34. -8 and -5 or 5 and 8   35. 12 cm wide and 16 cm long

36. 16 cm   37. 9 cm and 12 cm

38. 5 cm wide and 12 cm long   39. 5 hours   40. 8 m

Chapter Test 7

1(a) – 5 and 4   (b) – 6 and 7

2(a) $-\frac{1}{4}$ and 3   (b) $-1\frac{1}{2}$ and $2\frac{1}{2}$

3(a) - 1 .09 and 4.59   (b) – 2.26 and 0.59

4. 6 cm   5. 15 m long and 8 m wide

6. The height is 10 cm and the length of the base is 7 cm

7. 12 m   8. 3 seconds

Chapter 8

Try this

1. Domain = {- 1, 0, 1, 2, 3}   Range = {3, 5, 6, 7, 12}

2.

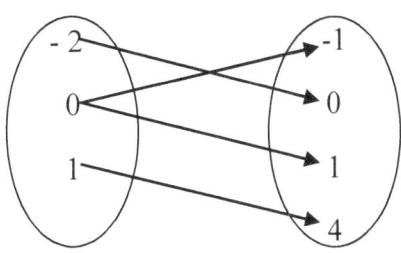

3. {(-3,1), (2,1), (5,2), (-4,3)}   4. $x \rightarrow 3x + 2$

5. x   0   1   2   3   4   5
   ↓   ↓   ↓   ↓   ↓   ↓   ↓
   y   - 4 - 1   2   5   8   11

6(a) $f(-2) = 4$, $f(-1) = 2$, $f(2) = -4$, $f(3) = -6$

(b) Range = {- 6 – 4, 2, 4}

7. $g(2) = 2$, $g(-1) = -7$ 8. $\sqrt{13}$ 9. (1, 3)   10. 1

11. $y = -2x + 3$ 12. $y = -2x + 1$ 13. $2x – 3y + 10 = 0$

14. $y = 2x – 4$ 15. $3x + 2y + 5 = 0$

Exercise 8.1

1(a) Domain = {-2, -1, 0, 3}   Range = {-2, -1, 1, 2}

(b) Domain = {2, 3, 5, 4}   Range = {1, 2, 3}

(c) Domain = {-3, 0, 5, 8}   Range = {0, 2, 4, 5, 6}

(d) Domain = {-2}   Range = {2, 5}

(e) Domain = {2, 4, 6, 8, 10}   Range = {1, 2, 3, 4, 5}

2(a)   (b)

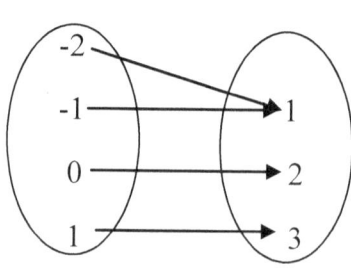
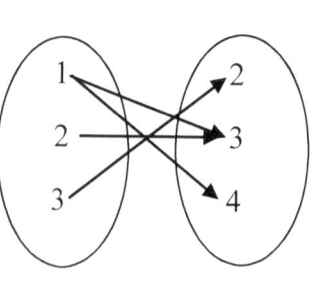

(c)

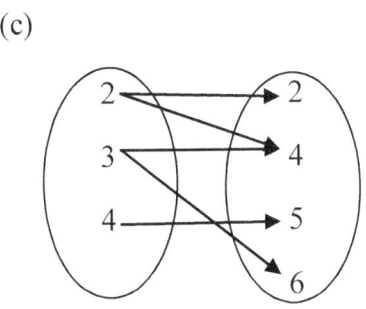

(d)

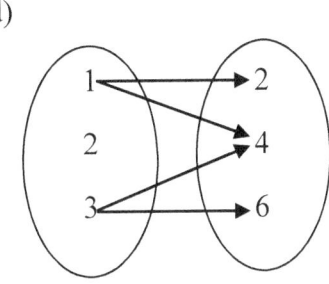

3. $\{(1,3),(1,5),(1,9),(2,3),(2,5),(2,9),(7,9)\}$

4. $\{(1,1),(2,8),(3,27),(4,64),(5,125)\}$

5. $\{(8,2),(9,3), 25,5),(49,7)\}$

6(a) $\{(1,1),(2,1),(5,2),(7,3)\}$ (b) $\{(-2,-3),(-1,0),(2,1),(2,2)\}$

 (c) $\{(1,2),(1,-1),(-2,3),(1,0)\}$ (d) $\{(0,0),(1,1),(2,-2),(-2,3),(1,0)\}$

7(a) 6   (b) $-\frac{1}{2}$ 8. (3,5), (-5,-1), (-2.5,0)

Exercise 8.2

1(a) many –to-one   (b) not a mapping   (c) not a mapping

(d) one-to-one   (e) one-to-many   (f) many-to-many

2(a) $x \rightarrow 3x + 1$ (b) $x \rightarrow -2x + 1$

 (c) $x \rightarrow 2x - 5$ (d) $x \rightarrow \frac{1}{2}x + 3$

 (e) $x \rightarrow x^2$ (f) $x \rightarrow 2^x$

 (g) $x \rightarrow 3x$ (h) $x \rightarrow x^3$

3(a) $\{- 5, - 2, 1, 4\}$ (b) $\{ -3, 1, 3, 5, 7\}$

 (c) $\{2, 8, 18\}$ (d) $\{ -5, -2, 1, 4, 10\}$

Exercise 8.3

1. (a), (b) and (c)   2. (a) and (d)

3(a) $f: x \rightarrow 3x + 4$ (b) $f: x \rightarrow 2x^2 - 3$ (c) $f: x \rightarrow \frac{x^2}{x-2}$, $x \neq 2$

4(a) 7   (b) – 8   (c) 9   (d) 15

5(a) 1   (b) – 2   (c) – 12   (d) – 1

6(a) 9   (b) – 5   (c) - 3   (d) 3

7. {- 5, -3, -1, 1, 3, 5}   8. – 2   9. – 1, 3

10(a) – 1   (b) 0   (c) – 2, 3   (d) – 2 , 2   (e) $x < 5$

Example 8.4(a)

1. 13   2. 5   3. 10   4. $5\sqrt{2}$ 5. 4   6. 10

7. 13   8. 17   9. 17   10. 20   11. $3\sqrt{5}$ 12. 13

Exercise 8.4(b)

1. (5, 3)   2. (2, 4)   3. (- 3, -4)   4, (- 2, 1)

5. (- 2, 3)   6. (1, 1)   7. $\left(2\frac{1}{2}, 3\right)$ 8. $\left(-\frac{1}{2}, -1\right)$

9. $\left(-1\frac{1}{2}, 0\right)$ 10. (- 1, 1)   11. (4, 3)   12. (- 3, -2)

Exercise 8.4(c)

1. – 2   2. 3   3. – 1   4. $\frac{1}{2}$ 5. $-\frac{3}{2}$ 6. 2

7. $\frac{3}{2}$ 8. $\frac{4}{3}$ 9. 2   10. $-\frac{2}{3}$ 11. $\frac{1}{3}$ 12. – 2

Exercise 8.4(d)

1. $y = 4x - 3$

2. $y = 2x + 3$

3. $y = -3x + 5$

4. $y = 4x + 6$

5. $y = -2x - 4$

6. $y = \frac{1}{2}x + 1$

7. $y = \frac{2}{3}x - 4$

8. $y = \frac{3}{4}x + 2$

Exercise 8.4(e)

1. $y = 2x + 1$

2. $y = 3x - 8$

3. $y = -3x + 5$

4. $y = -x + 5$

5. $y = -2x + 3$

6. $y = 4x - 3$

7. $y = 2x + 6$

8. $y = -2x - 7$

9. $y - 5x - 7$

10. $y = -4x - 2$

11. $y = -x - 1$

12. $y = 2x + 6$

Exercise 8.4(f)

1. $2x - 3y - 1 = 0$

2. $x - 4y + 5 = 0$

3. $3x - 4y - 5 = 0$

4. $3x - 2y - 18 = 0$

5. $3x + 4y + 1 - 0$

6. $3x + 5y - 61 = 0$

7. $4x - 3y - 16 = 0$

8. $5x - 3y + 7 = 0$

Exercise 8.4(g)

1. $y = -3x + 11$

2. $y = 2x + 3$

3. $y = 2x + 9$

4. $y = 2x - 4$

5. $y = -3x + 2$

6. $2x - 3y + 10 = 0$

7. $y = 5x + 14$

8. $8x - 7y - 10 = 0$

9. $y = -x + 12$ 10. $2x - 3y + 2 = 0$

11. $3x - 4y - 10 = 0$ 12. $3x + 2y - 18 = 0$

13. $4x + 3y - 22 = 0$ 14. $2x + 3y + 11 = 0$

15. $2x + 5y + 1 = 0$ 16. $5x + 2y - 20 = 0$

17. $5x - 4y - 12 = 0$ 18. $3x + 2y - 17 = 0$

19. $x + 3y + 8 = 0$ 20. $3x - 2y + 11 = 0$

Review exercise 8

1(a) Domain = {- 2, -1, 1, 2}   Range = {- 2, -1, 0, 3}

 (b) Domain = {2, 1, 3}   Range = {0, 2, 3, 4}

 (c) Domain = {0, 2, 4, 5, 6}   Range = {-1, 0, 5, 8}

 (d) Domain = {2, 5}   Range = {-3, -2}

2(a)

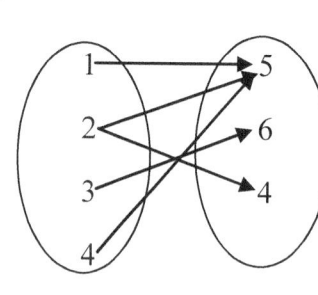

(b)

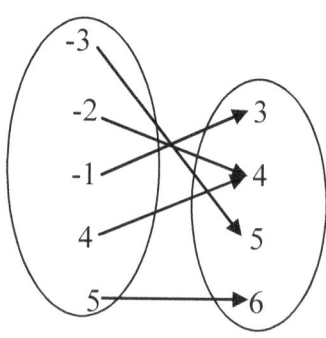

(c)

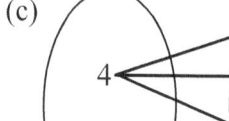

(d)

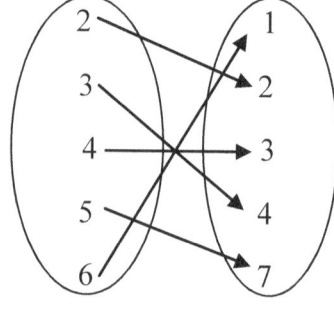

3(a) {(5, 2), (5, 3), (6, 3), (4, 2), (5, 5)}

(b) {(2, -1), (3, -2), (4, -2), (4, 3), (6, 5)}

(c) {(-2, 3), (1, 3), (-3, 4), (4,3)}

(d) {(3, 2), (5, 4), (1, 3), (2, 2), (3, 4)}

4(a) not a mapping   (b) one-to-many

(c) many-to-many   (d) one-to-one

5(a) $x \rightarrow 3x + 3$ (b) $x \rightarrow 5x$

6(a) Range = {-3, 2, 7, 12, 17, 22}   (b) Range = {2, 5, 14, 29}

7. (a) and (d)   8. (a), (c) and (d)

9(a) 4   (b) 7   (c) -11   (d) -2

10(a) 15   (b) - 1   (c) - 3   (d) 5

11. – 2 , 5   12(a) $\sqrt{5}$ (b) 5   (c) 13   (d) $\sqrt{13}$

13(a) (4, 1)   (b) (0, 2)   (c) $\left(4\frac{1}{2}, 3\frac{1}{2}\right)$ (d) $\left(3, 5\frac{1}{2}\right)$

14(a) – 2   (b) – 4   (c) $\frac{3}{2}$ (d) $\frac{5}{4}$

15(a) $y = 3x - 2$ (b) $y = 4x - 11$

(c) $y = -2x + 6$ (d) $3x - 4y + 18 = 0$

16(a) $x - 2y + 7 = 0$ (b) $5x + 2y - 22 = 0$

(c) $y = x + 1$ (d) $5x - 3y - 14 = 0$

(e) $y = 4$ (f) $y = -3x + 7$

(g) $4x + 3y + 17 = 0$ (h) $2x - 3y - 1 = 0$

Chapter Test 8

1(a) Domain = {0, 1, - 1, 2}   Range = {1, 2}

   (b) Domain = {3, 4}   Range = {1, 2, 3, 4}

2. (b) and (d)   3(a) many-to-many   (a) not a mapping

4(a) $x \rightarrow 2x + 7$ (b) $x \rightarrow -2x + 1$

5. (b)   6(a) 1   (b) 7   (c) 1

7(a) $a = 3$ $b = 5$ (b) 6   8. 1, $1\frac{1}{2}$ 9(a) $\frac{2}{3}$ (b) - 1, 3

10(a) $\sqrt{5}$ (b) 5   11(a) (0, - 2)   (b) (- 3, - 2)

12(a) 3   (b) 4   12. 4   13. 4

14(a) $2x - 3y - 9 = 0$ (b) $y = -3x - 3$

15(a) $y = 3x - 9$ (b) $3x + 2y + 7 = 0$

Chapter 9

Try this

1(a) 2 cm to 1 unit   (b) 2 cm to 2 units

   (c) 2 cm to 5 units   (d) 2 cm to 0.1 unit

   (e) 2 cm to 20 units   (f) 2 cm to 10 units

   (g) 2 cm to 0.5 unit   (h) 2 cm to 5 units

2.

x	-2	-1	0	1	2	3	4
y	4	-1	-4	-5	-4	-1	4

5. $x = 2$ $y = 1$ 7(a) – 1.2, 4.2   (b) – 1.5, 4.5

8. – 1, 3.5   9(a) – 1, 1.5   (b) - 1.6, 1.6   10. 5

Exercise 9.1

1.

x	-1	0	1	2	3	4
y	3	5	7	9	11	13

2.

x	-2	-1	0	1	2	3	4	5
y	7	5	3	1	-1	-3	-5	-7

3.

x	0	1	2	3	4	5	6
y	-7	-4	-1	2	5	8	11

4.

x	-2	-1	0	1	2	3	4
y	-4	-3.5	-3	-2.5	-2	-1.5	-1

5.

x	-1	0	1	2	3	4	5	6
y	12	6	2	0	0	2	6	12

6.

x	-2	-1	0	1	2	3	4	5	6
y	-7	0	5	8	9	8	5	0	-7

7.

x	-3	-2	-1	0	1	2
y	10	0	-4	-2	6	20

8.

x	-3	-2	-1	0	1	2	3	4	5
y	7	0	-5	-8	-9	-8	-5	0	7

9.

x	-3	-2	-1	0	1	2	3	4
y	-6	0	4	6	6	4	0	-6

10.

x	-3	-2	-1	0	1	2	3
y	-12	2	4	0	-4	-2	12

11.

x	-2	-1	0	1	2	3	4
y	-16	0	4	2	0	4	20

12.

x	0	1	2	3	4
y	-1.67	0.75	-0.20	0.17	0.43

Exercise 9.2(b)

1. $x = 3$, $y = 3$ 2. $x = -3$, $y = 8$ 3. $x = 0$, $y = 1$

4. $x = 3, \ y = -3$ 5. $x = 1, \ y = -2$ 6. $x = 1, \ y = 1$

Exercise 9.2(d)

1(a) – 3.4, 1.4   (b) $3 < x < 1$ (c) (- 1, - 4)

2(a) – 0.6, 4.6   (b) x = 1.5

3(a) - 1.3, 3.8   (b) – 6.1   1.25   (c) $-2 \leq x < 1.25$

4(a) - 1.4, 3.4   (b) 4

5(a)

x	-1	0	1	2	3	4	5	6	7
y	15	8	3	0	-1	0	3	8	15

(c) (i) 2, 4   (ii) 0.8, 5.2

6(a)

x	-3	-2	-1	0	1	2	3	4	5
y	-21	-8	1	6	7	4	-3	-14	-29

(c) (i) - 1.1, 2.6   (ii) - 1.4, 1.9

7(a)

x	-2	-1.5	-1	0	0.5	1	2	3	3.5	4
y	5	2.25	0	-3	-3.75	-4	-3	0	2.25	5

(c) (i) - 1.4, 3.4   (ii) -1.8, 2.8

8(a)

x	-1	-0.5	0	0.5	1	1.5	2	2.5	3
y	-1	0.75	2	2.75	3	2.75	2	0.75	-1

(c) (i) - 0.4, 2.4   (ii) 0.4, 2.6

9(a) -1.6,   1.6   (b) -1.3,   2.3

10. 3   11. -13   12. -1.8,

Review exercise 9

1.

x	-3	-2	-1	0	1	2
y	-11	-8	-5	-2	1	4

2.

x	-4	-3	-2	-1	0	1
y	-7	-4	-1	2	5	8

3.

x	-5	-4	-3	-2	-1	0	1	2
y	8	2	-2	-4	-4	-2	2	8

4.

x	-5	-4	-3	-2	-1	0	1	2
y	-22	-9	0	5	6	3	-4	-15

9. $x = 3$, $y = 2$ 10 $x = 2$, $y = 1$

11. $x = 3$, $y = 2$ 12. $x = -2$, $y = 3$

13(a) (2, -7)   (b) (-1, 4)   (c) (-1, -5)   (d) $\left(1\frac{1}{3}, -\frac{1}{3}\right)$

14(a) - 1.2,   4.2   (b) - 0.8,   2.4

15(a) – 1.2, 4.2   (b) 7.5   (c) $1.5 < x \le 5$

16(a)

x	-4	-3	-2	-1	0	1	2	3	4
y	29	14	3	-4	-7	-6	-1	8	21

(c) (i) – 1.6, 2.1   (ii) - 1.3, 3.3

17(a) – 1.4, 0   (b) – 1.9, 1.9

18. - 3   19. 8   20(a) – 2, 1

Chapter Test 9

1(a)

x	-2	-1	0	1	2	3	4	5	6
y	20	13	8	5	4	5	8	13	20

(c)(i) 1, 6   (ii) – 0.4, 4.4

2(a)

x	-2	-1	0	1	1.5	2	2.5	3	3.5	4
y	-5	0	3	4	3.75	3	1.75	0	-2.255	-5

(c) – 1.4, 3.4   (d) 4

3(b) (i) 196 m   (ii) 3.5 s   (iii) 2 s   (iv) 7 s

4(a) – 1.4, 2.4   (b) - 2.2, 2.2   5(b) – 2, 3

Chapter 10

Try this

1. 1(mod 5)

2.

+	0	1	2	3	4	5	6	7
0	0	1	2	3	4	5	6	7
1	1	2	3	4	5	6	7	0
2	2	3	4	5	6	7	0	1
3	3	4	5	6	7	0	1	2
4	4	5	6	7	0	1	2	3
5	5	6	7	0	1	2	3	4
6	6	7	0	1	2	3	4	5
7	7	0	1	2	3	4	5	6

3. 0 (mod)

4.

×	0	1	2	3	4	5	6
0	0	0	0	0	0	0	0
1	0	1	2	3	4	5	6
2	0	2	4	6	1	3	5
3	0	3	6	2	5	1	4
4	0	4	1	5	2	6	3
5	0	5	3	1	6	4	2
6	0	6	5	4	3	2	1

5. Tuesday

Exercise 10.1(a)

1(a) {0, 1, 2, 3, 4}   (b) {0, 1, 2, 3, 4, 5}

(c) {0, 1, 2, 3, 4, 5, 6, 7}   (d) {0, 1, 2, 3, 4, 5, 6, 7, 8}

2(a) 2(mod 5)   (b) 3(mod 4)   (c) 0(mod 6)

(d) 5(mod 12)   (e) 1(mod 8)   (f) 5(mod 9)

(g) 0(mod 7)   (h) 11(mod 13)   (i) 5(mod 12)

(j) 1(mod 3)   (k) 0(mod 5)   (l) 2(mod 13)

3(a) 0(mod 6)   (b) 1(mod 4)   (c) 1(mod 8)

(d) 0(mod 5)   (e) 3(mod 9)   (f) 3(mod 7)

(g) 2(mod 8)   (h) 3(mod 12)   (i) 1(mod 6)

(j) 2(mod 5)

Exercise 10.1(b)

1(a)

+	0	1	2	3
0	0	1	2	3
1	1	2	3	0
2	2	3	0	1
3	3	0	1	2

(b)

+	0	1	2	3	4	5
0	0	1	2	3	4	5
1	1	2	3	4	5	0
2	2	3	4	5	0	1
3	3	4	5	0	1	2
4	4	5	0	1	2	3
5	5	0	1	2	3	4

(c)

+	0	1	2	3	4	5	6	7	8
0	0	1	2	3	4	5	6	7	8
1	1	2	3	4	5	6	7	8	0
2	2	3	4	5	6	7	8	0	1
3	3	4	5	6	7	8	0	1	2
4	4	5	6	7	8	0	1	2	3
5	5	6	7	8	0	1	2	3	4
6	6	7	8	0	1	2	3	4	5
7	7	8	0	1	2	3	4	5	6
8	8	0	1	2	3	4	5	6	7

2.

+	4	7	9	11
4	8	11	1	3
7	11	2	4	6
9	1	4	6	8
11	3	6	8	10

3(a)

+	0	1	2	3	4	5	6
0	0	1	2	3	4	5	6
1	1	2	3	4	5	6	0
2	2	3	4	5	6	0	1
3	3	4	5	6	0	1	2
4	4	5	6	0	1	2	3
5	5	6	0	1	2	3	4
6	6	0	1	2	3	4	5

(a) 1   (b) 6

4.

+	1	3	5	7
1	2	4	6	0
3	4	6	0	2
5	6	0	2	4
7	0	2	4	6

(a) 3, 7   (b) 5   (c) 5

Exercise 10.2(a)

1. 2(mod 4)   2. 2(mod 5)   3. 3(mod 6)   4. 0(mod 8)

5. 5(mod 9)   6. 4(mod 12)   7. 2(mod 7)   8. 3(mod 5)

9. 2(mod 6)   10. 6(mod 7)   11. 5(mod 7)   12. 3(mod 6)

Exercise 10.2(b)

1(a)

×	0	1	2	3	4
0	0	0	0	0	0
1	0	1	2	3	4
2	0	2	4	1	3
3	0	3	1	4	2
4	0	4	3	2	1

(b)

×	0	1	2	3	4	5
0	0	0	0	0	0	0
1	0	1	2	3	4	5
2	0	2	4	0	2	4
3	0	3	0	3	0	3
4	0	4	2	0	4	2
5	0	5	4	3	2	1

(c)

×	0	1	2	3	4	5	6	7
0	0	0	0	0	0	0	0	0
1	0	1	2	3	4	5	6	7
2	0	2	4	6	0	2	4	6
3	0	3	6	1	4	7	2	5
4	0	4	0	4	0	4	0	4
5	0	5	2	7	4	1	6	3
6	0	6	4	2	0	6	4	2
7	0	7	6	5	4	3	2	1

2.

×	2	3	5	6
2	4	6	1	3
3	6	0	6	0
5	1	6	7	3
6	3	0	3	0

(a) 3   (b) 0

3.

×	3	5	7	9
3	9	3	9	3
5	3	1	11	9
7	9	11	1	3
9	3	9	3	9

(a) 3, 7   (b) 5, 7

4.

×	1	3	5	7
1	1	3	5	7
3	3	0	6	3
5	5	6	7	8
7	7	3	8	4

(a) 3   (b) 5

Exercise 10.3

1(a) Wednesday   (b) Tuesday   2(a) 10 a.m.   (b) 6 p.m.

3. Tuesday   4. Saturday   5(a) April (b) July

6(a) August   (b) September   7. Sunday   8. 3 p.m.

Review exercise 10

1(a) 2(mod 5)   (b) 0(mod 4)   (c) 0(mod 6)

(d) 9(mod 12)   (e) 7(mod 8)   (f) 6(mod 9)

2(a) 2(mod 6)   (b) 1(mod 4)   (c) 1(mod 8)

(d) 5(mod 9)   (e) 0(mod 7)   (f) 3(mod 12)

3(a) 0(mod 4)   (b) 4 (mod 5)   (c) 3(mod 6)

(d) 0(mod 9)   (e) 0(mod 12)   (f) 2(mod 7)

4.

+	0	1	2	3	4
0	0	1	2	3	4
1	1	2	3	4	0
2	2	3	4	0	1
3	3	4	0	1	2
4	4	0	1	2	3

×	0	1	2	3	4
0	0	0	0	0	0
1	0	1	2	3	4
2	0	2	4	1	3
3	0	3	1	4	2
4	0	4	3	2	1

(a) {2}   (b) {2, 3}

5.

+	3	7	9	11
3	6	10	12	1
7	10	1	3	5
9	12	3	5	7
11	1	5	7	9

×	3	7	9	11
3	9	8	1	7
7	8	10	11	12
9	1	11	3	8
11	7	12	8	4

(a) 5   (b) {9}

6.

×	2	3	5	6
2	4	6	2	4
3	6	1	7	2
5	2	7	1	6
6	4	2	6	4

(a) {3, 5}   (b) {2, 6}

7. 1 p.m.   8. 2 a.m

Chapter Test 10

1(a) 3(mod 5)   (b) 4(mod 6)   (c) 3(mod 8)

2(a) 3(mod 5)   (b) 1(mod 6)   (c) 4(mod 8)

3(a) 0(mod 6)   (b) 3(mod 9)   (c) 8(mod 12)

4.

+	1	5	7	11
1	2	6	8	0
5	6	10	0	4
7	8	0	2	6
11	0	4	6	10

×	1	5	7	11
1	1	5	7	11
5	5	1	11	7
7	7	11	1	5
11	11	7	5	1

(a) 0   (b) {1, 5, 7, 11}

5.

+	1	3	5	7
1	2	4	6	8
3	4	6	8	1
5	6	8	1	3
7	8	1	3	5

×	1	3	5	7
1	1	3	5	7
3	3	0	6	3
5	5	6	7	8
7	7	3	8	4

(a) 1   (b) {1, 7}

6. 2

Chapter 11

Try this

1(a) (i) $p \propto q$ (ii) $p \propto x^2$ (iii) $p \propto r^3$ (iv) $p \propto \sqrt[3]{q}$

(v) $y \propto \sqrt{x}$

(b) (i) $y = kx^3$ (ii) $p = k\sqrt[4]{q}$ (iii) $s = k\sqrt{t}$ (iv) $f = ka$

(v) $E = kv^2$ (vi) $l = kw$ (vii) $d = kt$

2. 9   3. 45   4. 30

5(a) (i) $p \propto \dfrac{1}{q}$ (ii) $y \propto \dfrac{1}{x^2}$ (ii) $s \propto \dfrac{1}{t^3}$

(iv) $z \propto \dfrac{1}{\sqrt{y}}$ (v) $p \propto \dfrac{1}{\sqrt[3]{r}}$

(b) (i) $p = \dfrac{k}{q^2}$ (ii) $s = \dfrac{k}{\sqrt[3]{r}}$ (iii) $p = \dfrac{k}{r^3}$ (iv) $F = \dfrac{k}{\sqrt{r}}$

6. 5   7. 16

8(a) $y = kx^2z$ (b) $p = kq\sqrt[3]{r}$ (c) $y = kx\sqrt{z}$ (d) $p = kqr^3$

9. 56   10. 35   11. 9

12(a) $y = a + bx$ (b) $p = a + \dfrac{b}{q^3}$ (c) $d = a + bt^2$

(d) $y = a + bxz^2$ (e) $p = a + b\dfrac{q}{\sqrt{r}}$

13. 57   14. 4

Exercise 11.1

1. $y = 5x$ 2. $p = 4r^2$ 3. $y = 25\sqrt{x}$ 4. $y = \dfrac{3}{4}x^3$

5. $p = 45\sqrt[3]{r}$ 6. 6   7. 30   8. 24   9. 5   10. 18

11. 2   12. 24   13. 9   14. 125   15. 3   16. 312 grams

17. GH¢ 2.88   18. 225 cm^3

Exercise 11.2

1. 5   2. 4   3. 4   4. $\dfrac{3}{5}$ 5. $\dfrac{2}{3}$

6. 25   7. 4   8. $20\dfrac{1}{2}$ 9. $1\dfrac{1}{2}$ 10. 6

Exercise 11.3

1. 30   2. 90   3. 10   4. 3   5. 25

6. 6   7. 4   8. 16

Exercise 11.4

1. 8   2. 8   3. 15   4. 4   5. 16

6. 49   7. 4   8. 6   9. 4   10. 60

Exercise 11.5

1. 18   2. 4   3(a) 18   (b) 16   4. 40   5. 58

6. 8   7. 8   8. 3   9. 32   10. $2\dfrac{1}{4}$

Review exercise 11

1. 2   2. $w = 9x$ 3. 30   4. $p = \frac{48}{\sqrt{r}}$ 5. 2

6. $w = 4x^2y$ 7. 12   8. $p = \frac{10q}{\sqrt{r}}$ 9. $w = \frac{15xy}{z}$ 10. 33

11. 50   12. $\frac{1}{3}$ 13. 16   14. 100   15. 4

16. 4   17. 9   18. 5   19. 16   20. 6

Chapter Test 11

1(a) 4   (b) 3   (c) 20

2(a) $y = 4x$ (b) $z = \frac{2y}{x}$ (c) $w = \frac{5xy}{z}$

3. 3   4. $\frac{4}{9}$ 5. $1\frac{1}{2}$ 6. 35   7. 24.5   8. 12

Chapter 12

Try this

1(a) 5^{12} (b) 7^7 (c) 3^{11} 2(a) 2^3 (b) 3^4 (c) 7^2

3(a) 7^{10} (b) 3^6 (c) 5^{12} 4(a) 10   (b) 4   (c) 3

5(a) 9   (b) 16   (c) 125

6(a) $\frac{1}{81}$ (b) $\frac{1}{4}$ (c) 125   (d) 8

7(a) 2   (b) $\frac{3}{2}$ 8(a) -5 (b) 3

9(a) 6.5×10^3 (b) 7.35×10^{-7}

10(a) 1.28×10^{-1} (b) 1.2×10^4

11(a) $\log_{10} 10000 = 5$ (b) $\log_2 \frac{1}{32} = -5$

12(a) $64 = 2^6$ (b) $\frac{1}{8} = 2^{-3}$ 13(a) 3 (b) -4

14(a) $\log_5 24$ (b) $\log_2 3$ 15(a) 3 (b) $\frac{5}{3}$

16(a) 1.3801 (b) 0.6990 17(a) $1\frac{1}{3}$ (b) 5

18(a) 0.7466 (b) 0.7657 (c) 0.7606 (d) 0.7851

19(a) 1.5563 (b) 3.3636 (c) 2.8668 (d) 5.8415

20(a) 5.396 (b) 71.29 (c) 4618 (d) 200.4

21(a) 1398 (b) 1.441 (c) 6.400 (d) 162.5

22. 11.52   23(a) $\bar{2}.2$ (b) 1.6 (c) $\bar{1}.4$ (d) $\bar{2}.9$

24(a) 0.01561 (b) 6.337 (c) 0.08112 (d) 0.0005434

25. 1.894

Exercise 12.1(a)

1(a) 2^7 (b) 3^5 (c) 5^8 (d) 6^9

2(a) 2^{-5} (b) 3^{-3} (c) 5^3 (d) 7^{-2}

3(a) 3^3 (b) 2^2 (c) 7^3 (d) 8^5

4(a) 2^{-2} (b) 3^5 (c) 5^2 (d) 6^{-8}

5(a) 4^6 (b) 2^{12} (c) 3^{10} (d) 7^{20}

6(a) 2^{-6} (b) 3^{-8} (c) 4^6 (d) 5^{-3}

7(a) 2^{-6} (b) 2^{-4} (c) 4^2 (d) 5^{-2}

8(a) 2   (b) 5   (c) 4   (d) 2

9(a) 27   (b) 4   (c) 125   (d) 32

10(a) $\frac{1}{2}$ (b) $\frac{1}{3}$ (c) $\frac{1}{25}$ (d) $\frac{1}{8}$

11(a) 1   (b) 16   (c) 2   (d) 1

12(a) 3^4 (b) 2^3 (c) 5^3 (d) $4^5 \times 3^2$ (e) 5^2 (f) 3^2

13(a) 3   (b) 16   (c) 48   (d) 972   (e) 12   (f) $\frac{1}{8}$

Example 12.1(b)

1. 5   2. 3   3. 3   4. $\frac{1}{5}$ 5. 3   6. $-\frac{8}{3}$

7. 2   8. -2 9. $-\frac{1}{2}$ 10. $\frac{4}{3}$ 11. 0   12. -2

Exercise 12.2(a)

1. 4.7×10^6 2. 8.6×10^4 3. 2.3×10^2

4. 7.6×10^1 5. 3.64×10^9 6. 1.24×10^5

7. 2.75×10^1 8. 5.725×10^3 9. 6.27×10^2

10. 9.5   11. 1.05×10^4 12. 4.156×10^3

13. 5.6×10^{-3} 14. 2.4×10^{-1} 15. 4.31×10^{-5}

16. 7.03×10^{-4} 17. 6.43×10^{-2} 18. 8.6×10^{-6}

19. 9×10^{-3} 20. 1.26×10^{-1} 21. 3.164×10^{-2}

22. 2×10^{-5} 23. 6.05×10^{-2} 24. 5.7×10^{-7}

Exercise 12.2(b)

1. 4×10^9 2. 4.05×10^{-6} 3. 2.25×10^2

4. 5.445×10^1 5. 2.448×10^3 6. 5.82×10^1

7. $6. \times 10^3$ 8. 2.4×10^7 9. 1×10^2

10. 2.8×10^2 11. 6.5×10^{-2} 12. 6×10^1

Exercise 12.3(a)

1. $log_2 16 = 4$ 2. $log_3 27 = 3$ 3. $log_3 81 = 4$

4. $log_5 25 = 2$ 5. $log_4 64 = 3$ 6. $log_2 \frac{1}{8} = -3$

7. $log_2 \sqrt[5]{32} = 1$ 8. $log_6 \frac{1}{36} = -2$ 9. $64 = 2^6$

10. $729 = 3^6$ 11. $625 = 5^4$ 12. $3 = 27^{\frac{1}{3}}$

13. $1 = 3^0$ 14. $5 = 5^1$ 15 3

16. 5   17. 2   18. 3   19. 4   20. 0

21. 4   22. $\frac{2}{3}$ 23. -5 24. $\frac{5}{3}$

Exercise 12.3(b)

1. $log_{10} 15$ 2. $log_5 16$ 3. $log_3 8$ 4. $log_8 25$

5. $log_3 \frac{1}{16}$ 6. $log_4 3$ 7. $log_2 3$ 8. $log_{10} 12$

9. $log_{10} 15$ 10. $log_{10} 18$ 11. 3   12. 5

13. 4   14. $\frac{2}{3}$ 15. $4\frac{1}{2}$ 16. $\frac{1}{3}$ 17. -4 18. $1\frac{1}{2}$

19. $\frac{2}{3}$ 20. $1\frac{1}{2}$ 21. $\frac{1}{2}$ 22. $1\frac{1}{3}$ 23. 0.7781

24. 1.2552   25. 1.3010   26. 1.7781   27. 1.6532

28. 1.3980   29. 1.9084   30. 1.8751   31. 0.1761

32. – 0.9030   33. 0.4771   34. – 0.1505

Exercise 12.3(c)

1. 25   2. 3   3. 16   4. 15   5. $1\frac{1}{2}$

6. 6   7. $\frac{1}{2}$ 8. 1   9. 6   10. 5

Exercise 12.4(a)

1. 0.6628   2. 0.9042   3. 0.7782

4. 0.3243   5. 0.0086   6. 0.7374

7. 0.7061   8. 0.6029   9. 0.4764

10. 0.0319   11. 1   12. 2   13. 3

14. 0   15. 4   16. 5   17. 3   18. 7

19. 4   20. 1   21. 1.4472

22. 1.7210   23. 3.6609   24. 2.8540

25. 1.6098   26. 4.8041   27. 5.5163

28. 4.8286   29. 4.0001   30. 2.8641

31. 1   32. 2   33. 3   34. 5

35. 4   36. 6   37. 7   38. 1

39. 2   40. 1   41. 108.6

42. 6.109   43. 1009   44. 83370

45. 625500   46. 30690000   47. 199.5

48. 10000   49. 61.24   50. 1085000

Exercise 12.4(b)

1. 2689   2. 6555   3. 29220

4. 233570000   5. 9.212   6. 1.8500

7. 75.06   8. 16.32   9. 5255

10. 1240000   11. 4064   12. 755.0

13. 6.530   14. 23.32   15. 5.4261

16. 3-712   17. 43.03   18. 112.0

19. 195.9   20. 3.139   21. 1.313

Exercise 12.4(c)

1. $\bar{1}$ 2. $\bar{2}$ 3. $\bar{3}$ 4. $\bar{1}$ 5. $\bar{5}$

6. $\bar{4}$ 7. $\bar{1}.6355$ 8. $\bar{2}.9212$

9. $\bar{3}.8261$ 10. $\bar{1}.3201$ 11. $\bar{4}.4771$

12. $\bar{3}.7173$ 13, 0.4808   14. 0.0005814

15. 0. 000006434   16. 0.00002590

17. 0.000001   18. 0.0000008185

20. 0.0001219   21. 0.00000002003

22. 0.0001021   23. 4.5   24, $\bar{2}.3$

25. $\bar{3}.7$ 26. $\bar{4}.8$ 27. $\bar{5}.2$

28. $\bar{4}.5$ 29. $\bar{1}.3$ 30. $\bar{2}.9$

Exercise 12.4(d)

1. 0.00005219   2. 0.000007975

3. 0.1383   4. 5.436   5. 0.0006924

6. 0.07464   7. 0.09221   8. 0.03903

9. 0,02689   10. 2.897   11. 0.0001337

12. 0.04801

Review exercise 12

1. 3^2 2. 2^{-1} 3. 5^{-2} 4. 5^{-4}

5. 3^2 6. 3^{-2} 7. 5^2 8. 2^6

9. 3^6 10. 7^{-2} 11. 1   12. 2^2

13. 2   14. 6   15. 25   16. 3

17. 4   18. 9   19. $\frac{1}{27}$ 20. $\frac{1}{16}$ 21. $2\frac{1}{2}$ 22. $\frac{1}{2}$

23. $\frac{2}{3}$ 24. 0

25. $-\frac{1}{2}$ 26. $-1\frac{1}{4}$ 27. 0   28. -2

29. 5.97×10^5 30. 6.40×10^2

31. 6.75×10^1 32. 7.582×10^3

33. 6.5×10^{-1} 34. 8.04×10^{-4}

35. 2.178×10^{-2} 36. 9×10^{-5}

37. 5.395×10^3 38. 4.863×10^{-3}

39. 2×10^6 40. 9×10^2 41. 2×10^1

42. 2.8×10^{-2} 43. 6.50×10^2

44. 6×10^{-1} 45. 2   46. $\frac{2}{3}$ 47. 0

48. $-\frac{1}{2}$ 49. 6^3 50. 2^{-5}

51. $9 = 81^{\frac{1}{2}}$ 52. $10{,}000 = 10^4$

53. $log_3 35$ 54. $log_2 5$ 55. $\log 5$

56. $log_4\left(\frac{9}{2}\right)$ 57. 7   58. 2   59. 3

60. $\frac{3}{2}$ 61. 0.0791   62. 1.8572

63. -0.6990 64. 1.6532   65. 4

66. 5   67. 5   68. 2   69. 973.1

70. 35110   71. 150.4   72. 3.173

73. 0.01456   74. 0.09301   75. 0.4034

76. 0.5065   77. 79.03   78. 5.953

79. 5.565   80. 16.20

Chapter Test 12

1. 3^2 2(a) 18   (b) 4   3(a) 3   (b) 6

4(a) 3.62×10^{-3} (b) $5{,}78 \times 10^1$ 5. 1.5×10^{-1}

6(a) 4   (b) - 5   (c) $\frac{2}{3}$ 7(a) 1.7781   (b) 2.1303

8(a) $log_3 6$ (b) $\log 12$ 9(a) 4   (b) 3

10(a) 0.2261   (b) 3.276   11. 0.8015

12(a) $x = 3, y = 2$ (b) $x = 2, y = 1$

Chapter 13

Try this

1(a) 201_4 (b) 1024_8 2. $4E1$

3(a) 43_5 (b) 111_2 4. 11424_5

5(a) 24321   (b) 13504   6. 196

7. 8   8. 11001_2 9. 1030_5

Exercise 13.1(a)

1(a) 1001   (b) 1000   (c) 1100

(d) 10101   (e) 11100   (f) 100001

2(a) 110   (b) 1102   (c) 1020

(d) 2110   (e) 1020   (f) 12021

3(a) 113   (b) 31   (c) 1200

(d) 1021   (e) 1310   (f) 1102

4(a) 114   (b) 100   (c) 421

(d) 2024   (e) 2420   (f) 2244

5(a) 123   (b) 121   (c) 215

(d) 10001   (e) 1445   (f) 5115

6(a) 123   (b) 60   (c) 1003

(d) 1311   (e) 1055   (f) 1265

7(a) 127   (b) 102   (c) 1237

(d) 753   (e) 1306   (f) 1143

8 (a) 57   (b) 137   (c) 808

(d) 780   (e) 514   (f) 2242

Exercise 13.1(b)

1. $T8$ 2. $24T$ 3. $E54$

4. $E6E$ 5. $T16$ 6. 2538

Exercise 13.1(c)

1(a) 11   (b) 11   (c) 1011

(d) 1   (e) 100000   (f) 1101

2(a) 101   (b) 112   (c) 20

(d) 202   (e) 12   (f) 1112

3(a) 212   (b) 23   (c) 123

(d) 22   (e) 313   (f) 2023

4(a) 11   (b) 234   (c) 24

(d) 331   (e) 2332   (f) 223

5(a) 22   (b) 35   (c) 132

(d) 1121   (e) 2103   (f) 412

6(a) 23   (b) 14   (c) 146

(d) 134   (e) 164   (f) 236

7(a) 22   (b) 14   (c) 127

(d) 175   (e) 326   (f) 556

8(a) 22   (b) 14   (c) 37

(d) 217   (e) 1271   (f) 838

9(a) 26   (b) 15   (c) 276

(d) 395   (e) $1T2$ (f) $T06$

Exercise 13.2(a)

1.

x	0	1	2
0	0	0	0
1	0	1	2
2	0	2	11

(a) 1022   (b) 2101   (c) 10122   (d) 11021

2.

x	0	1	2	3
0	0	0	0	0
1	0	1	2	3
2	0	2	10	12
3	0	3	12	21

(a) 201   (b) 20001   (c) 3302   (d) 101010

3.

x	0	1	2	3	4	5
0	0	0	0	0	0	0
1	0	1	2	3	4	5
2	0	2	4	10	12	14
3	0	3	10	13	20	23
4	0	4	12	20	24	32
5	0	5	14	23	32	41

(a) 212   (b) 10024   (c) 20053   (d) 154022

4.

x	1	2	3	4
1	1	2	3	4
2	2	4	6	10
3	3	6	11	14
4	4	10	14	20

(a) 146   (b) 5134   (c) 16212

5.

x	2	5	7	9
2	4	T	12	16
5	T	21	2E	39
7	12	2E	41	53
9	16	39	53	69

5(a) 14*E*   (b) 1343   (c) 17984

Exercise 13.2(b)

1(a) 100001   (b) 101101   (c) 10010110

  (d) 11001011   (e) 1101110   (f) 1011010

2(a) 112   (b) 2002   (c) 11011

  (d) 121121   (e) 1101221   (f) 1101111

3(a) 222   (b) 12300   (c) 20012

  (d) 311130   (e) 100323   (f) 232110

4(a) 1232   (b) 20004   (c) 14202

  (d) 1023202   (e) 101432   (f) 104210

5(a) 334   (b) 5354   (c) 25332

(d) 100200   (e) 120200   (f) 253530

6(a) 1336   (b) 13143   (c) 20033

(d) 1025046   (b) 1052433   (c) 631305

7(a) 2230   (b) 11253   (c) 41763

(d) 506766   (e) 272210   (f) 574665

8(a) 523   (b) 13556   (c) 73886

(d) 315322   (e) 155603   (f) 301363

9(a) 55E (b) 2T7E0   (c) 128E6

(d) 10588   (e) 2E0E67   (f) 36581T

Exercise 13.3(a)

1. 10   2. 27   3. 45   4. 16   5. 69   6. 61

7. 46   8. 121   9. 110   10. 73   11. 84   12. 738

13. 124   14. 176   15. 547   16. 249   17. 474   18. 731

19. 47   20. 309   21. 1623   22. 167   23. 3021   24. 764

25. 1707   26. 5   27. 8   28. 9   29. 6   30. 7

31. 5   32. 9   33. 8   34. 6   35. 4   36. 8

Exercise 13.3(b)

1(a) 10111   (b) 10001   (c) 1001   (d) 100010   (e) 11101

2(a) 111   (b) 1000   (c) 1012   (d) 1101   (e) 1111

3(a) 103   (b) 130   (c) 133   (d) 231   (e) 1323

4(a) 33   (b) 1012   (c) 144   (d) 441   (e) 412

5(a) 54   (b) 323   (c) 124   (d) 342   (e) 555

6(a) 54   (b) 122   (c) 266   (d) 642   (e) 1130

7(a) 108   (b) 363   (c) 254   (d) 33   (e) 524

8(a) 22   (b) 61   (c) 171   (d) 371   (e) 633

9(a) TE (b) 101   (c) 194   (d) 277   (e) $50T$

Exercise 13.3(c)

1. 214_5 2. 103_6 3. 351_7 4. 511_8 5. 1110_4

6. 10000_2 7. 520_9 8. 1200_3 9. $4TE_{12}$ 10. 602_9

Review exercise 13

1. 10000_2 2. 10000_2 3. 1110_3

4. 2020_3 5. 1120_4 6. 2021_4

7. 1014_5 8. 2001_5 9. 523_6

10. 10100_6 11. 1023_7 12. 1126_7

13. 530_8 14. 1041_8 15. 1001_9

16. 1084_9 17. 1030_{12} 18. $T0T_{12}$

19. 10_2 20. 10001_2 21. 112_3

22. 220_3 23. 231_4 24. 2122_4

25. 41_5 26. 114_5 27. 1312_6

28. 1344_6 29. 65_7 30. 553_7

31. 345_8 32. 2256_8 33. 168_9

34. 445_9 35. $TT8_{12}$ 34. $12E4_{12}$

37. 1000001_2 38. 111111_2

39. 11011_3 40. 11122_3

41. 12012_4 42. 10132_4

43. 22322_5 44. 33222_5

45. 10202_6 46. 43532_6

47. 31646_7 48. 10605_7

49. 10227_8 50. 6350_8

51. 4336_9 52. 7672_9

53. $2E184_{12}$ 54. 2676_{12}

55. 29   56. 11   57. 17   58. 68

59. 54   60. 107   61. 117   62. 49

63. 164   64. 196   65. 141   66. 527

67. 302   68. 748   69. 176   70. 1578

71. 1606   72. 3197   73. 5   74. 9

75. 7   76. 5   77. 5   78. 5

79. 1111101_2 80. 100101_2

81. 212_3 82. 2111_3 83. 212_4

84. 1323_4 85. 1014_5 86. 322_5

87. 340_6 88. 535_6 89. 133_7

90. 236_7 91. 101_8 92. 206_8

93. 63_9 94. 181_9 95. EE_{12}

96. $3TT_{12}$ 97. 210_5 98. 130_6

99. 1241_7 100. 466_8 101. 278_9

102. 1392_{12} 103. 1154_6 104. 11001_2

105. 1143_8

Chapter Test 13

1(a) 1000100_2 (b) 403_5 (c) 31052_6

2(a) 956_{12} (b) $ET7$ (c) $25E28$ 3. 735_8

4(a)

×	2	3	4	5
2	4	6	8	11
3	6	10	13	16
4	8	13	17	22
5	11	16	22	27

 (b) (i) 862_9 (ii) 13263_9

5. 164   6. 1404_7 7. 3112_5 8. 13021_5

9. 66_8 10(a) 7   (b) 5

Chapter 14

Try this

1(a) $3\sqrt{2}$ (b) $2\sqrt{5}$ 2(a) $7\sqrt{5}$ (b) $3\sqrt{7}$ (c) 0

3(a) $6\sqrt{2}$ (b) $\sqrt{3}$ 4(a) $15\sqrt{6}$ (b) $4\sqrt{15}$

5(a) $\frac{4}{3}\sqrt{3}$ (b) $2\sqrt{5}$ 6(a) $30 - 3\sqrt{10}$ (b) $-6 + 5\sqrt{3}$

7(a) -1 (b) $13 - 4\sqrt{3}$ 8. 3.535 9. $3\sqrt{5} - 6$

Exercise 14.1

1. $2\sqrt{6}$ 2. $5\sqrt{2}$ 3. $2\sqrt{2}$ 4. $4\sqrt{2}$

5. $6\sqrt{3}$ 6. $2\sqrt{7}$ 7. $6\sqrt{2}$ 8 $3\sqrt{5}$

9. $5\sqrt{3}$ 10. $3\sqrt{10}$ 11. $4\sqrt{5}$ 12. $6\sqrt{3}$

13. $8\sqrt{2}$ 14. $7\sqrt{3}$ 15. $6\sqrt{5}$ 16. $4\sqrt{7}$

17. $10\sqrt{3}$ 18. $9\sqrt{5}$ 19. $4\sqrt{30}$ 20. $8\sqrt{10}$

Exercise 14.2(a)

1. $10\sqrt{5}$ 2. $5\sqrt{7}$ 3. $6\sqrt{3}$ 4. $3\sqrt{7}$ 5. $2\sqrt{3}$ 6. $-3\sqrt{2}$

7. 0 8. $5\sqrt{3}$ 9. $2\sqrt{5}$ 10. $7\sqrt{3}$ 11. $5\sqrt{2}$ 12. $5\sqrt{7}$

13. $\sqrt{2}$ 14. $\sqrt{3}$ 15. $\sqrt{5}$ 16. $-4\sqrt{2}$ 17. $15\sqrt{2}$ 18. $\sqrt{5}$

19. $3\sqrt{2}$ 20, $4\sqrt{2}$ 21. $4\sqrt{3}$

Exercise 14.2(b)

1. 18 2. $8\sqrt{10}$ 3. $4\sqrt{30}$ 4. $7\sqrt{30}$ 5. $9\sqrt{5}$ 6. $10\sqrt{15}$

7. $15\sqrt{6}$ 8. $6\sqrt{6}$ 9. $18\sqrt{6}$ 10. $18\sqrt{5}$ 11. $10\sqrt{3}$ 12. $180\sqrt{3}$

Exercise 14.2(c)

1. $3\sqrt{3}$ 2. $\frac{1}{2}\sqrt{2}$ 3. $3\sqrt{3}$ 4. 1 5. $\sqrt{10}$

6. $2\sqrt{7}$ 7. $4\sqrt{2}$ 8. $\frac{3\sqrt{2}}{8}$ 9. $\frac{1}{2}\sqrt{10}$ 10. $\sqrt{2}$

11. 2   12. $\frac{1}{2}\sqrt{10}$ 13. $\frac{1}{3}\sqrt{6}$ 14. $\frac{3}{5}\sqrt{3}$ 15. $\frac{1}{4}\sqrt{30}$

Exercise 14.3(a)

1. $3+\sqrt{15}$ 2. $2\sqrt{3}-2$ 3. $30-10\sqrt{6}$

4. $12+4\sqrt{6}$ 5. 7   6. 3   7. -13

8. -10 9. $11-6\sqrt{2}$ 10. $5+2\sqrt{6}$

11. $11-4\sqrt{6}$ 12. $18-12\sqrt{2}$ 13. $22+12\sqrt{2}$

14. $30-12\sqrt{6}$ 15. $6\sqrt{6}+6-6\sqrt{2}-2\sqrt{3}$ 16. $24-7\sqrt{6}$

17. $2\sqrt{6}-2$ 18. $2+3\sqrt{2}-2\sqrt{3}-3\sqrt{6}$ 19. 8   20. 30

21. 9.928   22. 5.758   23. 8.083   24. 19.484   25. 6.07

26. -0.672 27. 3.554   28. 0.707   29. 2.866   30. 5.464

Exercise 14.3(b)

1. $\sqrt{2}+1$ 2. $6+3\sqrt{3}$ 3. $\sqrt{5}+1$ 4. $4\sqrt{3}+6$

5. $-8-4\sqrt{6}$ 6. $\frac{-1-2\sqrt{3}}{11}$ 7. $7-4\sqrt{3}$ 8. $\frac{7-3\sqrt{3}}{11}$

9. $-4-\sqrt{5}$ 10. $\frac{-1-\sqrt{15}}{2}$

Review exercise 14

1. $2\sqrt{15}$ 2. $3\sqrt{5}$ 3. $3\sqrt{2}$ 4. $5\sqrt{6}$ 5. $8\sqrt{3}$ 6. $7\sqrt{5}$

7. $6\sqrt{6}$ 8. $9\sqrt{3}$ 9. $8\sqrt{6}$ 10. $21\sqrt{2}$ 11. $6\sqrt{2}$ 12. $8\sqrt{3}$

13. $2\sqrt{3}$ 14. $-\sqrt{5}$ 15. $5\sqrt{5}$ 16. $-\sqrt{2}$ 17. $6\sqrt{3}$ 18. $\sqrt{3}$

19. $7\sqrt{5}$ 20. $6\sqrt{3}$ 21. $6\sqrt{3}$ 22. $4\sqrt{5}$ 23. $2\sqrt{30}$ 24. $6\sqrt{15}$

25. $25\sqrt{6}$ 26. $6\sqrt{15}$ 27. $16\sqrt{10}$ 28. $18\sqrt{6}$ 29. $6\sqrt{6}$

30. $60\sqrt{5}$ 31. $\frac{2}{3}\sqrt{6}$ 32. $\frac{6}{5}\sqrt{10}$ 33. $\frac{4}{3}\sqrt{3}$ 34. $\sqrt{3}$ 35. $\sqrt{5}$

36. $\frac{2}{3}\sqrt{3}$ 37. $\frac{\sqrt{30}}{15}$ 38. $\frac{2}{3}$ 39. $\frac{2}{3}\sqrt{3}$ 40. $\frac{5}{4}\sqrt{2}$

41. $8+\sqrt{6}$ 42. $10\sqrt{6}-6\sqrt{2}$ 43. $6\sqrt{2}-3$ 44. $-6+\sqrt{6}$

45. $14-6\sqrt{5}$ 46. $8-5\sqrt{21}$ 47. 3 48. 1

49. $29+12\sqrt{5}$ 50. $57-9\sqrt{2}$ 51. 11.708 52. -3.816

53. 3.464 54. 38.888 55. 22.168 56. $-6+3\sqrt{5}$

57. $\frac{6+4\sqrt{3}}{-3}$ 58. $2\sqrt{3}+3$ 59. $\frac{5-\sqrt{15}}{2}$ 60. $4+\sqrt{15}$

Chapter Test 14

1(a) $4\sqrt{6}$ (b) $8\sqrt{5}$ 2(a) $-2\sqrt{5}$ (b) $4\sqrt{2}$

3(a) $12\sqrt{6}$ (b) $15\sqrt{2}$ 4(a) $\frac{3\sqrt{6}}{8}$ (b) $\sqrt{3}$

5. $\frac{2}{3}\sqrt{3}+1$ 6(a) $8\sqrt{10}$ (b) $3\sqrt{15}-5\sqrt{6}+6\sqrt{10}-20$

7(a) 2 (b) $79-20\sqrt{5}$ 8(a) $-\frac{1}{2}$ (b) - 2, 2

9(a) 5.07 (b) 6.3507 10(a) $4+3\sqrt{2}$ (b) $\frac{6+5\sqrt{2}}{14}$

Chapter 15

Try this

1. 3, - 2, - 7   2. 59   3. 15   4. 10, 7, 4   5. 680   6. − 80

7. GH¢ 23.25   8. $r = \frac{1}{3}, \frac{4}{3}, \frac{4}{9}$ 9. 256   10. $a = 2,\ r = 3$

11. $40\frac{1}{3}$ 12. GH¢ 3166.39

Chapter 15.1(a)

1(a) 3   (b) 4   (c) − 6   (d) − 12   (e) 4   (f) − 2

2(a) 21, 24, 27   (b) 27, 33, 39   (c) − 10, - 15, - 20

(d) 31, 37, 39   (e) − 29, - 38, - 47   (f) 27, 31, 35

Exercise 15.1(b)

1(a) 115   (b) − 59   (c) $27\frac{1}{2}$ (d) 3   (e) − 21   (f) − 99

2(a) $3n + 4,\ 64$ (b) $4n,\ 100$ (c) $3n + 2,\ 38$

(d) $7n − 5,\ 142$ (e) $−2n + 19,\ −1$ (f) $2n + 3,\ 35$

3(a) 23   (b) 15   (c) 13   (d) 41   (e) 32   (f) 16

4(a) 5, 8, 11   (b) 9, 13, 17   (c) 13, 11, 9

(d) 9, 12, 15   (e) - 35, - 23, -11   (f) 25, 23, 21

Exercise 15.1(c)

1(a) 2800   (b) 808   (c) 2173   (d) 483   (e) 352   (f) − 483

2(a) 504   (b) 0   (c) 1625   (d) − 568   (e) 135   (f) $−157\frac{1}{2}$

3. 2, 195   4. 1, 3, 590   5. GH¢ 460   6. 42   7. 126

8. GH¢ 1150   9. GH¢ 29.45   10. GH¢ 116.25

Exercise 15.2(a)

1(a) 2   (b) 3   (c) 0.1   (d) $\frac{1}{3}$ (e) $-\frac{1}{2}$ (f) 8

2(a) 40, 80, 160   (b) 1, $\frac{1}{2}$, $\frac{1}{4}$ (c) $\frac{2}{9}$, $-\frac{2}{27}$, $\frac{2}{81}$

    (d) 0.027, 0.0081, 0.00243   (e) $\frac{1}{8}$, $\frac{1}{16}$, $\frac{1}{32}$ (f) $\frac{5}{4}$, $\frac{5}{8}$, $\frac{5}{16}$

3(a) 384   (b) $\frac{2}{243}$ (c) $\frac{1}{32}$ (d) -2048 (e) $\frac{1}{81}$ (f) 4

Exercise 15.2(b)

1(a) 765   (b) 364   (c) $15\frac{1}{2}$ (d) $6\frac{20}{27}$ (e) $7\frac{127}{128}$ (f) 255

2. 3, 6560   3. $5\frac{1}{3}$, $70\frac{1}{3}$ 4. 7300   5. GH¢ 7577.60

6. 12   7. GH¢ 5318.84   8. 8

Review exercise 15

1(a) arithmetic sequence   (b) not arithmetic sequence

    (c) arithmetic sequence   (d) not arithmetic sequence

    (e) not arithmetic sequence   (f) arithmetic sequence

2(a) geometric sequence   (b) not geometric sequence

    (c) geometric sequence   (d) not geometric sequence

    (e) geometric sequence   (f) geometric sequence

3(a) 17, 21, 25   (b) $-3, -8, -13$ (c) 11, 14, 17

    (d) 1, $-1, -3$ (e) 3, 8, 13   (f) 5, 3, 1

4(a) 24, 48, 96   (b) $\frac{1}{5}, \frac{1}{25}, \frac{1}{125}$ (c) 32, -64, 128

(d) -2 , $1-\frac{1}{2}$ (e) 2, 8, 32   (f) $\frac{1}{3}$, $\frac{1}{9}$, $\frac{1}{27}$

5(a) 93   (b) -61 (c) -37 (d) 75   (e) 24   (f) 67

6(a) 729   (b) $\frac{9}{64}$ (c) $\frac{1}{32}$ (d) 27   (e) 512   (f) $\frac{1}{9}$

7(a) 23   (b) 13   (c) 12   (d) 15   (e) 31   (f) 20

8(a) 8   (b) 8   (c) 11   (d) 10   (e) 8   (f) 7

9(a) 2600   (b) 728   (c) 2542   (d) 546   (e) 384   (f) -310

10(a) 480   (b) 460   (c) -120 (d) -520 (e) 225   (f) 660

11(a) 255   (b) $40\frac{4}{9}$ (c) $86\frac{7}{9}$ (d) $5\frac{1}{4}$ (e) $193\frac{1}{32}$ (f) -257

12. 3, 12, 255   13. 13, -120 14. 7450   15. $\frac{1}{3}$, 81, $121\frac{1}{3}$

16. $\frac{1}{2}$, 16 $-\frac{1}{2}$, -16 17(a) GH¢ 12,500   (b) GH¢ 143,700

18(a) 48   (b) 900   19. GH¢ 15,853.02   20. GH¢ 10,413.49

Chapter Test 15

1(a) 5, 0, -5 -75 (b) 5, $\frac{5}{2}$, $\frac{5}{4}$ $\frac{5}{32}$

2(a) $2n+1$ (b) n^2 3(a) 1, 7, 17, 31, 49   (b) 199

4(a) 17   (b) 8   5(a) 1683   (b) -900

6(a) 5115   (b) $17\frac{79}{81}$ 7. 114   8. 3   9. 300   10. 2740

11. GH¢ 12,100   GH¢ 131,700   12(a) GH¢ 905.49   (b) 8 years

Chapter 16

Try this

1(a) 5 : 4   (b) 5 : 12   (c) 8 : 15   2. 1 : 3   3. $\frac{5}{3}$: 1   4. 9

5. GH¢ 22.50   6. 30 people   7. 1680, 1200, 720   8. GH¢ 360

9. GH¢ 3020, GH¢ 6040, GH¢ 18,120   10. $\frac{1}{200}$ 11. 10 m

12. GH ¢ 93.60   13. GH¢ 122.50   14. GH¢ 61.50

15(a) 10 am   (b) 30 km   (c) 12 noon to 1 pm, 4 pm to 5 pm

  (d) (i) 10 km h^{-1} (ii) 10 km h^{-1} (e) 8 pm

16(a) $66\frac{2}{5}$%   (b) 70 %   17(a) $\frac{3}{25}$ (b) $\frac{13}{20}$

18(a) 48 %   (b) 3.7 %   19(a) 0.32   (b) 0.184

20. 20 %   21. 240   22. 900   23. 1456

24. 3390   25. $2\frac{1}{2}$%   26. 540

Exercise 16.1(a)

1(a) 4 : 5   (b) 7 : 9   (c) 4 : 3   (d) 6 : 5

2(a) 3 : 4   (b) 3 : 5   (c) 5 : 9   (d) 3 : 2

3(a) 3 : 50   (b) 9 : 40   (c) 9 : 20 (d) 5 : 3

4. 2 : 3   5. 3 : 2   6. 3 : 5   7. 11 : 16

8. 2 : 5   9. 2 : 5   10. 15 : 16

Exercise 16.1(b)

1(a) 1 : 3   (b) 1 : $\frac{3}{8}$ (c) 1 : $\frac{9}{2}$ (d) 1 : $\frac{3}{5}$ (e) 1 : $\frac{9}{7}$

2(a) 2 : 1   (b) $\frac{5}{2}$: 1   (c) $\frac{4}{3}$: 1   (d) $\frac{5}{2}$: 1   (e) $\frac{4}{5}$: 1

Exercise 16.1(c)

1(a) 14   (b) 32   (c) 20   (d) $9\frac{1}{2}$ (e) 4   (f) 6

(g) 80   (h) 30   (i) 17   (j) $3\frac{3}{4}$ (k) 5   (l) 10

2. 25.7   3. GH¢ 56.80   4. GH¢ 120   5. 9600   6. 25

7. 7   8. 336   9. 15   10. 2.47 Kg   11. 15

12. 10,000   13. GH¢ 1,000   14. 18   15. GH¢ 9375

16. 75 books   17. 20 h 50 min   18. 6000 bricks   19. 48 h

20. 750 litres   21. 1,000 trees   22. 7 h 30 min   23. 40 km/h

24. 2 h 15 min   25. 6 h   26. 12   27. 3600 copies

Exercise 16.1(d)

1. GH¢ 1,000,   GH¢ 1,500   2. GH¢ 2,250,   GH¢ 1,500

3. GH¢ 6,300   GH¢ 8,400   GH¢ 10,500   4. 360, 450

5. 28, 24   6. GH¢ 12,400, GH¢ 18,600   GH¢ 6,200

7. GH¢ 1,000   8. GH¢ 8,640   9. GH¢ 15,750   10. 6

11. GH¢ 13,440   12. GH¢ 640

Exercise 16.1(e)

1. 14   2. 56   3. GH¢ 13,680   4. 6,000   5. GH¢ 1,250

6. GH¢ 1,500   7. GH¢ 104   GH¢ 156

8. GH¢ 180   GH¢ 360   GH¢ 540

9. GH¢ 468   10. 160   240   120

Exercise 16.1(f)

1. $\dfrac{1}{25}$ 2. $\dfrac{1}{25,000}$ 3. 3.5 km   4. 12 km   5. 25 cm

6. 20 m   12.5 m   7. 5 m   4 m   8. 1 m^2 9. 12 km^2 10. 4 cm^2

11. 64 cm^2 12. 50 m   13.14.5 cm   14. 1.8 m   15. 2.7 m

Exercise 16.2(a)

1(a) 12 km^{-1} (b) GH¢ 250 per ticket   (c) 75 patients per doctor

  (d) 40 students per class

2. GH¢ 500 per night   3. GH¢ 245

4. GH¢ 72   5. 1.5 cm^2 s^{-1} 6. 270 cm^3 7. 25 m^2

8. GH¢ 62.55   9. 6 cm^3 s^{-1} 10. 50 m   11. GH¢ 181.20

12. $ 200   13. £ 5245.90   14. GH¢ 1.19 per litre

15. GH¢ 2490.90   16. 1620   17. 2 h 15 min   18. 18,000

19. 2 h   20. 20 min.

Exercise 16.2(b)

1. GH¢ 14.90   2 (b) GH¢ 54   12   3. GH¢ 18.62

4. GH¢ 120.30   5. GH¢ 130.81

Exercise 16.2(c)

5(a) 2 h   (b) 9 km   6(a) 1 h   (b) 3 h 30 min   (c) 48.75 m

7(a) 1 s   (b) 16 m/s^2 (c) 54 m

8(a) He stopped   (b) 8 km/h   (c) 1 h   (d) 4 pm

Exercise 16.3(a)

1(a) 20 %   (b) 80 %   (c) 6 %   (d) 3 %   (e) 25 %

2(a) 75 %   (b) 80 %   (c) 32 %   (d) $16\frac{2}{3}$%

 (e) 120 %   (f) 350 %

3(a) $\frac{17}{20}$ (b) $\frac{3}{20}$ (c) $\frac{3}{50}$ (d) $2\frac{1}{2}$ (e) $\frac{3}{9}$ (f) $\frac{9}{200}$

4(a) 65 %   (b) 5 %   (c) 23 %

 (d) 136 %   (e) 207 %   (f) 4.8 %

5(a)0.56   (b) 0.45   (c) 0.125

 (d) 1.78   (e) 2.06   (f) 0.002

Exercise 16.3(b)

1(a) 20 %   (b) 30 %   (c) 50 %

 (d) 25 %   (e) 50 %   (f) 25 %

2. 75 %   3. $12\frac{1}{2}$%   4. 20 %   5. 2 %   6. 80%

7. $33\frac{1}{3}$%   8. 40 %   9. 40 %   10. $66\frac{2}{3}$%

Exercise 16.3(c)

1(a) 40   (b) 1110   (c) 800   (d) 60   (e) 351   (f) 27

 2.GH¢ 68   3. 144   4. 375   5. GH¢ 37.50

 6. 19   7. 420   8. 50

Exercise 16.3(d)

1. 128   2. 2,000   3. 22,400   4. 240

5. GH¢ 5,000   6. 40   7. GH¢ 56,000

8. GH¢ 1,200   9. GH¢ 1,600   10. GH¢ 5,000

Exercise 16.3(e)

1. 17 %   2. $6\frac{2}{3}\%$ 3. 60 %

4. 20 %   5. 25 %   6. 24 %

7(a) 10,800   (b) 936   (c) 305.5   (d) 3516.8

8(a) 450   (b) 1,350   (c) 288   (d) 999

9. GH¢ 4.20   10. GH¢ 95.06   11. 200

12. GH¢ 1,456   13. GH¢ 437   14. GH¢ 4,440   15. GH¢ 14.55

Exercise 16.3(f)

1.1,152   2. 1,575   3. 4,116   4. 5,040   5. GH¢ 912

6.GH¢ 3,780   7. 860   8(a) 12.6 %   (b) GH¢ 3.93

Review exercise 16

1(a) 4 : 1   (b) 1 : 3   (c) 5 : 6   (d) 10 : 13

2(a) 2 : 3   (b) 9 : 20   (c) 3 : 1   (d) 2 : 1   (e) 7 : 3   (f) 40 : 3

3(a) $1:\frac{3}{2}$ (b) 1 : 3   (c) 1 : 8   (d) $1:\frac{3}{2}$

4(a) $\frac{2}{3}:1$ (b) $\frac{5}{4}:1$ (c) $\frac{4}{3}:1$ (d) $\frac{2}{3}:1$

5(a) 40   (b) 27   (c) 6   (d) 8.4   (e) 45   (f) 4

6. 140   7. 7h 30 min   8. GH¢ 11.97   9. 280 km

10. GH¢ 13.44   11. 225 kg   12. 1750 minutes

13(a) 240   400   (b) 520   728   (c) 240   840   (d) 2416 2114

14. GH¢ 780   GH¢ 1,170   15. GH¢ 42   GH¢ 30   16. 30

17. GH¢ 10.50   GH¢ 31.50   GH¢ 21.00

18. GH¢ 30   GH¢ 45   GH¢ 135

19(a) $\frac{1}{250}$ (b) $\frac{1}{20}$ (c) $\frac{1}{4,000}$ (d) $\frac{1}{200}$ (e) $\frac{1}{250,000}$

20. 1 : 200,000   21. 1.2 km   22. 500 m^2 23. 3750 km^2

24. GH¢ 12   25. GH¢ 120   26. GH¢ 60   27. GH¢ 810

28(a) GH¢ 22   (b) GH¢ 34   (c) GH¢ 46   29. GH¢ 253.30

30. $ 838.93   31. GH¢ 357   32. GH¢ 106.50

33(a) 10 km   (b) 14 km   (c) 21 km   (d) 3 hr   (e) 7 km h^{-1}

34(a) $37\frac{1}{2}\%$ (b) 76 %   (c) $62\frac{1}{2}\%$ (d) 85 %

35(a) $\frac{3}{20}$ (b) $\frac{29}{40}$ (c) $1\frac{1}{2}$ (d) $1\frac{1}{8}$

36(a) 21 %   (b) 73.5 %   (c) 136 %   (d) 5.4 %

37(a) 314   (b) GH¢ 100   (c) 2405   (d) GH¢ 780

38(a) 20 %   (b) $33\frac{1}{3}\%$ (c) $37\frac{1}{2}\%$ (d) $7\frac{1}{2}\%$

39(a) 840   (b) 80   (c) 5760   (d) 36,00

40(a) 230   (b) 2175   (c) 697.5   (d) 1332

41(a) 540   (b) 461.7   (c) 1,144   (d) 860

42. GH¢ 1,380   43. GH¢ 120   44. GH¢ 63   45. 12.8 % 46. 75 %

47. GH¢ 187.50   48. GH¢ 1250   49. GH¢ 218.40   50. 38 %

Chapter Test 16

1(a) 1 : 3   (b) 5 : 7   (c) 3 : 10   (d) 4 : 1

2(a) 8 : 7   (b) GH¢ 700   3. GH¢ 1000, GH¢ 3000, GH¢ 4500

4(a) $\dfrac{1}{200}$ (b) 24 km^2 5(a) GH¢ 18 per hour   (b) GH¢ 648

7. $12,416   8. GH¢ 30   9(a) $22\dfrac{1}{2}\%$ (b) 20 %

10(a) GH¢ 56.00   (b) GH¢ 57.60

Chapter 17

Try this

1. GH¢ 60   2. GH¢ 15,300   3. GH¢ 572

4. GH¢ 127.06   5. GH¢ 600   6. GH¢ 600

7. GH¢ 9030   8. GH¢ 603.42   9. GH¢ 156.89

10. GH¢ 98.92   11. GH¢ 1,100   12. GH¢ 50

13. 18.5 %   14. 19.2 %

15. Esi had GH¢ 1,200   Kojo had GH¢ 1,440

16. GH¢ 506.25   17. GH¢ 1,325   18. GH¢ 595

19(a) GH¢ 2,900 (b) GH¢ 9,600 (c) GH¢ 186.67 (d) GH¢ 855

Exercise 17.1(a)

1. GH¢ 2,000   2. GH¢ 3,600   3. 7.5 %   4. 4 %

5. GH¢ 1,785   6. GH¢ 270   7. GH¢ 25,785   8. GH¢ 28,500

9. GH¢ 12,000   10. GH¢ 14,500   11. GH¢ 5,600

12. GH¢ 20,500   13. GH¢ 248,500   14. GH¢ 3,000

15. GH¢ 716.70   16. GH¢ 1,220   17. 8 %   18. GH¢ 4,296

Exercise 17.1(b)

1. GH¢ 108   2. GH¢ 6   3. GH¢ 1.50   4. GH¢ 637.50

5. GH¢ 6,175   6. GH¢ 51.00   7. GH¢ 8,000

8. GH¢ 4,000   9. GH¢ 5,525   10. GH¢ 5,400

Exercise 17.1(c)

1. GH¢ 765   2. GH¢ 1,665   3. GH¢ 16,256.25   4. GH¢ 491.52

5. GH¢ 11,880   6. GH¢ 20,000   7. GH¢ 50,000

8. GH¢ 50,000   9. GH¢ 5,500   10. 10 %

Exercise 17.2(a)

1(a) GH¢ 750   (b) GH¢ 6,000   (c) GH¢ 3,600   (d) GH¢ 483.75

2(a) GH¢ 29,760   (b) GH¢ 7,920   (c) GH¢ 1,040   (d) GH¢ 10,810

3(a) GH¢ 1,080   (b) GH¢ 29,666.67

  (c) GH¢ 4,375   (d) GH¢ 8,500

4(a) 6 %   (b) $3\frac{1}{2}$%   (c) $3\frac{3}{4}$%   (d) 6 %

5(a) $8\frac{1}{3}$%   (b) 3 years   (c) 5 years   (d) 3 years

6(a) 5 %   (b) $2\frac{1}{2}\%$ (c) $2\frac{1}{3}\%$ (d) 8 %

Exercise 17.2(b)

1(a) GH¢ 4,977.79   (b) GH¢ 982,282.25

(c) GH¢ 33,162.75   (d) GH¢ 5,700.47

2(a) GH¢ 820   (b) GH¢ 3,972

(c) GH¢ 824.32   (d) GH¢ 2,602.02

3. GH¢ 2,789.19   4. GH¢ 3,851.36

5. GH¢ 11,821.88   6. GH¢ 4,193.36

Exercise 17.3

1. GH¢ 1,800   2. GH¢ 280   3. GH¢ 600

4. GH¢ 1,266.67   5. GH¢ 240 6 . GH¢ 30

7. GH¢ 50   8. GH¢ 80   9. GH¢ 36

10. 6   11. 19.9 %   12. 26.7 %

13. GH¢ 240   14. GH¢ 188.40   15. GH¢ 5796

Exercise 17,4

1. GH¢ 1,050   GH¢ 630   2. GH¢ 2,400   GH¢ 4,000

3. GH¢ 7,200   GH¢ 8,400   4. GH¢ 7,825   GH¢ 7,775

5. GH¢ 12,500   GH¢ 7,500   GH¢ 5,000   6. GH¢ 6,000

7. GH¢ 15,600   GH¢ 10,400   8. GH¢ 29,154.10

Exercise 17.5(a)

1. GH¢ 94.50   2. GH¢ 91.88   3. GH¢ 3,189.38

4. GH¢ 2,193.75   5. GH¢ 17.55   6. GH¢ 520   7. GH¢ 950

8. GH¢ 178.89   9. GH¢ 60   10. GH¢ 1,675.04

Exercise 17.5(b)

1. GH¢ 33   2. GH¢ 81.25   3. GH¢ 746.75   4. GH¢ 406.35

5. GH¢ 150   6. GH¢ 1,250   7. GH¢ 136   GH¢ 141.44

8. GH¢ 145.33   GH¢ 149.69   9. 5 %   10. 2 %

11. GH¢580.13   12. GH¢1200

Exercise 17.5(c)

1. GH¢ 358   2. GH¢ 4,200   3. GH¢ 812.50   4. GH¢ 14,940

5. GH¢ 25,225   6 GH¢ 149.50   7(a) GH¢ 700   (b) GH¢ 9,300

8(a) GH¢ 157.50   (b) GH¢ 1,592.50

9(a) GH¢ 9,400   (b) GH¢ 139.17   (c) GH¢ 902.50

10(a) GH¢ 554   (b) GH¢ 11,500

Review exercise 17

1. GH¢ 900   2. GH¢ 1,000   3. 15 %   4. GH¢ 8,260

5. GH¢ 1731   6. GH¢ 279   7. GH¢ 187.50

8. GH¢ 600   9. GH¢ 5,152   10. GH¢ 13,083.98

11. GH¢ 25,700   12. GH¢ 1,080   13. GH¢ 9,683

14. GH¢ 2,586.08   15. GH¢ 3137.73   16. GH¢ 470

17. GH¢ 35,000   18. GH¢ 15   19. 67 %

20. GH¢ 2,608   GH¢ 3,912   21. GH¢ 1,680   GH¢ 1,995

22. GH¢ 192.50   23. GH¢ 213.75   24. GH¢ 388

25. GH¢ 500   27. GH¢ 245   28. GH¢ 1,950

29. GH¢ 172   30(a) GH¢ 113.33   (b) GH¢ 790.8

Chapter Test 17

1. GH¢ 463.25   2. GH¢ 2880   3. GH¢ 11,840

4. GH¢ 3,500   5. GH¢ 4,320   6. GH¢ 7,024.64

7. GH¢ 148.50   8. GH¢ 313.88

9. GH¢ 7966.63   GH¢ 9559.95   10. GH¢ 160.40

Chapter 18

Try this

1(a) $\angle ABC$ (b) $\angle PQR$ (c) $\angle STR$

2(a) complementary   (b) supplementary   (c) neither

(d) complementary   (e) supplementary   (f) neither

(g) supplementary   (h) complementary

3. 25   4. 10   5. $x = 40°$ $y = 120°$

6(a) 66⁰   (b) 116⁰   7. 17 cm   8. 24 m   9. 50⁰

10(a) congruent   (b) congruent   (c) not congruent   (d) congruent

(e) not congruent   (f) congruent

11. AB = 15   BC = 12   12. 320 m   13. 147.3⁰

14. 8   15. 24   16. 150⁰

17(a) $\angle$ AEB, $\angle$ ADB   (b) $\angle$ AFC (c) $\angle$ EAD, $\angle$ EBD

(d) $\angle AOB$ (e) $\angle BOC$ (f) $\angle AOC$

18(a) $\angle FAC$, $\angle FBC$ (b) $\angle FEC$, $\angle FDC$

19(a) 54°   (b) 156°   20. 107^0 21. 58^0

22(a) $x = 75°, y = 150°$ 23(a) $x = 63°, y = 117°$

24. $x = 68°, y = 136°$ 25. $x = 35°, y = 35°$ 26. 20^0

Exercise 18.1(a)

1(a) $\angle$ ABC   (b) $\angle$ PQR

(c) $x° = \angle RQS$ $y° = \angle QSR$ $z° = \angle$ QRS

(d) $a° = \angle FGI$ $b° = \angle GFH$ $c° = \angle FHI$ $d° = \angle GIH$

3(a) 67^0 (b) 112^0 (c) 50^0 (d) 258^0

Exercise 18.1(b)

1. 105^0 2. 9   3. 24^0 4. $x = 11$ $y = 125$

5. 30   6.$x = 10, y = 18$ 7. 40   8. 39

9.$x = 10, y = 30$ 10. $x = 30, y = 125$

11. $x = 15, y = 120$ 12. $x = 11, y = 114$

Exercise 18.1(c)

1(a) $p = 120°$ $q = 60°$ $r = 120°$ $t = 120°$

$s = 60°$ $v = 60°$ $u = 120°$

(b) $p = 45°$ $q = 135°$ $r = 45°$ $s = 135°$

$t = 45°$ $u = 135°$ $v = 45°$

(c) $p = 110°$ $q = 110°$ $r = 110°$

(d) $p = 65°$ $q = 115°$ $r = 65°$ $s = 65°$ $t = 115°$

2(a) 12   (b) $x = 124$, $y = 14$ (c) 34

(d) $x = 112°$ $y = 28$ (e) $x = 12$, $y = 10$

(f) $x = 28$, $y = 19$

3(a) 60^0 (b) 60^0 (c) 75^0 (d) 70^0

Exercise 18.2(a)

1. 104^0 2. 23   3. 21   4. 30   5. 53   6. 30

7. 67   8. 21   9. 26   10. 42   11. 18   12. 20

13. 24   14. 56   15. 24   16. 10

Exercise 18.2(b)

1(a) 15   (b) 26   (c) 7.81   (d) 12.04

2(a) 24   (b) 8   (c) 36   (d) 7

3(a) 9   (b) 16   (c) 25   (d) 9

(e) 15   (f) 4   (g) 12   (h) 2

4. 72 cm   5. 8.6 m   6. 2.6 km   7. 44.7 km   8. 161.6 km

9. 13.8 m   10. 12.7 m   11. 2.5 m   12. 90 km

Exercise 18.2(c)

1. 55^0 2. 63^0 3.86^0 4.40^0 5. 30^0

6. 124^0 7. 45^0 8. 36^0 9. 32^0 10. 34^0

11. 24^0 12. 55^0 13. 22.5^0 14. 94^0 15. 40^0

16. 114⁰   17. 120⁰   18. 45⁰   19. 72⁰   20. 55⁰

Exercise 18.3

1. (a) and (g) ASA   (b) and (h) SAS

   (c) and (e) hypotenuse, side   (d) and (f) SSS

2(a) ΔBAC ΔDBC (b) ΔRQS ΔRPT

   (c) ΔEDA ΔCBA

3(a) ΔCBA ΔEDA 9   (b) ΔPQS ΔTRS 1.4

   (c) ΔABC ΔEDC 3

4(a) ΔABC ΔADE 7.5   (b) ΔCDA ΔBEA 5

   (c) ΔEAC ΔBDC 3

5(a) ΔABC ΔEDC AC = 7   CE = 5

   (b) ΔPQR ΔTUS QR = 5   US = 8

   (c) ΔDCA ΔBEA AC = 9   AE = 6

6. 13.5 m   7. 1.6 m   8. 58 cm   9. 427.5 m

Exercise 18.4(a)

1(a) 2160⁰   (b) 2520⁰   (c) 2880⁰   (d) 3240⁰   (e) 6120⁰

2(a) 108⁰   (b) 135⁰   (c) 150⁰   (d) 156⁰   (e) 165.6⁰

3(a) 9   (b) 6   (c) 18   (d) 10   (e) 15

4(a) 90⁰   (b) 75⁰   (c) 72⁰   (d) 45⁰   (e) 60⁰   (f) 36⁰

5. 144⁰   6. 75⁰, 150⁰

Exercise 18.4(b)

1(a) 12   (b) 8   (c) 5   (d) 6   (e) 10

2(a) 45^0 (b) 51.4^0 (c) 72^0 (d) 30^0

3(a) 15   (b) 6   (c) 12   (d) 8

4(a) 40^0 (b) 25^0 (c) 30^0 (d) 36^0 (e) 41^0 (f) 42^0

Exercise 18.5(a)

1(a) 66^0 (b) 76^0 (c) $x = 54°$ $y = 54°$

(d) $x = 98°$ $y = 98°$ (e) $x = 80°$ $y = 160°$

(f) $x = 65°$ $y = 65°$ (g) $x = 20°$ $y = 70°$

(h) $x = 56°$ $y = 35°$

2(a) 55^0 (b) $48°$ (c) $x = 38°$ $y = 38°$

(d) $x = 42°$ $y = 60°$

3(a) $x = 70°$ $y = 110°$ (b) $x = 115°$ $y = 230°$

(c) $x = 38°$ $y = 90°$ (d) $x = 72°$ $x = 144°$

(e) $x = 72°$ $y = 73°$ (f) $x = 65°$ $y = 105°$

(g) $x = 70°$ $y = 140°$ (h) $x = 66°$ $y = 36°$

Exercise 18.5(b)

4(a) $x = 70°$ $y = 70°$ (b) $x = 52°$ $y = 52°$

(c) $x = 34°$ $y = 34°$ (d) $x = 70°$ $y = 70°$

2(a) 26   (b) 38   (c) 60   (d) 35   (e) 58   (f) 37

3(a) 15   (b) 16   (c) 12   (d) 6

Review exercise 18

2(a) 115^0 (b) 43^0 (c) $x = 75°$ $y = 105°$ (d) 91^0

3 (a) $x = 4°$ $y = 26°$ (b) $x = 63°$ $y = 63°$ $z = 117°$

(c) $x = 68°$ (d) 50^0

4(a) 40^0 (b) $x = 55°$ $y = 50°$ (c) 16^0 (d) 23^0

5(a) 17 cm   (b) 9 cm   (c). 25 cm   (d) 15 cm

6. 15.6 m   7. 8.9 m   8(a) 1620^0 (b) 2520^0 (c) 4140^0

9(a) 120^0 (b) 140^0 (c) 157.5^0 10(a) 12   (b) 20

11(a) 60^0 (b) 32.7^0 12. 125 m   13. 0.9 cm

14. (a) and (e),   (b) and (d)   (c) and (f)

15(a) $\triangle AEB$ $\triangle CDB$ $x = 6$ AB = 12   BC = 10

(b) $\triangle CDE$ $\triangle CAB$ $x = 6$ DE = 16

(c) $\triangle CBA$ $\triangle CED$ $x = 9$ AB = 18   ED = 15

16(a) 110^0 (b) 30^0

17(a) 74^0 (b) $x = 53°$ $y = 106°$ (c) 102^0

(d) $x = 54°$ $y = 54°$

18(a) $x = 64°$ $y = 118°$ (b) $x = 85°$ $y = 40°$

19(a) $x = 35°$ $y = 35°$ (b) $x = 65°$ $y = 130°$

20(a) 36   (b) 35   (c) 12 cm   (d) 9 cm

Chapter Test 18

1(a) 25   (b) $x = 60°$, $y = 30°$

2(a) $x = 75°$ $y = 117°$ (b) $x = 64°$ $y = 130°$

3(a) 8   (b) $x = 55°$ $y = 110°$

4(a) 4.58 cm   (b) $x = 5$ $y = 12$ (c) 400 km

5(a) $\triangle PTR$, $\triangle SQR$, 3   (b) $\triangle PQR$, $\triangle PTS$, 5

6. 400 m   7(a) $40°$ (b) $35°$ 8. $x = 30°$, $y = 60°$

9. $x = 37°$ $y = 37°$ 10(a) $18°$ (b) $72°$

www.ingramcontent.com/pod-product-compliance
Lightning Source LLC
Chambersburg PA
CBHW071408180526
45170CB00001B/8